Introduction to
ANTENNAS & PROPAGATION

Introduction to
ANTENNAS & PROPAGATION

James R. Wait

Peter Peregrinus Ltd on behalf of the Institution of Electrical Engineers

Published by Peter Peregrinus Ltd., London, United Kingdom
© 1986: Peter Peregrinus Ltd.

All rights reserved. No part of this publication may be reproduced, stored in a retrieval system or transmitted in any form or by any means—electronic, mechanical, photocopying, recording or otherwise—without the prior written permission of the publisher.

While the author and the publishers believe that the information and guidance given in this work is correct, all parties must rely upon their own skill and judgment when making use of it. Neither the author nor the publishers assume any liability to anyone for any loss or damage caused by any error or omission in the work, whether such error or omission is the result of negligence or any other cause. Any and all such liability is disclaimed.

British Library Cataloguing in Publication Data

Introduction to antennas and propagation.
   1. Electromagnetic fields
  I. Wait, James R.
     530.1'41   QC665.E4

ISBN: 0-86341-054-5

Printed in England by Short Run Press Ltd., Exeter

Contents

	Preface	ix
	Useful data and glossary	xi
1	**Review of phasors and vectors**	**1**
1.1	Phasor concept	1
1.2	Addition and subtraction	2
1.3	Multiplication and division	3
1.4	Square root	4
1.5	Complex conjugate	4
1.6	Real vectors	4
1.7	Addition and subtraction of vectors	5
1.8	Multiplication of vectors	6
1.9	Complex vectors	7
1.10	Multiplication of complex vectors	8
1.11	Time averaging	8
1.12	Rotating vectors	10
1.13	Gradient operator	10
1.14	Divergence operator	11
1.15	Curl operator	11
1.16	Two theorems	11
1.17	Miscellaneous formulas from vector calculus	12
2	**Quasi-static fields**	**17**
2.1	Current flow in conductive media and Ohm's law	17
2.2	Lossy capacitor	18
2.3	Laplace's equation	19
2.4	Current point source	20
2.5	Transition to electrostatics	26
2.6	Magnetostatics	27
2.7	Bibliography	36
3	**Dynamic fields**	**37**
3.1	Ampère's and Faraday's laws: Maxwell's equations	37
3.2	Plane wave transmission	39
3.3	Plane wave solutions	40
3.4	Transmission in material media	42
3.5	Excitation of plane waves	44
3.6	Oblique excitation of plane waves	47
3.7	Further exercises	51
3.8	Review problems	56
3.9	Bibliography	63

4	**Reflection and refraction**	**64**
	4.1 Perfect reflector	64
	4.2 Reflection from a dielectric half-space	66
	4.3 Critical reflection	73
	4.4 Reflection from a lossy half-space	75
	4.5 Further exercises	78
	4.6 Bibliography	85
5	**Electromagnetic fields of current distributions**	**86**
	5.1 Vector potential	86
	5.2 Complete dipole field expressions	88
	5.3 Free space fields and radiation patterns	90
	5.4 Fields of a linear antenna	92
	5.5 Radiation of linear antenna in free space	94
	5.6 Antenna arrays	97
	5.7 Power considerations	100
	5.8 Further exercises	103
	5.9 Bibliography	119
6	**Guided waves**	**121**
	6.1 Coaxial cable and wire lines	121
	6.2 Simple transmission line theory	123
	6.3 Rectangular waveguides	127
	6.4 Wave circuit parameters	133
	6.5 General slab model	136
	6.6 Limit of perfect wall conductivity	138
	6.7 Effect of finite wall conductivity	139
	6.8 Earth–ionosphere waveguide model	140
	6.9 Surface waves	144
	6.10 Further exercises	147
	6.11 References	151
7	**Cylindrical waves and scatter**	**152**
	7.1 General formulation	152
	7.2 Fields of line sources	154
	7.3 Internal fields of the cylindrical conductor	157
	7.4 Scattering from a thin cylinder	159
	7.5 Oblique incidence scattering from a thin wire	161
	7.6 Plane wave scattering from a cylinder of any radius	162
	7.7 Scattering from a semi-cylindrical boss on ground plane	167
	7.8 Application to scattering of a radio ground wave by a ridge	169
	7.9 Scattering by two parallel wires	172
	7.10 Scattering from a wire located over a plane conductor	175
	7.11 Line source excitation of a homogeneous cylinder	178
	7.12 Application to subsurface radio	179
	7.13 Scattering of a plane wave from a cylinder and parallel wire combination	181
	7.14 Scattering by two parallel wires for oblique incidence	183
	7.15 Scattering at oblique incidence from cylindrical structures	185
	7.16 Scattering from conducting cylinder of arbitrary cross-section	188
	7.17 Further exercises	195
	7.18 Appendix: cylindrical Bessel functions – an outline	199
	7.19 References	207

Contents vii

8 Electromagnetic TM and TE spherical waves — **210**
- 8.1 Introduction — 210
- 8.2 The Debye potentials — 210
- 8.3 External excitation of sphere — 214
- 8.4 Concentrically layered model — 216
- 8.5 The iterative solution — 218
- 8.6 Scattering of a plane wave from a sphere: radar cross-sections — 221
- 8.7 References — 227

9 Antennas mounted on smooth convex surfaces — **229**
- 9.1 Introduction — 229
- 9.2 Formulation — 230
- 9.3 Formal solution — 231
- 9.4 Complex integral representation — 232
- 9.5 Discussion of the cutback factor — 235
- 9.6 Extension to E-parallel polarisation — 239
- 9.7 Application to moderately large cylinders — 243
- 9.8 Concluding remarks — 245
- 9.9 Appendix: mathematical details — 245
- 9.10 References — 252

Index — 255

Preface

Chapters 1–6 of this book are based on courses on electromagnetic fields at the undergraduate level at the University of Arizona. An attempt has been made to provide a concise introduction in a self-contained format. The stress is on basic concepts. There are numerous exercises, many with solutions.

A unique feature of the presentation is the introduction of quasi-static theory at an early stage via Ohm's law for conducting media (Chapter 2). Electrostatics and magnetostatics are recovered as special cases. Then, with a suitable generalisation of Ampère's and Faraday's laws, Maxwell's equations are set out in a logical fashion without appealing *ad hoc* postulates (Chapter 3).

Another novel feature is the use of a uniform current sheet model to illustrate the excitation of plane waves in material media (Chapters 3 and 4). Impedance concepts are also used extensively, and they help the student relate to his earlier ideas from circuit courses.

Antenna theory is introduced via a simple treatment of vector potential without the need to solve inhomogeneous wave equations (Chapter 5). Various illustrations of radiation patterns are given, and antenna arrays are discussed very briefly.

Guided waves are introduced by a quasi-static analysis of the familiar coaxial cable (Chapter 6). Then transmission line and waveguide theory are outlined, with attention drawn to related circuit equivalences. The author's earlier research on the transmission of long electromagnetic waves in the earth–ionosphere waveguide is used as a graphic example. Dielectric waveguides are also discussed briefly, including the basic mechanism in fibre optic transmission.

The mathematical background needed includes calculus and elementary differential equations. Some familiarisation with circuit theory would also be helpful. In any case, some of the analytical tools, such as vector calculus, are reviewed in Chapter 1. An effort is made here to remove the common confusion between phasors and vectors.

Chapters 7, 8 and 9 contain slightly more advanced material with numerous engineering applications. The mathematics needed is covered in a fairly concise manner, including a much needed outline of cylindrical Bessel functions (Section 7.18). The corresponding spherical wave functions are covered in Chapter 8. Some applications to antennas on the earth's surface are included in the final chapter.

Acknowledgements

I am grateful to the many useful and cogent comments that I have received from my friends in the academic community particularly Prof. Donald G. Dudley. I am also indebted to Mikaya Lumori, Thomas Gruszka, Steven Webster and Manuel Nelson at the University of

Arizona for their important comments. Paul Teschan and Diane King have been particularly helpful in checking the many solutions to the worked exercises. It has been a real pleasure working with Nick Bliss of PPL during the production of this book.

James R. Wait
Tucson, July 1985

Useful data and glossary

ε_0	$= 8.854 \times 10^{-12}\,\text{F/m}$, permittivity of free space
μ_0	$= 4\pi \times 10^{-7}\,\text{H/m}$, magnetic permeability of free space
c	$= (\varepsilon_0 \mu_0)^{-1/2} = 2.9979 \times 10^8\,\text{m/s}$, velocity of light in free space
η_0	$= (\mu_0/\varepsilon_0)^{1/2} \simeq 120\pi \simeq 377\,\Omega$, intrinsic or characteristic impedance of free space
k	$= (\mu_0 \varepsilon_0)^{1/2} \omega$
$\varepsilon/\varepsilon_0$	dielectric constant or permittivity relative to free space
μ/μ_0	magnetic permeability relative to free space
η	$= [j\mu\omega/(\sigma + j\varepsilon\omega)]^{1/2}$, characteristic impedance in the medium
v_0	$= (\varepsilon\mu)^{-1/2}$, phase velocity in lossless medium ($\sigma = 0$)
λ_0	$= c/f$, free space wavelength; f is frequency (Hz)
σ	$=$ conductivity (S/m)
ω	$= 2\pi f$, radian frequency (rad/s)
γ	$= [j\mu\omega(\sigma + j\varepsilon\omega)]^{1/2}$, (complex) propagation constant for harmonic time factor $\exp(+j\omega t)$
α	$= \text{Re}\,\gamma$, attenuation rate (Np/m) (Re $=$ real part of)
β	$= \text{Im}\,\gamma$, phase factor (rad/m) (Im $=$ imaginary part of)
v	$= \omega/\beta$, phase velocity in medium
λ	$= 2\pi/\beta$, wavelength in medium
dB	$= (8.686)^{-1}\,\text{Np}$ (dB is decibels)

ELF	extremely low frequency (1 to 3000 Hz)
VLF	very low frequency (3 to 30 kHz)
LF	low frequency (30 to 300 kHz)
MF	medium frequency (300 to 3000 kHz)
HF	high frequency (3 to 30 MHz)
VHF	very high frequency (30 to 300 MHz)
UHF	ultra high frequency (300 to 3000 MHz)
SHF	super high frequency (3 to 30 GHz)
EHF	extra high frequency (30 to 300 GHz)

$\sigma + j\varepsilon\omega =$ (complex) admittivity
$j\mu\omega \quad\;\; =$ (complex) impedivity

Chapter 1

Review of phasors and vectors

1.1 Phasor concept A time-harmonic real physical quantity may be designated by

$$v(t) = V_0 \cos(\omega t + \phi_0) \tag{1.1a}$$

Here V_0 is the amplitude, ω is the frequency, t is the time and ϕ_0 is the phase. Note that $\omega = 2\pi f$, where f is the frequency in hertz. We may plot $v(t)$ as a function of time t as indicated in Fig. 1.1.

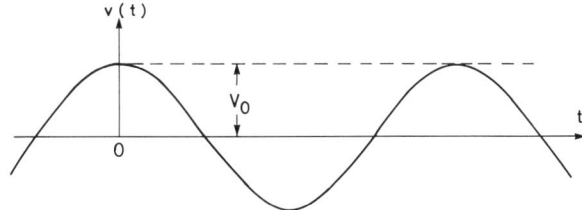

Fig. 1.1 Plot of $v(t)$ shown for $\varphi_0 = 0$

Now we can also write $v(t)$ in terms of a *complex phasor* V in the manner

$$v(t) = \text{Re}\,(V\,e^{j\omega t}) \tag{1.1b}$$

where Re means 'real part of'. In the present example, we would choose

$$V = V_0\,e^{j\phi_0} \tag{1.2}$$

where V_0 is the amplitude or magnitude of the complex quantity or phasor V. The phase angle ϕ_0 is the phase or argument of V. Here $j = (-1)^{1/2}$.

The manipulation of complex quantities (or phasors) follows well defined rules. For example, we may write

$$V = \text{Re}\,V + j\,\text{Im}\,V \tag{1.3}$$

where Re V is the real part and Im V is the imaginary part (Fig. 1.2). Then, of course,

$$V = |V|\,e^{j\phi_0} = |V|\cos\phi_0 + j|V|\sin\phi_0 \tag{1.4}$$

so that

$$\text{Re}\,V = |V|\cos\phi_0 = V_0 \cos\phi_0 \tag{1.5}$$

$$\text{Im } V = |V| \sin \phi_0 = V_0 \sin \phi_0 \tag{1.6}$$

We also note that

$$|V| = V_0 = [(\text{Re } V)^2 + (\text{Im } V)^2]^{1/2} \tag{1.7}$$

and

$$\tan \phi_0 = \text{Im } V/\text{Re } V \tag{1.8}$$

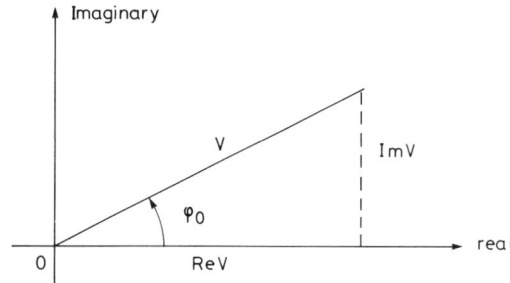

Fig. 1.2 Complex V-plane

1.2 Addition and subtraction

Consider two voltages at the same frequency:

$$v_1(t) = V_1 \cos(\omega t + \phi_1) \tag{1.9}$$

$$v_2(t) = V_2 \cos(\omega t + \phi_2) \tag{1.10}$$

But let us not use phasors for the moment. Instead, we add the two equations by writing

$$\begin{aligned}v_1(t) + v_2(t) &= V_1 \cos \omega t \cos \phi_1 - V_1 \sin \omega t \sin \phi_1 \\ &\quad + V_2 \cos \omega t \cos \phi_2 - V_2 \sin \omega t \sin \phi_2 \\ &= (V_1 \cos \phi_1 + V_2 \cos \phi_2) \cos \omega t \\ &\quad - (V_1 \sin \phi_1 + V_2 \sin \phi_2) \sin \omega t\end{aligned} \tag{1.11}$$

Now our desired form of the combined signal is

$$v_1(t) + v_2(t) = V_c \cos(\omega t + \phi_c) \tag{1.12}$$

$$= V_c \cos \omega t \cos \phi_c - V_c \sin \omega t \sin \phi_c \tag{1.13}$$

Thus

$$V_c \cos \phi_c = V_1 \cos \phi_1 + V_2 \cos \phi_2 \tag{1.14}$$

and

$$V_c \sin \phi_c = V_1 \sin \phi_1 + V_2 \sin \phi_2 \tag{1.15}$$

Squaring and adding, we get

$$\begin{aligned}V_c^2 &= (V_1 \cos \phi_1 + V_s \cos \phi_2)^2 \\ &\quad + (V_1 \sin \phi_1 + V_2 \sin \phi_2)^2\end{aligned} \tag{1.16}$$

Review of phasors and vectors

To obtain ϕ_c we take the ratio of eqns. 1.15 and 1.14; thus

$$\tan \phi_c = \frac{V_1 \sin \phi_1 + V_2 \sin \phi_2}{V_1 \cos \phi_1 + V_2 \cos \phi_2} \tag{1.17}$$

Now, working with phasors, we would first write

$$v_1(t) = \mathrm{Re}\, V_1 \, e^{j\phi_1} \, e^{j\omega t} \tag{1.18}$$

$$v_2(t) = \mathrm{Re}\, V_2 \, e^{j\phi_2} \, e^{j\omega t} \tag{1.19}$$

The resultant phasor is simply given by (Fig. 1.3)

$$V_c \, e^{j\phi_c} = V_1 \, e^{j\phi_1} + V_2 \, e^{j\phi_2} \tag{1.20}$$

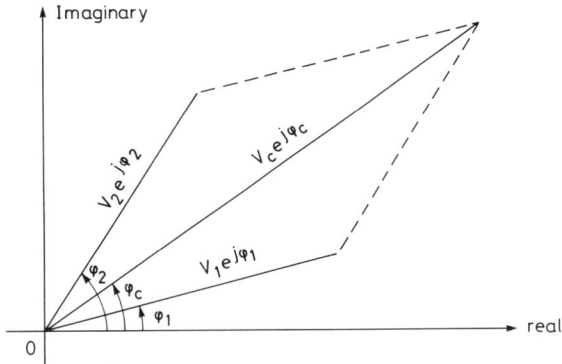

Fig. 1.3 Graphic representation of the addition of two complex numbers

Then, of course the desired signal is

$$v_c(t) = \mathrm{Re}\, (V_c \, e^{j\phi_c} \, e^{j\omega t})$$

$$= \mathrm{Re}\, (V_1 \, e^{j\phi_1} + V_2 \, e^{j\phi_2}) \, e^{j\omega t} \tag{1.21}$$

In the case where the signals are being subtracted, the same reasoning applies. In using phasors at the outset we merely write down that

$$v_c(t) = v_1(t) - v_2(t)$$

$$= \mathrm{Re}\, (V_1 \, e^{j\phi_1} - V_2 \, e^{j\phi_2}) \, e^{j\omega t} \tag{1.22}$$

1.3 Multiplication and division

In summary if we have two complex numbers $c = a \pm jb$ and $h = f \pm jg$, then

$$c \pm h = (a \pm f) + j(b \pm g) \tag{1.23}$$

Also, we can note that the multiplication rule is

$$ch = (a + jb)(f + jg)$$

$$= (af - bg) + j(bf + ag) \tag{1.24}$$

where we have used the identity $j^2 = -1$. The corresponding division rule is

$$\frac{c}{h} = \frac{a + jb}{f + jg} = \frac{(a + jb)(f - jg)}{f^2 + g^2}$$

$$= \frac{af + bg}{f^2 + g^2} + j\frac{bf - ag}{f^2 + g^2} \tag{1.25}$$

If we write $c = |c| e^{j\phi_1}$ and $h = |h| e^{j\phi_2}$, the multiplication and division can be done more directly. Thus

$$ch = |ch| e^{j(\phi_1 + \phi_2)} \tag{1.26}$$

$$\frac{c}{h} = \left|\frac{c}{h}\right| e^{j(\phi_1 - \phi_2)} \tag{1.27}$$

1.4 Square root

Say we wish to find the square root of the complex number $c = a + jb$, where a and b are real. To this end we first write the number in the polar form

$$c = |c| e^{j\phi}$$

where $\tan \phi = b/a$. Thus the solution is

$$c^{1/2} = |c|^{1/2} e^{j\phi/2} \tag{1.28}$$

which is readily verified to be correct. However, it is evident that

$$c^{1/2} = -|c|^{1/2} e^{j\phi/2} = |c|^{1/2} e^{j[\pi + (\phi/2)]} \tag{1.29}$$

is also a solution.

1.5 Complex conjugate

For $c = a + jb$ with a and b real, the complex conjugate is

$$c^* = a - jb \tag{1.30}$$

Here we note that

$$cc^* = a^2 + b^2 = |c|^2 \tag{1.31}$$

Also $c = |c| e^{+j\phi}$, the complex conjugate in polar form, is simply

$$c^* = |c| e^{-j\phi} \tag{1.32}$$

1.6 Real vectors

A real quantity that is characterised by both magnitude and direction is a real vector. The velocity of a parcel of air in the atmosphere is an example of a real vector. In this book, vectors will be printed in bold italic.

With reference to a Cartesian or rectangular co-ordinate system (x, y, z), we can define a real vector $\boldsymbol{v}$ as follows:

$$\boldsymbol{v} = v_x \boldsymbol{i}_x + v_y \boldsymbol{i}_y + v_z \boldsymbol{i}_z \tag{1.33}$$

Here v_x, v_y and v_z are the x, y and z components of the vector $\boldsymbol{v}$; $\boldsymbol{i}_x$, $\boldsymbol{i}_y$ and $\boldsymbol{i}_z$ are *unit vectors*, that is, they have unit magnitude and they are oriented in the x, y, z directions, respectively (Fig. 1.4). Actually v_x, v_y and v_z may be positive or negative real quantities, so that the direction

Review of phasors and vectors

of v can be up, down, or sideways. The magnitude of v is clearly given by

$$v = (v_x^2 + v_y^2 + v_z^2)^{1/2} \tag{1.34}$$

Note that we designate the magnitude of the vector by v rather than $|v|$. Also we prefer the designations i_x, i_y, i_z, for the unit vectors rather than the forms such as $\hat{x}$, $\hat{y}$, $\hat{z}$ or i, j, k as found in many text books.

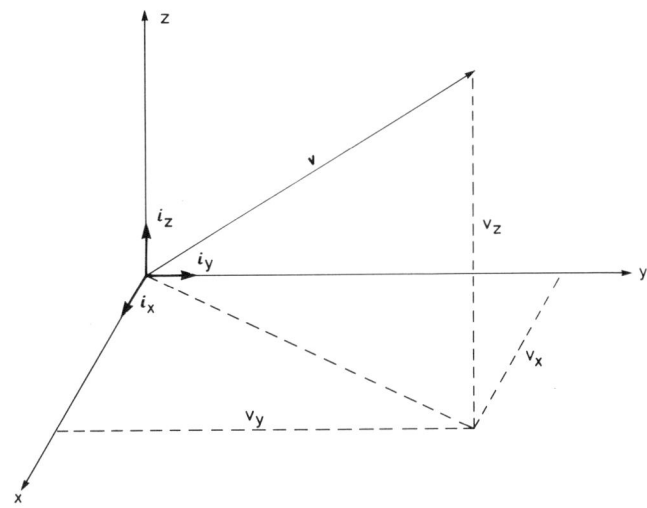

Fig. 1.4 Vector v, rectangular components v_x, v_y, v_z and unit vectors i_x, i_y, i_z

1.7 Addition and subtraction of vectors

Consider two separate real vectors:

$$v = v_x i_x + v_y i_y + v_z i_z \tag{1.35}$$

$$u = u_x i_x + u_y i_y + u_z i_z \tag{1.36}$$

By addition we form the new vector

$$v + u = (v_x + u_x) i_x + (v_y + u_y) i_y + (v_z + u_z) i_z \tag{1.37}$$

Clearly the x, y and z components are $v_x + u_x$, $v_y + u_y$ and $v_z + u_z$, respectively (see Fig. 1.5).

To deal with the subtraction of two vectors we note that

$$v - u = (v_x - u_x) i_x + (v_y - u_y) i_y + (v_z - u_z) i_z \tag{1.38}$$

Exercise

An aeroplane flies from Tucson to Las Vegas in a roughly north-west direction at a velocity $v = 707$ km/h. There is a westerly wind $w = 100$ km/h (i.e. the wind blows from the west). How fast and in what direction would the plane travel with no wind?

Solution

Let i_x be east and i_y be north. Then

$$v = -500 i_x + 500 i_y$$

and

$$w = 100\boldsymbol{i}_x$$

Then, clearly, the velocity with no wind is

$$\boldsymbol{v} - \boldsymbol{w} = -600\boldsymbol{i}_x + 500\boldsymbol{i}_y$$

In other words the air speed, which is a scalar, is

$$[(600)^2 + (500)^2]^{1/2} \simeq 780 \text{ km/h}$$

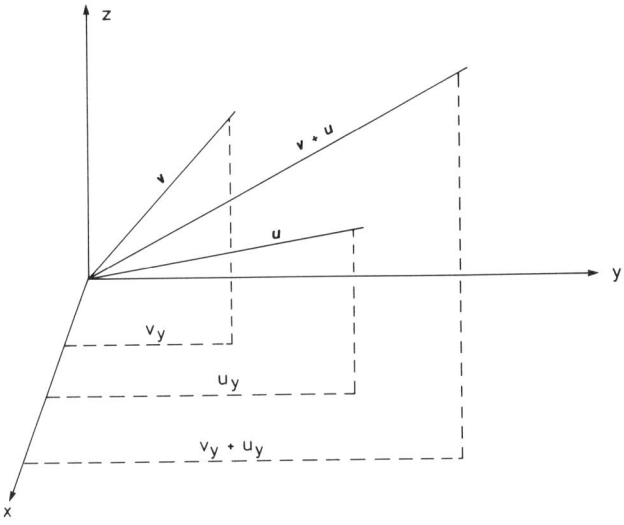

Fig. 1.5 Addition of two vectors **v** and **u** and their resultant vector **v** + **u**

(As a matter of interest, deduce the ground speed, or velocity relative to the ground, for the straight line return journey, assuming you still have money for fuel.)

The rules for the addition and subtraction of complex numbers and those for two-dimensional vectors are very similar. Unfortunately this can be a source of confusion, particularly when a phasor is referred to as a 'vector'.

1.8 Multiplication of vectors

There are two kinds of vector multiplication. The first is the *dot* or *scalar product*. It is defined by

$$\boldsymbol{v} \cdot \boldsymbol{u} = v_x u_x + v_y u_y + v_z u_z \tag{1.39}$$

where clearly the right-hand side is a scalar quantity. An equivalent definition is

$$\boldsymbol{v} \cdot \boldsymbol{u} = vu \cos\theta \tag{1.40}$$

where θ is the angle subtended by $\boldsymbol{v}$ and $\boldsymbol{u}$, and v and u are the respective magnitudes.

The second kind of vector multiplication is the *cross* or *vector product*. It is defined by

$$\boldsymbol{v} \times \boldsymbol{u} = \boldsymbol{i}_x(v_y u_y - v_z u_y) + \boldsymbol{i}_y(v_z u_x - v_x u_z)$$

$$+ \boldsymbol{i}_z(v_x u_y - v_y u_x) \tag{1.41}$$

Another way to express this definition is the determinant form

$$\bm{v} \times \bm{u} = \begin{vmatrix} \bm{i}_x & \bm{i}_y & \bm{i}_z \\ v_x & v_y & v_z \\ u_x & u_y & u_z \end{vmatrix} \quad (1.42)$$

An equivalent geometrical definition is

$$\bm{v} \times \bm{u} = (vu \sin \theta)\bm{i}_n \quad (1.43)$$

where θ is the angle subtended by $\bm{v}$ and $\bm{u}$, and $\bm{i}_n$ is a unit vector perpendicular to the plane containing $\bm{v}$ and $\bm{u}$. The geometry of the situation is shown in Fig. 1.6. Here it is important to note that $\bm{i}_n$ is drawn in the direction of a right-handed screw rotation from $\bm{v}$ to $\bm{u}$.

Exercise

Show that

$$\bm{v} \times \bm{u} = -\bm{u} \times \bm{v} \quad (1.44)$$

and explain change in Fig. 1.6.

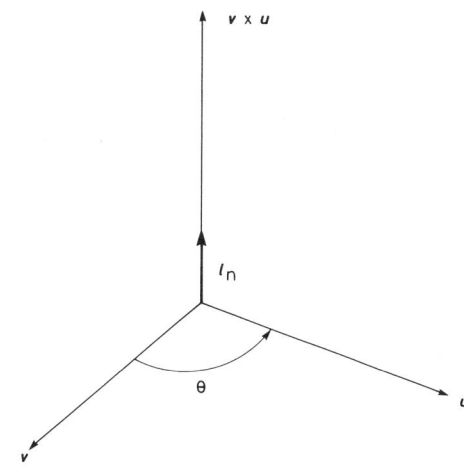

Fig. 1.6 Illustrating the right-hand rule

1.9 Complex vectors

Hopefully we have clarified the properties of (complex) phasors and real vectors. Now we introduce complex vectors by first noting that a vector that varies time harmonically can be written

$$\bm{e}(t) = \bm{i}_x e_x(t) + \bm{i}_y e_y(t) + \bm{i}_z e_z(t) \quad (1.45)$$

or

$$\bm{e}(t) = \bm{i}_x E_x \cos(\omega t + \phi_x) + \bm{i}_y E_y \cos(\omega t + \phi_y)$$
$$+ \bm{i}_z E_z \cos(\omega t + \phi_z) \quad (1.46)$$

or

$$\bm{e}(t) = \mathrm{Re}\,[(\bm{i}_x E_x\, e^{j\phi_x} + \bm{i}_y E_y\, e^{j\phi_y} + \bm{i}_z E_z\, e^{j\phi_z})\, e^{j\omega t}] \quad (1.47)$$

where, for example, E_x and ϕ_x are the amplitude and phase of the x component. Thus we can summarise the above sequence by the compact statement

$$e(t) = \text{Re}\,(\boldsymbol{E}\,e^{j\omega t}) \tag{1.48}$$

where $\boldsymbol{E}$ is a complex vector. The cartesian or rectangular components of $\boldsymbol{E}$ are the (complex) phasors $E_x\,e^{j\phi_x}$, $E_y\,e^{j\phi_y}$ and $E_z\,e^{j\phi_z}$.

1.10 Multiplication of complex vectors

The rules for the dot product of two complex vectors are the same as for two real vectors, i.e.

$$\boldsymbol{e} \cdot \boldsymbol{d} = e_x d_x + e_y d_y + e_z d_z \tag{1.49}$$

where e_x and d_x, for example, are the x components of $\boldsymbol{e}$ and $\boldsymbol{d}$ respectively. Either or both of e_x and d_x can be complex phasors.

The cross product of two complex vectors is

$$\boldsymbol{h} \times \boldsymbol{e} = \boldsymbol{i}_x(h_y e_z - h_z e_y) + \boldsymbol{i}_y(h_z e_x - h_x e_z) + \boldsymbol{i}_z(h_x e_y - h_y e_x) \tag{1.50}$$

where again the components $(h_y e_z - h_z e_y)$ etc. are complex phasors.

We note that the geometrical depictions in Fig. 1.5 and Fig. 1.6 are not strictly applicable when dealing with complex vectors. The subtending angle would need to be complex.

1.11 Time averaging

An important concept in signal analysis is the intensity or time average of the square of the real signal. For example, if our scalar voltage, as a function of time, is

$$v(t) = V_0 \cos(\omega t + \phi_0) \tag{1.51}$$

then

$$\langle v^2(t) \rangle = \frac{1}{T}\int_0^T v^2(t)\,dt \tag{1.52}$$

is the average over one period, where $T = 1/f = 2\pi/\omega$. As an exercise the reader may show that

$$\langle v^2(t) \rangle = V_0^2/2 \tag{1.53}$$

for any value of ϕ_0. Here we note that $V_0/\sqrt{2}$ is the RMS (root mean square) amplitude of the signal.

Exercise

Show that $\langle v(t) \rangle = 0$.

Exercise

Show that the average over one half cycle is

$$\frac{2}{T}\int_0^{T/2} v(t)\,dt = \begin{array}{l} = 2V_0/\pi \text{ for } \phi_0 = -\pi/2 \\ = 0 \quad\quad \text{ for } \phi_0 = 0 \end{array}$$

Another useful concept is illustrated as follows. Consider two time-varying real vectors

$$\boldsymbol{a}(t) = \text{Re}\,\boldsymbol{A}\,e^{j\omega t} \tag{1.54}$$

$$\boldsymbol{b}(t) = \text{Re}\,\boldsymbol{B}\,e^{j\omega t} \tag{1.55}$$

where **A** and **B** are the corresponding complex vectors. Now we can write

$$a(t) = \text{Re } A \cos \omega t - \text{Im } A \sin \omega t \qquad (1.56)$$

$$b(t) = \text{Re } B \cos \omega t - \text{Im } B \sin \omega t \qquad (1.57)$$

The cross product is given by

$$\begin{aligned} a(t) \times b(t) &= (\text{Re } A \times \text{Re } B) \cos^2 \omega t \\ &+ (\text{Im } A \times \text{Im } B) \sin^2 \omega t \\ &- \tfrac{1}{2} (\text{Re } A \times \text{Im } B + \text{Im } A \times \text{Re } B) \sin 2\omega t \end{aligned} \qquad (1.58)$$

Then clearly the corresponding time average is

$$\langle a(t) \times b(t) \rangle = \tfrac{1}{2} (\text{Re } A \times \text{Re } B + \text{Im } A \times \text{Im } B) \qquad (1.59)$$

because the average of the $\sin 2\omega t$ term in eqn. 1.58 is zero. We can readily verify that

$$\langle a(t) \times b(t) \rangle = \tfrac{1}{2} \text{Re } (A \times B^*) \qquad (1.60)$$

where $B^* = \text{Re } B - j \text{ Im } B$ is the complex conjugate of the complex vector **B**.

Exercise

We represent a general periodic function $v(t)$ in the form

$$v(t) = \sum_{m=0}^{\infty} a_m \cos m\omega t + b_m \sin m\omega t$$

where $\omega = 2\pi f$, $T = 1/f$ is the fundamental period, and a_m and b_m are coefficients. Show that $\langle v(t) \rangle = a_0$

Solution

We note that

$$\begin{aligned} \frac{1}{T} \int_0^T v(t) \, dt &= \frac{1}{T} \left(a_0 \int_0^T dT + \sum_{m=1}^{\infty} a_m \int_0^T \cos m\omega T \, dt \right. \\ &\quad \left. + \sum_{m=1}^{\infty} b_m \int_0^T \sin m\omega t \, dt \right) \\ &= a_0 + \frac{1}{T} \sum_{m=1}^{\infty} a_m \left. \frac{\sin m\omega t}{m\omega} \right|_0^T - \frac{1}{T} \sum_{m=1}^{\infty} b_m \left. \frac{\cos m\omega t}{m\omega} \right|_0^T \\ &= a_0 + \frac{1}{T} \sum_{m=1}^{\infty} a_m \frac{\sin m\omega T}{m\omega} - \frac{1}{T} \sum_{m=1}^{\infty} b_m \frac{\cos (m\omega T) - 1}{m\omega} \end{aligned}$$

Note that

$$\sin m\omega T = \sin (2m\pi) = 0$$

$$\cos (m\omega T) - 1 = \cos (2m\pi) - 1 = 0$$

Thus only a_0 (i.e. the DC term) survives.

Exercise

Verify that

$$\langle a(t) \times b(t) \rangle = \tfrac{1}{2} \operatorname{Re}(A^* \times B) = -\tfrac{1}{2} \operatorname{Re}(B^* \times A) \tag{1.61}$$

1.12 Rotating vectors

Consider the real time-varying vector

$$e(t) = E_0(i_x \cos \omega t + i_y \sin \omega t) \tag{1.62}$$

where E_0 is a real constant. This rotating vector is sketched in Fig. 1.7, where its position is shown at the times ωt given by 0, $\pi/2$, π and $3\pi/2$. The sense of rotation (i.e. anti-clockwise) is also shown. The phasor form of eqn. 1.62 is the complex vector

$$E = E_0(i_x - j i_y) \tag{1.63}$$

For example we readily deduce that

$$e(t) = \operatorname{Re}(E e^{j\omega t}) \tag{1.64}$$

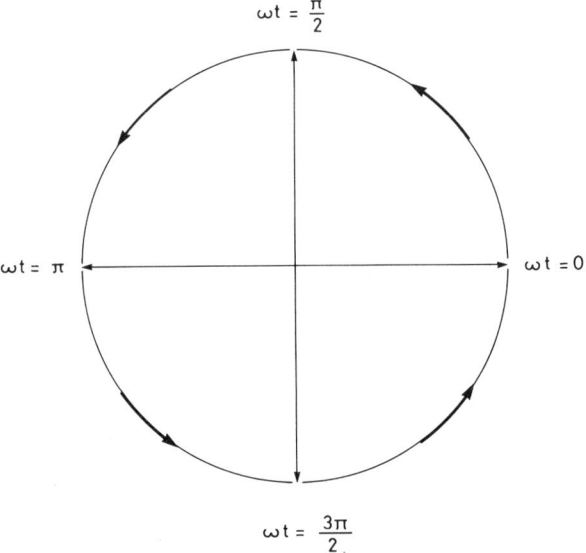

Fig. 1.7 Rotating vector

Exercise

Show that the phasor or complex vector form of the real time-varying vector

$$i_x \sin \omega t - i_y z \cos(\omega t + \phi) \tag{1.65}$$

is

$$-i_x j - i_y z e^{j\phi} \tag{1.66}$$

1.13 Gradient operator

In *potential theory* one needs to deal with the spatial rate of change of a scalar quantity h. The latter may be complex. Thus the *gradient* of h is defined by

Review of phasors and vectors

$$\text{grad } l = \mathbf{i}_x \frac{\partial h}{\partial x} + \mathbf{i}_y \frac{\partial h}{\partial y} + \mathbf{i}_z \frac{\partial h}{\partial z} \tag{1.67}$$

The gradient, which operates on a scalar h, is a vector with components $\partial h/\partial x$, $\partial h/\partial y$ and $\partial h/\partial z$. We also can write

$$\text{grad } h = \nabla h \tag{1.68}$$

where

$$\nabla = \mathbf{i}_x \frac{\partial}{\partial x} + \mathbf{i}_y \frac{\partial}{\partial y} + \mathbf{i}_z \frac{\partial}{\partial z} \tag{1.69}$$

is the vector operator *del*. Its components are $\partial/\partial x$, $\partial/\partial y$ and $\partial/\partial z$. It is important to remember that grad and ∇ are vector operators and have the form of vectors.

1.14 Divergence operator

The *flux* of a physical quantity flowing in or out of a small volume is characterised by the divergence. The *divergence* of a vector is defined by

$$\text{div } \mathbf{v} = \frac{\partial v_x}{\partial x} + \frac{\partial v_y}{\partial y} + \frac{\partial v_z}{\partial z} \tag{1.70}$$

which is a scalar. An equivalent statement is the dot product form

$$\text{div } \mathbf{v} = \nabla \cdot \mathbf{v} \tag{1.71}$$

where ∇ is the gradient operator defined above.

1.15 Curl operator

The vorticity or rotation of the physical field quantity is characterised by the *curl* of a vector $\mathbf{a}$. The curl operator is defined by

$$\text{curl } \mathbf{a} = \mathbf{i}_x \left(\frac{\partial a_z}{\partial y} - \frac{\partial a_y}{\partial z} \right) + \mathbf{i}_y \left(\frac{\partial a_x}{\partial z} - \frac{\partial a_z}{\partial x} \right)$$

$$+ \mathbf{i}_z \left(\frac{\partial a_y}{\partial x} - \frac{\partial a_x}{\partial y} \right) \tag{1.72}$$

An equivalent statement is

$$\text{curl } \mathbf{a} = \nabla \times \mathbf{a} \tag{1.73}$$

where ∇ is the gradient operator. For example, note that we can recover (1.72) by performing the operation

$$\nabla \times \mathbf{a} = \begin{vmatrix} \mathbf{i}_x & \mathbf{i}_y & \mathbf{i}_z \\ \frac{\partial}{\partial x} & \frac{\partial}{\partial y} & \frac{\partial}{\partial z} \\ a_x & a_y & a_z \end{vmatrix} \tag{1.74}$$

1.16 Two theorems

The *divergence theorem* says that for *any* vector A

$$\iiint_V \nabla \cdot \mathbf{A} \, dv = \iint_S \mathbf{A} \cdot \mathbf{i}_n \, da \tag{1.75}$$

where V is a volume bounded by a closed surface S. Here dv is an element of the volume, da is an element of area of the surface S and $\mathbf{i}_n$ is an outward-pointing normal unit vector.

Stokes's theorem takes the form

$$\iint_S (\nabla \times A) \cdot i_n \, ds = \oint_C A \cdot i_l \, dl \qquad (1.76)$$

where now the closed contour is capped by *any* surface S. Here i_l is a unit vector at the element dl on the contour C and ds is an element of area of the capping surface. The right-hand rule can be used to specify the direction of the contour (i.e. the fingers are in the direction of C while the thumb is in the direction of the normal i_n to the surface S).

Exercise

Consider the two-dimensional vector

$$v(t) = v_x(t) \, i_x + v_y(t) \, i_y$$

Represent the phasor form in the usual manner

$$V = V_x i_x + V_y i_y$$

where

$$v(t) = \text{Re } V \, e^{j\omega t}$$

For the five cases indicated below, plot the time history of the function $v(t)$ for one complete cycle for ωt from 0 to 2π:

(a) $V = i_x + j \, i_y$
(b) $V = i_x + 3 \, i_y$
(c) $V = i_x + 3j \, i_y$
(d) $V = 2 \exp(j\pi/4) \, i_x + 2 \exp(-j\pi/4) \, i_y$
(e) $V = i_x + 3 \exp(-j\pi/4) \, i_y$

Solutions and plots

It is perhaps useful to note that these plots would correspond to the trace on the face of a cathode ray tube if the signal $v_x(t)$ was applied to the horizontal plates and $v_y(t)$ was applied to the vertical plates. This is a concrete way to visualise the nature of a rotating vector.

On the plots, the arrows indicate the direction of increasing t.
(a) $v(t) = \cos \omega t \, i_x - \sin \omega t \, i_y$ See Fig. 1.8.
(b) $v(t) = \cos \omega t \, i_x + 3 \cos \omega t \, i_y$ See Fig. 1.9.
(c) $v(t) = \cos \omega t \, i_x - 3 \sin \omega t \, i_y$ See Fig. 1.10.
(d) $v(t) = 2[\cos(\omega t + \pi/4) \, i_x + \cos(\omega t - \pi/4) \, i_y]$ See Fig. 1.11.
(e) $v(t) = \cos \omega t \, i_x + 3 \cos(\omega t - \pi/4) \, i_y$ See Fig. 1.12.

1.17 Miscellaneous formulas from vector calculus

We list below explicit expressions for gradient, divergence, curl and Laplacian operators in the three most common co-ordinate systems.

In the rectangular (x, y, z) system we have the following:

$$\text{grad } V = i_x \frac{\partial V}{\partial x} + i_y \frac{\partial V}{\partial y} + i_z \frac{\partial V}{\partial z}$$

$$\text{div } J = \frac{\partial J_x}{\partial x} + \frac{\partial J_y}{\partial y} + \frac{\partial J_z}{\partial z}$$

$$\text{curl } A = i_x\left(\frac{\partial A_z}{\partial y} - \frac{\partial A_y}{\partial z}\right) + i_y\left(\frac{\partial A_x}{\partial z} - \frac{\partial A_z}{\partial x}\right)$$

$$+ i_z\left(\frac{\partial A_y}{\partial x} - \frac{\partial A_x}{\partial y}\right)$$

$$\nabla^2 \Phi = \frac{\partial^2 \Phi}{\partial x^2} + \frac{\partial^2 \Phi}{\partial y^2} + \frac{\partial^2 \Phi}{\partial z^2}$$

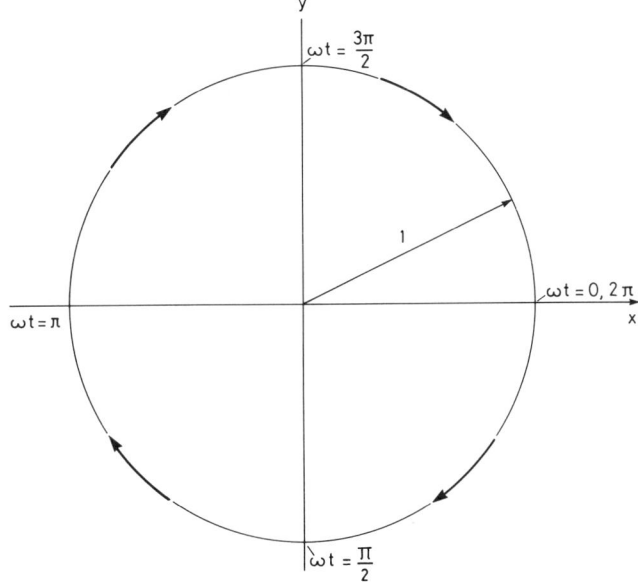

Fig. 1.8

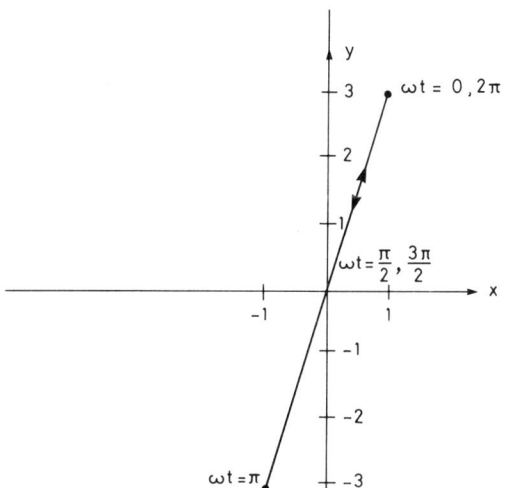

Fig. 1.9

14 Review of phasors and vectors

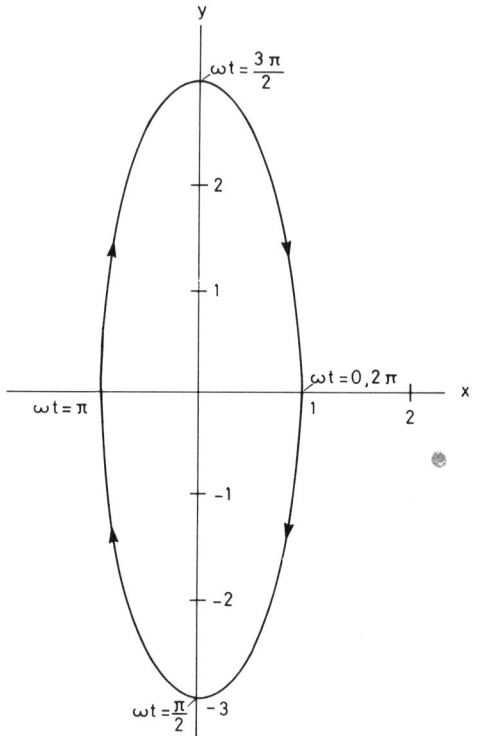

Fig. 1.10

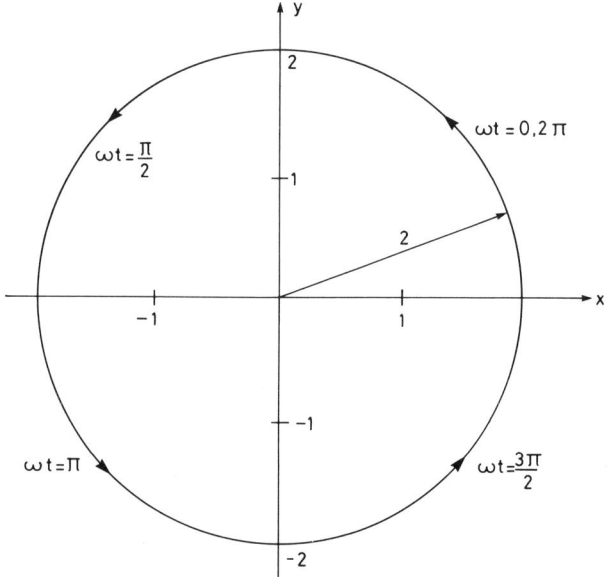

Fig. 1.11

In cylindrical co-ordinates (ϱ, ϕ, z) we have the following:

$$\text{grad } V = i_\varrho \frac{\partial V}{\partial \varrho} + i_\phi \frac{\partial V}{\varrho \partial \phi} + i_z \frac{\partial V}{\partial z}$$

$$\text{div } J = \frac{1}{\varrho} \frac{\partial}{\partial \varrho}(\varrho J_\varrho) + \frac{1}{\varrho} \frac{\partial J_\phi}{\partial \phi} + \frac{\partial J_z}{\partial z}$$

$$\text{curl } A = i_\varrho \left(\frac{1}{\varrho} \frac{\partial A_z}{\partial \phi} - \frac{\partial A_\phi}{\partial z} \right) + i_\phi \left(\frac{\partial A_\varrho}{\partial z} - \frac{\partial A_z}{\partial \varrho} \right)$$

$$+ i_z \left[\frac{1}{\varrho} \frac{\partial}{\partial \varrho}(\varrho A_\phi) - \frac{1}{\varrho} \frac{\partial A_\varrho}{\partial \phi} \right]$$

$$\nabla^2 \Phi = \frac{1}{\varrho} \left[\frac{\partial}{\partial \varrho} \left(\varrho \frac{\partial \Phi}{\partial \varrho} \right) \right] + \frac{1}{\varrho} \frac{\partial \Phi}{\partial \phi^2} + \frac{\partial^2 \Phi}{\partial z^2}$$

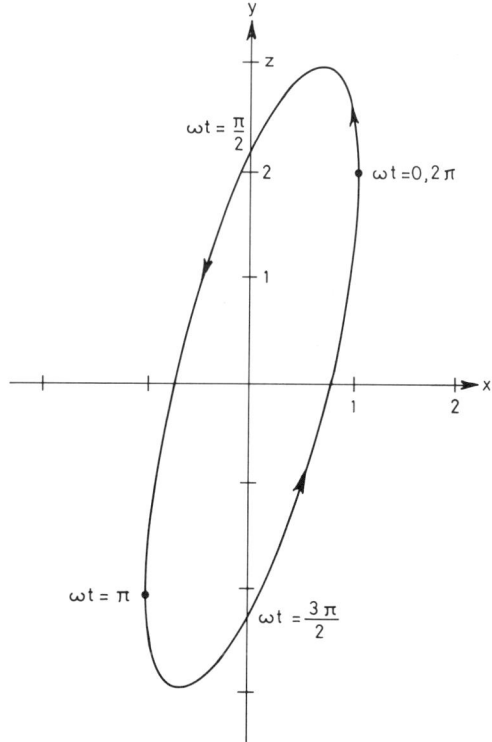

Fig. 1.12

In spherical co-ordinates (r, θ, ϕ) we have the following:

$$\text{grad } V = i_r \frac{\partial V}{\partial r} + i_\theta \frac{1}{r} \frac{\partial V}{\partial \theta} + i_\phi \frac{1}{r \sin \theta} \frac{\partial V}{\partial \phi}$$

$$\text{div } J = \frac{1}{r^2 \sin \theta} \left[\sin \theta \frac{\partial}{\partial r}(r^2 J_r) + r \frac{\partial}{\partial \theta}(\sin \theta J_\theta) + r \frac{\partial J_\phi}{\partial \phi} \right]$$

$$\text{curl } \mathbf{A} = \mathbf{i}_r \frac{1}{r \sin \theta} \left[\frac{\partial}{\partial \theta} (\sin \theta \, A_\phi) - \frac{\partial}{\partial \phi} A_\theta \right]$$

$$+ \mathbf{i}_\theta \frac{1}{r \sin \theta} \left[\frac{\partial}{\partial \phi} A_r - \sin \theta \frac{\partial}{\partial r} (r A_\phi) \right]$$

$$+ \mathbf{i}_\phi \frac{1}{r} \left[\frac{\partial}{\partial r} (r A_\theta) - \frac{\partial}{\partial \theta} A_r \right]$$

$$\nabla^2 \Phi = \frac{1}{r^2 \sin \theta} \left[\sin \theta \frac{\partial}{\partial r} \left(r^2 \frac{\partial \Phi}{\partial r} \right) + \frac{\partial}{\partial \theta} \left(\sin \theta \frac{\partial \Phi}{\partial \theta} \right) \right.$$

$$\left. + \frac{1}{\sin \theta} \frac{\partial^2 \Phi}{\partial \phi^2} \right]$$

Some useful relationships are:

$$\text{div curl } \mathbf{A} = 0$$

$$\text{curl grad } V = 0$$

$$\text{div } V\mathbf{J} = V \text{ div } \mathbf{J} + \mathbf{J} \cdot \text{grad } V$$

$$\text{curl } V\mathbf{A} = V \text{ curl } \mathbf{A} - \mathbf{A} \times \text{grad } V$$

$$\text{div } (\mathbf{A} \times \mathbf{J}) = \mathbf{J} \cdot \text{curl } \mathbf{A} - \mathbf{A} \cdot \text{curl } \mathbf{J}$$

If $\mathbf{A} = \mathbf{i}_z A_z$, we have

$$\text{curl curl } \mathbf{A} = \text{grad } \frac{\partial A_z}{\partial z} - \mathbf{i}_z \left(\frac{\partial^2 A_z}{\partial z^2} + \frac{\partial^2 A_z}{\partial y^2} + \frac{\partial^2 A_z}{\partial x^2} \right)$$

Chapter 2

Quasi-static fields

2.1. Current flow in conductive media and Ohm's law

As a starting point for our treatment of electromagnetic fields, we will review certain concepts in static electricity. Let us deal with a region that has an electrical *conductivity* σ in siemens per metre (S/m). At some point in this region we can invoke Ohm's law, which says

$$e_0 = (1/\sigma) j_0 \tag{2.1}$$

where e_0 is the *electric field* vector and j_0 is the *current density* vector. Under truly static conditions e_0 and j_0 are real vectors that do not vary with time.

We now wish to generalize eqn. 2.1 to allow us to characterise time-varying fields. Without loss of generality we will deal with harmonically varying quantitites at frequency $f = \omega/2\pi$. In accordance with our discussion in Chapter 1, we note that

$$e(t) = \text{Re}\,(E\,e^{j\omega t}) \tag{2.2}$$

and

$$j(t) = \text{Re}\,(J\,e^{j\omega t}) \tag{2.3}$$

A generalisation of Ohm's law from the static theory is simply

$$J = (\sigma + j\varepsilon\omega)\,E \tag{2.4}$$

where $\sigma + j\varepsilon\omega$ can be described as a *complex conductivity* or *admittivity*. Here, of course, σ and ε are the *conductivity* and *permittivity* of the medium. In the case where σ and ε do not vary with frequency, we note that Ohm's Law in the time domain is

$$j(t) = \left(\sigma + \varepsilon\frac{\partial}{\partial t}\right) e(t) \tag{2.5}$$

The right-hand side may be decomposed into

$$j(t) = j_c(t) + j_d(t) \tag{2.6}$$

where

$$j_c(t) = \sigma\,e(t) \tag{2.7}$$

is the *conduction current density* and

$$j_d(t) = \varepsilon\frac{\partial}{\partial t} e(t) \tag{2.8}$$

is the *displacement current density*. The corresponding decomposition of the complex vector is

$$\boldsymbol{J} = \boldsymbol{J}_c + \boldsymbol{J}_d \qquad (2.9)$$

where $\boldsymbol{J}_c = \sigma \boldsymbol{E}$ and $\boldsymbol{J}_d = j\varepsilon\omega \boldsymbol{E}$. An equivalent statement for the latter is $\boldsymbol{J}_d = j\omega \boldsymbol{D}$, where $\boldsymbol{D} = \varepsilon \boldsymbol{E}$ is the displacement vector.

Here we should note that since $\nabla \cdot (\boldsymbol{J}_c + \boldsymbol{J}_d) = 0, \nabla \cdot \boldsymbol{J}_d = -\nabla \cdot \boldsymbol{J}_c \neq 0$. In fact $\nabla \cdot \boldsymbol{J}_c = -j\omega \varrho_e$, where ϱ_e is the *induced charge density*. An equivalent statement is $\nabla \cdot \boldsymbol{D} = \varrho_e$, which is *Poisson's equation*.

2.2 The lossy capacitor

To provide a physical understanding of the process, consider the lossy capacitor or impedor depicted in Fig. 2.1. Essentially here we have selected a rectangular slice of the medium of cross-sectional area A and thickness d. We insert this block into an ideal capacitor with metal plates of area A. Then, neglecting any fringing effects, the impedance Z is given by

$$Z = \frac{V}{I} = \frac{1}{\sigma + j\varepsilon\omega} \frac{d}{A} \qquad (2.10)$$

where the phasors V and I are the voltage and current at the terminals. (We have also ignored any polarisation effects at the metal plates and in the contained region.)

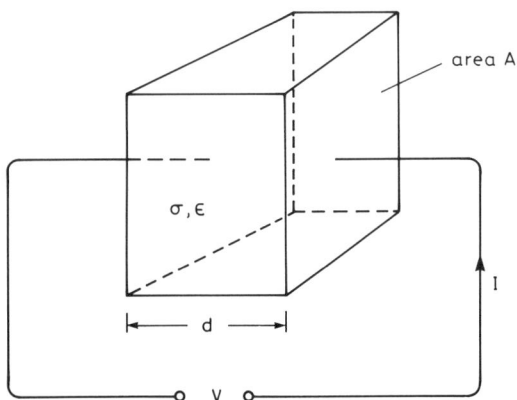

Fig. 2.1 An ideal impedor or lossy capacitor

Actually we can write eqn. 2.10 more conveniently in terms of the admittance Y of the device:

$$Y = \frac{I}{V} = G + j\omega C \qquad (2.11)$$

where

$$G = \sigma A/d \text{ siemens (S)} \qquad (2.12)$$

$$C = \varepsilon A/d \text{ farads (F)} \qquad (2.13)$$

Also note that if we orient the x axis to be perpendicular to the plates, we see that

$$I = J_x A \qquad (2.14)$$

Quasi-static fields

and
$$V = E_x d \tag{2.15}$$

Thus it is clear that eqns. 2.10 or 2.11 are equivalent to
$$J_x = (\sigma + j\varepsilon\omega)E_x \tag{2.16}$$
which is consistent with eqn. 2.4.

It is important to note that in the above example we are dealing with potential concepts. We have assumed that all significant dimensions of the problem are small compared with the wavelength. Obviously this restriction is less stringent as the frequency is lowered. This is precisely what is meant by the term *quasi-static* in the title of the chapter.

2.3 Laplace's equation

We now return to the general volume conductor and deal with a time-harmonic electric field $e(t)$. We assume that this vector can be derived from the gradient of a scalar function $\phi(t)$. Thus, with a prefixed minus sign, we write
$$e(t) = -\operatorname{grad} \phi(t) \tag{2.17}$$
In terms of a complex vector E and a phasor scalar Φ, we have
$$E = -\operatorname{grad} \Phi = -\nabla\Phi \tag{2.18}$$
where Φ is a phasor defined such that
$$\phi(t) = \operatorname{Re}(\Phi\, e^{j\omega t})$$
In terms of the potential Φ the current density J is obtained from Ohm's Law, as given by eqn. 2.16. Thus
$$J = -(\sigma + j\varepsilon\omega)\nabla\Phi \tag{2.19}$$
To obtain the governing equation for the potential Φ we now invoke a fundamental property of current flow in regions where there are *no sources*. Specifically we say that
$$\operatorname{div} J = 0 \text{ or } \nabla \cdot J = 0 \tag{2.20}$$
which is a statement that the current has no divergence (i.e. what flows into a small volume, flows out). Thus on using eqn. (2.19) in eqn. 2.20 we find that
$$\nabla \cdot [(\sigma + j\varepsilon\omega)\nabla\Phi] = 0 \tag{2.21}$$
This equation is vastly simplified when we assume that σ and ε do not depend on x, y, z. Then, of course, the potential satisfies
$$\nabla \cdot \nabla\Phi = 0$$
This is usually written
$$\nabla^2 \Phi = 0 \tag{2.22}$$
and is called *Laplace's equation*. ∇^2 is the *Laplacian operator*. In rectangular co-ordinates,
$$\left(\frac{\partial^2}{\partial x^2} + \frac{\partial^2}{\partial y^2} + \frac{\partial^2}{\partial z^2}\right)\Phi = 0 \tag{2.23}$$

Exercise

Let
$$j(t) = i_x j_x(t) \tag{2.24}$$
where $j_x(t) = j_0 u(t)$ and $u(t)$ is the unit step function at $t = 0$. Then, assuming that eqn. 2.16 is valid, show that
$$e_x(t) = (j_0/\sigma)\left[1 - \exp\left(-\frac{\sigma}{\varepsilon}t\right)\right]u(t) \tag{2.25}$$
Here $e_x(0) = 0$ and $e_x(\infty) = j_0/\sigma$.

Solution

We are given
$$j_x(t) = j_0 u(t) \qquad u(t) = 1 \text{ for } t > 0$$
$$= 0 \text{ for } t < 0$$

Then
$$\mathscr{L} j_x(t) = j_0/s = J_x(s)$$
where $\mathscr{L}$ is the *Laplace transform operator*. But
$$\mathscr{L} e_x(t) = E_x(s) = \int_0^\infty e^{-st} e_x(t)\,dt$$
and Ohm's law says that
$$E_x(s) = (\sigma + \varepsilon s)^{-1} J_x(s)$$

Therefore
$$e_x(t) = \mathscr{L}^{-1} E_x(s) = j_0 \mathscr{L}^{-1} \frac{1}{s(\sigma + \varepsilon s)}$$
$$= \frac{j_0}{\sigma} \mathscr{L}^{-1}\left[\frac{1}{s} - \frac{1}{s + (\sigma/\varepsilon)}\right] = \frac{j_0}{\sigma}(1 - e^{-\sigma t/\varepsilon})u(t)$$
where $\mathscr{L}^{-1}$ is the inverse Laplace transform operator.

2.4 Current point source

We now consider a very simple but important source problem (see Fig. 2.2). We locate a point source of current I amperes with the usual time factor $\exp(j\omega t)$ in an unbounded homogeneous region of complex conductivity $\sigma + j\varepsilon\omega$. The rectangular co-ordinate system (x, y, z) is chosen with the origin at the point source. Now the current density emanating from the source is symmetric and, at a distance r, it is given by
$$\mathbf{J} = J_r \mathbf{i}_r \tag{2.26}$$
where $\mathbf{i}_r$ is a unit vector in the r direction. Now clearly
$$J_r = I/4\pi r^2 \tag{2.27}$$
Then the corresponding electric field is given by
$$\mathbf{E} = E_r \mathbf{i}_r \tag{2.28}$$
where, in accordance with Ohm's law,
$$E_r = \frac{J_r}{\sigma + j\varepsilon\omega} = \frac{I}{4\pi(\sigma + j\varepsilon\omega)r^2} \tag{2.29}$$

with
$$r^2 = x^2 + y^2 + z^2$$

The reader may now readily verify that

$$E_x = \frac{Ix}{4\pi(\sigma + j\varepsilon\omega)r^3} \qquad (2.30)$$

$$E_y = \frac{Iy}{4\pi(\sigma + j\varepsilon\omega)r^3} \qquad (2.31)$$

$$E_z = \frac{Iz}{4\pi(\sigma + j\varepsilon\omega)r^3} \qquad (2.32)$$

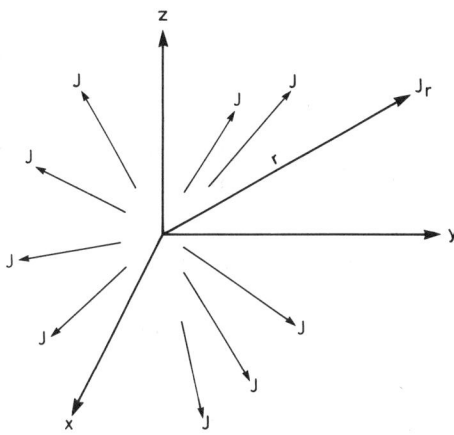

Fig. 2.2 Current emanating radially from a point source of I amperes in an unbounded homogeneous medium

The potential Φ that generates these field components is simply

$$\Phi = \frac{I}{4\pi(\sigma + j\varepsilon\omega)r} \qquad (2.33)$$

For example, note that
$$E_x = -\partial\Phi/\partial x$$
and
$$\frac{\partial}{\partial x}\frac{1}{r} = -\frac{x}{r^3}$$

Exercise Show that div $J = 0$ in a homogeneous medium for a current point source of current I. Assume a distance from observer to source of $r > 0$ and work in rectangular co-ordinates.

Exercise Locate the current point source on the surface of a homogeneous flat earth (i.e. a half-space), as indicated in Fig. 2.3. Show that the corresponding potential at any point in, or on the surface of, the half-space is given by

$$\Phi = I/2\pi(\sigma + j\varepsilon\omega)r \qquad (2.34)$$

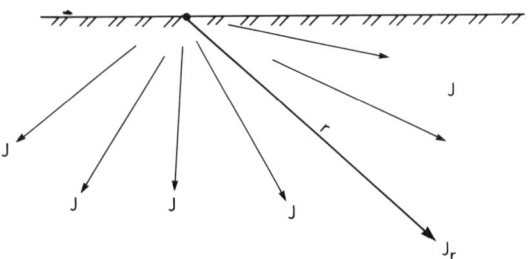

Fig. 2.3 Current emanating radially from a point source of I amperes on the surface of a homogeneous half-space. The upper half-space is an insulator

Then show that the normal current density in the conductor, as we approach the interface, vanishes. Assume that the current density in the air is zero.

Solution

For a given current I, the current density J_r in the lower half-space is doubled. Thus $J_r = I/2\pi r^2$, where $2\pi r^2$ is the area of a hemispherical bowl of radius r. Thus $E_r = (\sigma + j\omega\varepsilon)^{-1} J_r$. Then it follows that the required potential is

$$\Phi = I/2\pi(\sigma + j\omega\varepsilon)r$$

as can be verified by noting that $E_r = -\partial\Phi/\partial r$. Furthermore, the normal current density is given by

$$J_z = (\sigma + j\omega\varepsilon)E_z = -(\sigma + j\omega\varepsilon)\frac{\partial\Phi}{\partial z} = \frac{Iz}{2\pi r^3},$$

Now clearly $J_z = 0$ as $z \to 0$.

Exercise

With respect to a rectangular co-ordinate system (x, y, z), locate a current point source I amperes at $z = +l/2$ and a current point sink at $z = -l/2$ (Fig. 2.4). Then assuming that the medium is unbounded and homogeneous with an admittivity $\sigma + j\varepsilon\omega$, show that the resultant potential of the source and sink (i.e. a bipole) is given by

$$\psi = \frac{I}{4\pi(\sigma + j\varepsilon\omega)}\left(\frac{1}{r_+} - \frac{1}{r_-}\right) \qquad (2.35)$$

where

$$r_+ = \left[x^2 + y^2 + \left(z - \frac{l}{2}\right)^2\right]^{1/2} \qquad (2.36)$$

$$r_- = \left[x^2 + y^2 + \left(z + \frac{l}{2}\right)^2\right]^{1/2} \qquad (2.37)$$

Solution   The potential of the source I is

$$\psi_+ = I/4\pi(\sigma + j\omega\varepsilon)r_+$$

and the potential of the sink is

$$\psi_- = -I/4\pi(\sigma + j\omega\varepsilon)r_-$$

Thus

$$\psi = \psi_+ + \psi_-$$

which is the result given by eqn. 2.35.

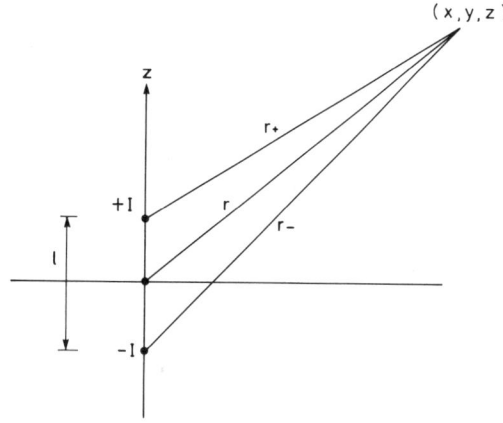

Fig. 2.4 Current point source at $z = l/z$ and a current point sink at $z = -l/z$, both on the z axis. The combination is called a bipole. The medium is homogeneous and unbounded

Exercise   In the previous exercise, let $l \to 0$ and $I \to \infty$ such that Il remains finite. Then show that the potential of this dipole, as in Fig. 2.5, is

$$\psi = (Il)z/4\pi(\sigma + j\varepsilon\omega)r^3 \tag{2.38}$$

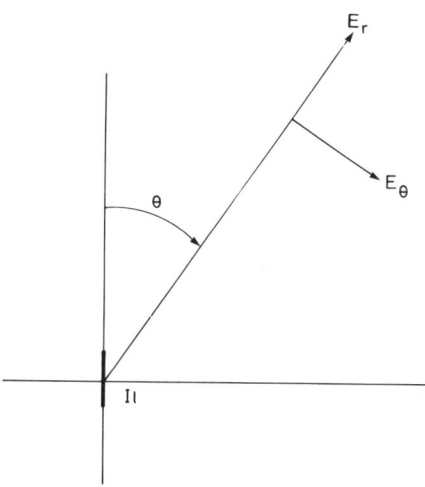

Fig. 2.5 Limiting case of the bipole when the distance becomes small compared with r. We now have a dipole

Hint: note that

$$1/r_+ = [x^2 + y^2 + z^2 - zl + (l/2)^2]^{-1/2}$$

$$\simeq \frac{1}{(r^2 - zl)^{1/2}} = \frac{1}{r}\left(1 - \frac{zl}{r^2}\right)^{-1/2}$$

$$\simeq \frac{1}{r}\left(1 + \frac{zl/2}{r^2}\right) = \frac{1}{r} + \frac{zl/2}{r^3} \qquad (2.39)$$

Solution

In analogy to eqn. 2.39, we have

$$1/r_- \simeq 1/r - zl/2r^3$$

Thus

$$\psi = -\frac{I}{4\pi(\sigma + j\omega\varepsilon)}\left(\frac{1}{r_+} - \frac{1}{r_-}\right)$$

$$\simeq \frac{I}{4\pi(\sigma + j\omega\varepsilon)}\left(\frac{1}{r} + \frac{zl/2}{r^3} - \frac{1}{r} + \frac{zl/2}{r^3}\right)$$

$$\simeq \frac{Ilz}{4\pi(\sigma + j\omega\varepsilon)r^3}$$

Exercise

Derive expressions for the field components using $E_x = -\partial\psi/\partial x$, $E_y = -\partial\psi/\partial y$, $E_z = -\partial\psi/\partial z$.

Solution

The expressions follow directly from eqn. 2.38. We make use of

$$\frac{\partial}{\partial x}\frac{1}{r} = \frac{\partial}{\partial r}\left(\frac{1}{r}\right)\frac{\partial r}{\partial x} = -\frac{1}{r^2}\frac{x}{r} = -\frac{x}{r^3}$$

Similarly

$$\frac{\partial}{\partial y}\frac{1}{r} = -\frac{y}{r^3} \quad \text{and} \quad \frac{\partial}{\partial z}\frac{1}{r} = -\frac{z}{r^3}$$

Then also note that

$$\frac{\partial^2}{\partial z^2}\frac{1}{r} = \frac{\partial}{\partial z}\left(-\frac{z}{r^3}\right) = -\frac{1}{r^3} + \frac{3z}{r^4}\frac{\partial r}{\partial z} = -\frac{1}{r^3} + \frac{3z^2}{r^5}$$

These lead to

$$E_x = \frac{(Il)3xz}{4\pi(\sigma + j\varepsilon\omega)r^5} \qquad (2.40)$$

$$E_y = \frac{(Il)3\,yz}{4\pi(\sigma + j\varepsilon\omega)r^5} \qquad (2.41)$$

$$E_z = \frac{Il}{4\pi(\sigma + j\varepsilon\omega)}\left(\frac{3z^2}{r^5} - \frac{1}{r^3}\right) \qquad (2.42)$$

Exercise

Choosing spherical co-ordinates (r, θ, ϕ) such that $x = r \sin \theta \cos \phi$, $y = r \sin \theta \sin \phi$ and $z = r \cos \theta$, show that the field components of the dipole are

$$E_r = Il \cos \theta / 2\pi(\sigma + j\varepsilon\omega)r^3 \tag{2.43}$$

$$E_\theta = Il \sin \theta / 4\pi(\sigma + j\varepsilon\omega)r^3 \tag{2.44}$$

Hint: as indicated in Fig. 2.6, note that $\cos \theta = z/r$. In cylindrical co-ordinates (ϱ, ϕ, z) we would have $E_\varrho^2 = E_x^2 + E_y^2$, so that

$$E_\varrho = Il\, 3\varrho z / 4\pi(\sigma + j\varepsilon\omega)r^5 \tag{2.45}$$

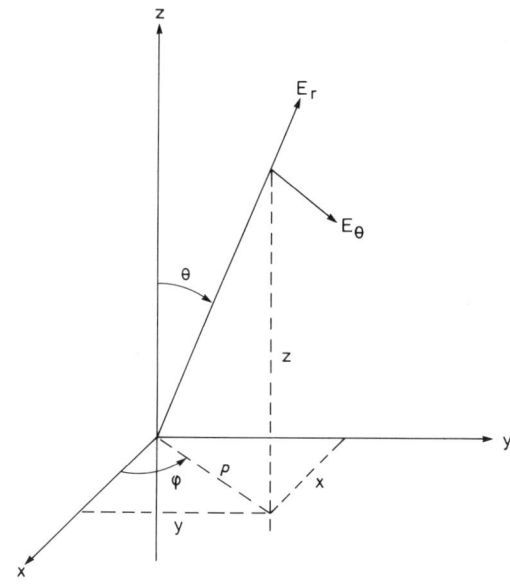

Fig. 2.6 Co-ordinate systems (x, y, z), (ρ, φ, z) and (r, θ, φ), and the electric field components in the r and θ directions

where $r^2 = (\varrho^2 + z^2)$. Then

$$E_r = E_\varrho \sin \theta + E_z \cos \theta \tag{2.46}$$

and

$$E_\theta = E_\varrho \cos \theta - E_z \sin \theta \tag{2.47}$$

as depicted in Figs. 7a and 7b.

Solution

Here we take the expressions for E_x, E_y and E_z in rectangular co-ordinates given by eqns. 2.40, 2.41 and 2.42 to get forms for E_ϱ and E_z in cylindrical co-ordinates. Of course, E_z is the same in both rectangular and cylindrical co-ordinates. Also we note that $E_\phi = 0$. The final step is to insert E_ϱ and E_z into eqns. 2.46 and 2.47 to yield E_r and E_θ in the spherical system. Here we make use of the trigonometric identities $\varrho = r \sin \theta$, $z = r \cos \theta$ and $\sin^2 \theta + \cos^2 \theta = 1$.

2.5 Transition to electrostatics

We have been dealing with harmonically varying fields in a medium of conductivity σ and permittivity ε. We now consider the limit where $\sigma \to 0$ and, at the same time, we recognise that I is really the rate of change of charge q, that is $I = j\omega q$. Then, for example, the potential Φ of the point charge from eqn. 2.33 becomes

$$\Phi = q/4\pi r\varepsilon \tag{2.48}$$

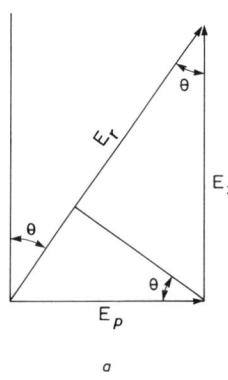
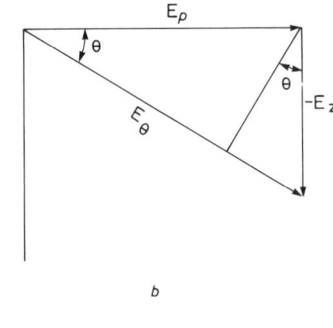

Fig. 2.7

The corresponding expressions for the electric field components are

$$E_x = qx/4\pi\varepsilon r^3 \tag{2.49}$$

$$E_y = qy/4\pi\varepsilon r^3 \tag{2.50}$$

$$E_z = qz/4\pi\varepsilon r^3 \tag{2.51}$$

The radial component is radially symmetric and given by

$$E_r = q/4\pi\varepsilon r^2 \tag{2.52}$$

The dielectric displacement vector, defined by $\mathbf{D} = \varepsilon \mathbf{E}$ (see Section 2.1), has only a radial component and, for this case,

$$D_r = q/4\pi r^2 \tag{2.53}$$

We will not pursue the subject of electrostatics at this stage, except to note that it can be regarded as the low-frequency limit of the time-harmonic formulation.

Exercise

Show that the electrostatic potential ϕ of two charges $+q$ and $-q$ separated by an infinitesimal distance l is given by

$$\phi = ql \cos\theta / 4\pi\varepsilon r^2 \tag{2.54}$$

Here (r, θ) are the spherical co-ordinates of the observer when the dipole is located at the origin and oriented in the $\theta = 0$ direction.

Solution

This is precisely the same situation as in one of the exercises in Section 2.4. Thus, in eqn. 2.35, replace I by $j\omega q$ and $\sigma + j\omega\varepsilon$ by $j\omega\varepsilon$, where the $j\omega$ cancel. The result is given by eqn. 2.38 with the same substitutions, noting of course that $z = r \cos\theta$.

Quasi-static fields

Exercise

If $E = -\text{grad } \Phi$, where Φ is a potential, show that curl $E = 0$ by expanding the result in rectangular co-ordinates.

Solution

Here Φ can be any scalar potential point function. Then, noting that $E_x = -\partial\Phi/\partial x$, $E_y = -\partial\Phi/\partial y$ and $E_z = -\partial\Phi/\partial z$,

$$\text{curl } E = \nabla \times E = \begin{vmatrix} i_x & i_y & i_z \\ \dfrac{\partial}{\partial x} & \dfrac{\partial}{\partial y} & \dfrac{\partial}{\partial z} \\ -\dfrac{\partial\Phi}{\partial x} & -\dfrac{\partial\Phi}{\partial y} & -\dfrac{\partial\Phi}{\partial z} \end{vmatrix}$$

$$= i_x\left(-\frac{\partial^2\Phi}{\partial y \partial z} + \frac{\partial^2\Phi}{\partial y \partial z}\right) + \cdots$$

$$= i_x(0) + i_y(0) + i_z(0) = 0$$

Exercise

Using known vector identities, show that eqn. 2.21 can be written generally in the form

$$\nabla\Phi \cdot \nabla(\sigma + j\varepsilon\omega) + (\sigma + j\varepsilon\omega)\nabla^2\Phi = 0$$

This result can also be proved by expanding eqn. 2.21 in rectangular co-ordinates.

Solution

Now, starting with eqn. 2.21, we write

$$\nabla \cdot [(\sigma + j\omega\varepsilon)\nabla\Phi] = \nabla \cdot (\hat{\sigma}\nabla\Phi)$$

$$= \left(i_x \frac{\partial}{\partial x} + i_y \frac{\partial}{\partial y} + i_z \frac{\partial}{\partial z}\right) \cdot \left(\hat{\sigma}\, i_x \frac{\partial\Phi}{\partial x} + \hat{\sigma}\, i_y \frac{\partial\Phi}{\partial y} + \hat{\sigma}\, i_z \frac{\partial\Phi}{\partial z}\right)$$

$$= \frac{\partial}{\partial x}\left(\hat{\sigma}\frac{\partial\Phi}{\partial x}\right) + \frac{\partial}{\partial y}\left(\hat{\sigma}\frac{\partial\Phi}{\partial y}\right) + \frac{\partial}{\partial z}\left(\hat{\sigma}\frac{\partial\Phi}{\partial z}\right)$$

$$= \frac{\partial\hat{\sigma}}{\partial x}\frac{\partial\Phi}{\partial x} + \hat{\sigma}\frac{\partial^2\Phi}{\partial x^2} + \frac{\partial\hat{\sigma}}{\partial y}\frac{\partial\Phi}{\partial y} + \hat{\sigma}\frac{\partial^2\Phi}{\partial y^2} + \frac{\partial\hat{\sigma}}{\partial z}\frac{\partial\Phi}{\partial z} + \hat{\sigma}\frac{\partial^2\Phi}{\partial z^2}$$

$$= \frac{\partial\hat{\sigma}}{\partial x}\frac{\partial\Phi}{\partial x} + \frac{\partial\hat{\sigma}}{\partial y}\frac{\partial\Phi}{\partial y} + \frac{\partial\hat{\sigma}}{\partial z}\frac{\partial\Phi}{\partial z} + \hat{\sigma}\left(\frac{\partial^2\Phi}{\partial x^2} + \frac{\partial^2\Phi}{\partial y^2} + \frac{\partial^2\Phi}{\partial z^2}\right)$$

$$= \nabla\hat{\sigma} \cdot \nabla\Phi + \hat{\sigma}\nabla^2\Phi$$

where $\hat{\sigma} = \sigma + j\omega\varepsilon$.

Since this is true in rectangular co-ordinates, it must be true in any co-ordinate system.

2.6 Magnetostatics

As a prelude to discussing Maxwell's equations, we introduce magnetic induction for time-harmonic fields. Thus, in analogy to Ohm's law as given in eqn. 2.4 for phasors, we have

$$j\omega B = j\mu\omega H \tag{2.55}$$

(or simply $\boldsymbol{B} = \mu \boldsymbol{H}$), where $\boldsymbol{B}$ is the *magnetic induction*, $\boldsymbol{H}$ is the *magnetic field strength* and μ is the *magnetic permeability*. Our implicit assumption here is that μ is a constant of proportionality between $\boldsymbol{B}$ and $\boldsymbol{H}$ which is independent of the strength and direction of the field.

Another basic property that we invoke is

$$\text{div } \boldsymbol{B} = 0 \tag{2.56}$$

at any point in the region (except possibly at the source). This non-divergence property is analogous to its counterpart div $\boldsymbol{J} = 0$ (eqn. 2.20).

We now again assume that the frequency is sufficiently low that quasistatic conditions prevail. Thus we assert that $\boldsymbol{H}$ can be derived from a gradient of a potential U, that is

$$\boldsymbol{H} = -\text{grad } U \tag{2.57}$$

or

$$\boldsymbol{H} = -\nabla U \tag{2.58}$$

By combining eqns. 2.55, 2.56 and 2.58, we deduce that

$$\nabla^2 U = 0 \tag{2.59}$$

where ∇^2 is the Laplacian operator defined by eqn. 2.22. Here we have also assumed that μ does not depend on position (i.e. we are dealing with a homogeneous region).

Exercise

If μ is position dependent, show that

$$\nabla U \cdot \nabla \mu + \mu \nabla^2 U = 0 \tag{2.60}$$

Solution

The logic is identical to the last exercise in Sec. 2.5. Now we start with $\nabla \cdot \boldsymbol{B} = 0$. Since $\boldsymbol{B} = \mu \boldsymbol{H}$ and $\boldsymbol{H} = \nabla U$, we can see immediately that eqn. 2.60 follows.

As we may observe, there is a close analogy between static-like electric fields and static-like magnetic fields. This is manifested in the close correspondence of eqns. 2.4, 2.18 and 2.22 with eqns. 2.55, 2.57 and 2.59, respectively.

In fact we could postulate the existence of a single magnetic pole and deduce the radial flux B_r in the same manner as for the point source of electric current. Such isolated magnetic poles are artificial, but a pair of such poles with equal and opposite sign is a realisable configuration. In the case where the charge separation is infinitesimal, we have the following expression for the magnetic potential ψ_m:

$$\psi_m = \frac{(j\omega q_m)lz}{4\pi j \mu \omega r^3} = \frac{q_m lz}{4\pi \mu r^3} \tag{2.61}$$

Here $j\omega q_m$ is the rate of change of the magnetic charge. This particular configuration is called the *magnetic dipole* and has *magnetic moment* $q_m l$. We have omitted the explicit derivation of eqn. 2.61, but it is clear that the result is fully analogous to the electrostatic counterpart given by eqn. 2.38.

From a practical viewpoint we are interested in the magnetic fields of distributions of electric current such as carried by a wire or cable. We can get at this in a simple manner by working with the known electric field expressions of an electric dipole of infinitesimal *current element Il* located at the origin. Thus, on using eqn. 2.43 we see that the radial current density is given by

$$J_r = (\sigma + j\varepsilon\omega)E_r = Il\cos\theta'/2\pi r^3 \tag{2.62}$$

where θ' denotes the polar angle. Now we apply *Ampère's law* to the circular path of radius $r\sin\theta$ about the vertical axis of the current element. The situation is illustrated in Fig. 2.8. To be specific, Ampère's law states that

$$\oint H_\phi \, ds = \iint_C J_r \, da \tag{2.63}$$

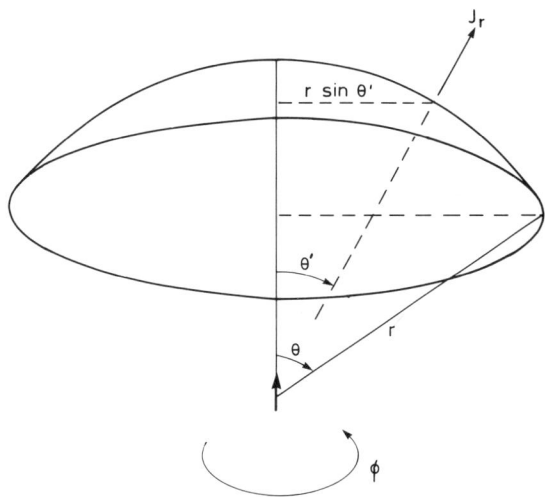

Fig. 2.8 Illustrating Ampère's law applied to a spherical capping surface of radius r for a current element at the origin

where the quantity on the left-hand side is the line integral of the magnetic field around the closed contour. The quantity on the right-hand side is the surface integral of the radial current density over the capping surface C. Now in terms of the spherical co-ordinates (r, θ, ϕ) shown in Fig. 2.8, we see that eqn. 2.63 is equivalent to

$$\int_0^{2\pi} H_\phi r \sin\theta \, d\phi = \int_{\phi=0}^{2\pi}\int_0^\theta J_r r^2 \sin\theta' \, d\theta' \, d\phi \tag{2.64}$$

where J_r is given by eqn. 2.62.

Because of obvious ϕ symmetry, the ϕ integration is trivial and amounts to multiplying both sides by 2π. The θ integration can be carried out by using

$$\int_0^\theta \cos\theta' \sin\theta' \, d\theta' = \frac{\sin^2\theta}{2} \tag{2.65}$$

thus we find that

$$H_\phi = Il\sin\theta/4\pi r^2 \tag{2.66}$$

In terms of rectangular co-ordinates,

$$\boldsymbol{H} = -\boldsymbol{i}_x H_\phi \sin \phi + \boldsymbol{i}_y H_\phi \cos \phi \qquad (2.67)$$

$$= \frac{Il\varrho}{4\pi r}\left(-\boldsymbol{i}_x \frac{1}{r^2}\frac{y}{\varrho} + \boldsymbol{i}_y \frac{1}{r^2}\frac{x}{\varrho}\right) \qquad (2.68)$$

This result can be reduced to the form

$$\boldsymbol{H} = \frac{I}{4\pi}\frac{\boldsymbol{l} \times \boldsymbol{r}}{r^3} \qquad (2.69)$$

where
$$\boldsymbol{l} = \boldsymbol{i}_z l$$
$$\boldsymbol{r} = \boldsymbol{i}_x x + \boldsymbol{i}_y y + \boldsymbol{i}_z z$$

Equation 2.69 is really another form of Ampère's law; it is independent of the co-ordinate system when written in terms of vectors. Note that here $\boldsymbol{l}$ is a vector of length l and $\boldsymbol{r}$ is a vector of length r. The tip of $\boldsymbol{r}$ is (x, y, z), being the co-ordinates of the observer. Also we should remember that $r \gg l$ for the dipole idealisation (i.e. l is effectively infinitesimal).

Exercise

Instead of an infinitesimal current element or dipole source, consider a linear current of I amperes that flows from $z' = -L/2$ to $L/2$ along the z axis, as indicated in Fig. 2.9. For convenience we have used the primed co-ordinate: z' to distinguish it from the axial co-ordinate z of the

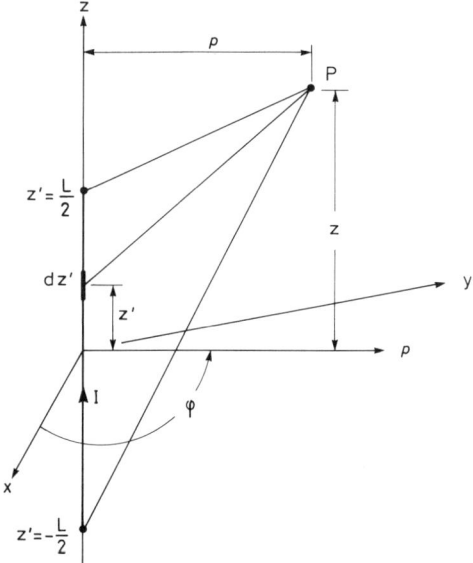

Fig. 2.9 Geometry for calculating the magnetic field of a linear wire of length L carrying a uniform current I

observer. Show that the magnetic field for this configuration at P is given by

$$\boldsymbol{H} = \boldsymbol{i}_\phi H_\phi$$

Quasi-static fields

where

$$H_\phi = \frac{I\varrho}{4\pi} \int_{-L/2}^{L/2} \frac{dz'}{[x^2 + y^2 + (z - z')^2]^{3/2}} \qquad (2.70)$$

Carry out the integration to get

$$H_\phi = \frac{I}{4\pi\varrho} \left(\frac{z + (L/2)}{\{\varrho^2 + [z + (L/2)]^2\}^{1/2}} - \frac{z - L/2}{\{\varrho^2 + [z - (L/2)]^2\}^{1/2}} \right) \qquad (2.71)$$

Hint:

$$\frac{\partial}{\partial z} \frac{z}{(\varrho^2 + z^2)^{1/2}} = \frac{\varrho^2}{(\varrho^2 + z^2)^{3/2}}$$

Now let the length of the current-carrying wire be infinite (i.e. $L \to \infty$), and show that

$$H_\phi]_{L \to \infty} = \frac{I}{4\pi\varrho}[+1 - (-1)] = \frac{I}{2\pi\varrho} \qquad (2.72)$$

This result is consistent with a statement of Ampère's law; $2\pi\varrho H_\phi$, the integral of H_ϕ around the closed circuit $2\pi\varrho$, is equal to I. Finally, show that in the limit $L \ll (\varrho^2 + z^2)^{1/2}$, our general result given by eqn. 2.71 reduces to the dipole form

$$H_\phi \simeq IL\varrho/4\pi r^3 \qquad (2.73)$$

Solution

With reference to Fig. 2.9, we merely integrate the individual dipole elements $I\,dz'$ over the length of the antenna from $z' = -L/2$ to $+L/2$ to give eqn. 2.71. The limit given by eqn. 2.72 follows directly from eqn. 2.71 when we let L tend to infinity in both numerator and denominator. To obtain the dipole limit, we deal with eqn. 2.71 as follows:

$$\frac{z + L/2}{\left[\varrho^2 + \left(z + \frac{L}{2}\right)^2\right]^{1/2}} = \frac{z + L/2}{\left[\varrho^2 + z^2 + zL + \left(\frac{L}{2}\right)^2\right]^{1/2}}$$

$$\simeq \frac{z + L/2}{(r^2 + zL)^{1/2}} = \frac{z + L/2}{r\left(1 + \frac{zL}{r^2}\right)^{1/2}} = \frac{z + L/2}{r}\left(1 + \frac{zL}{r^2}\right)^{-1/2}$$

$$\simeq \frac{z + L/2}{r}\left(1 - \frac{zL}{2r^2}\right) = \frac{z + L/2}{r} - \frac{(z + L/2)zL}{2r^3}$$

$$= \frac{z}{r} + \frac{L}{2r} - \frac{z^2 L}{2r^3} - \frac{zL^2}{2r^3} \qquad (A)$$

Similarly, we can show that

$$\frac{z - L/2}{\left[\varrho^2 + \left(z - \frac{L}{2}\right)^2\right]^{1/2}} \simeq \frac{z}{r} - \frac{L}{2r} + \frac{z^2 L}{2r^3} - \frac{zL^2}{2r^3} \qquad (B)$$

The L^2 terms in the right-hand sides of eqns (A) and (B) can be discarded. Then we see that the difference (A) − (B) reduces to

$$\frac{L}{r} - \frac{z^2 L}{r^3} = L\frac{r^2 - z^2}{r^3} = L\frac{\varrho^2}{r^3}$$

Then clearly

$$H_\phi \simeq \frac{I}{4\pi\varrho} L \frac{\varrho^2}{r^3} \simeq \frac{IL\varrho}{4\pi r^3}$$

as desired.

Here and elsewhere we have used the binomial theorem:

$$(1 + x)^n = 1 + nx + \frac{n(n-1)}{2!}x^2 + \frac{n(n-1)(n-2)}{3!}x^3 + \cdots$$

$$\simeq 1 + nx \quad \text{when} \quad x \ll 1$$

Exercise

Show that another form of eqn. 2.69 is the vector relationship

$$H = \text{curl } A = \nabla \times A \tag{2.74}$$

where

$$A = Il/4\pi r \tag{2.75}$$

Here, as we shall note later, A is the vector potential at distance r from the current element Il. However, bear in mind that we are dealing here with static-like fields that are strictly valid only when the frequency approaches zero. (Note that sometimes A is defined such that $\mu H = B = \nabla \times A$.)

Solution

Here we note that, if I is z directed,

$$\nabla \times \frac{Il}{4\pi r} = \nabla \times \frac{Il}{4\pi r} i_z = \begin{vmatrix} i_x & i_y & i_z \\ \frac{\partial}{\partial x} & \frac{\partial}{\partial y} & \frac{\partial}{\partial z} \\ 0 & 0 & Il/4\pi r \end{vmatrix}$$

$$= \frac{Il}{4\pi}\left[\left(\frac{\partial}{\partial y}\frac{1}{r}\right)i_x - \left(\frac{\partial}{\partial x}\frac{1}{r}\right)i_y\right] = \frac{Il}{4\pi}\left[-\frac{yi_x}{r^3} + \frac{xi_y}{r^3}\right]$$

but

$$\frac{I}{4\pi}\frac{l \times r}{r^3} = \frac{I}{4\pi r^3}\begin{vmatrix} i_x & i_y & i_z \\ 0 & 0 & l \\ x & y & z \end{vmatrix} = \frac{I}{4\pi r^3}[-ly\, i_x + lx\, i_y]$$

which is the same!

Exercise

Show that eqn. 2.71 can be obtained from a direct application of Ampère's law using a geometry similar to Fig. 2.4, where we must now develop an expression for J_r or J_z for a point source I at $z = L/2$ and a point sink at $z = -L/2$.

Solution

The situation is illustrated in Fig. 2.10. Now we apply Ampère's law to the circular path of radius ϱ for a current source at $z = +L/2$ and a

Quasi-static fields

sink at $z = -L/2$. That is,

$$\iint J_z \, da = \oint H_\phi \, ds$$

On the left-hand side we are integrating J_z over the area of the circle of radius ϱ, and on the right-hand side we are integrating H_ϕ around the circumference. Now the element of area $da = \varrho' \, d\varrho' \, d\phi'$, where ϱ' is the radial co-ordinate of da. Also we have the element of length $ds = \varrho \, d\phi$ at the outer radius ϱ. The vertical component of the current

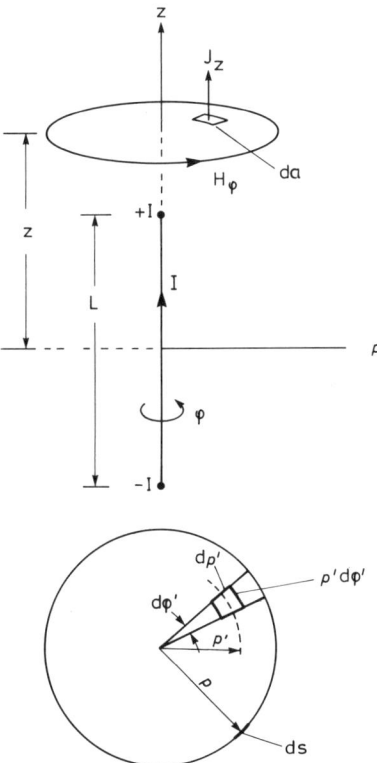

Fig. 2.10 Geometry for applying Ampère's law in cylindrical co-ordinates to deduce the magnetic field of a straight current-carrying wire of length L

density is given by

$$J_z = -(\sigma + j\omega\varepsilon) \frac{\partial \psi}{\partial z}$$

where

$$\psi = \frac{I}{4\pi(\sigma + j\omega\varepsilon)} \left(\frac{1}{r^+} - \frac{1}{r^-} \right)$$

and

$$r^+ = \left[\varrho + \left(z - \frac{L}{2} \right)^2 \right]^{1/2}$$

$$r^- = \left[\varrho + \left(z + \frac{L}{2} \right)^2 \right]^{1/2}$$

Thus

$$J_z = -\frac{I}{4\pi}\frac{\partial}{\partial z}\left[\frac{1}{r^+} - \frac{1}{r^-}\right]$$

$$= \frac{I}{4\pi}\left[\frac{z - \frac{L}{2}}{(r^+)^3} - \frac{z + \frac{L}{2}}{(r^-)^3}\right]$$

The area integration is thus effected as follows:

$$\iint J_z\, da = \int_{\varrho'=0}^{\varrho}\int_0^{2\pi} J_z\, \varrho'\, d\varrho'\, d\phi = 2\pi \int_{\varrho'=0}^{\varrho} J_z\, \varrho'\, d\varrho'$$

$$= \frac{I}{2}\left\{\int_0^{\varrho}\frac{\left(z - \frac{L}{2}\right)\varrho'}{(r^+)^3}\, d\varrho' - \int_0^{\varrho}\frac{\left(z + \frac{L}{2}\right)\varrho'}{(r^-)^3}\, d\varrho'\right\}$$

$$= \frac{I}{2}\left\{\left(z - \frac{L}{2}\right)\int_0^{\varrho}\frac{\partial}{\partial\varrho'}\left(-\frac{1}{r^+}\right)d\varrho' - \left(z + \frac{L}{2}\right)\int_0^{\varrho}\frac{\partial}{\partial\varrho'}\left(-\frac{1}{r^-}\right)d\varrho'\right\}$$

$$= \frac{I}{2}\left\{\left(z - \frac{L}{2}\right)\left(-\frac{1}{r^+}\right)\Big|_{\varrho'=0}^{\varrho'=\varrho} - \left(z + \frac{L}{2}\right)\left(-\frac{1}{r^-}\right)\Big|_{\varrho'=0}^{\varrho'=\varrho}\right\}$$

$$= \frac{I}{2}\left\{\frac{z + \frac{L}{2}}{\left[\varrho^2 + \left(z + \frac{L}{2}\right)^2\right]^{1/2}} - \frac{z - \frac{L}{2}}{\left[\varrho^2 + \left(z - \frac{L}{2}\right)^2\right]^{1/2}}\right\}$$

the last expression is now equated to

$$\oint H_\phi\, ds = \int_0^{2\pi} H_\phi \varrho\, d\phi = 2\pi\varrho H_\phi$$

which immediately gives eqn. 2.71.

Perhaps it is useful to note that the preceding derivation for H_ϕ is strictly true only if $z > L/2$. However, the result as derived, is valid for all z. For example, if $z < L/2$, we see that in the limit $\varrho' \to 0$

$$\frac{I}{2}\left(z - \frac{L}{2}\right)\left(-\frac{1}{r^+}\right) = \frac{I}{2}\left(z - \frac{L}{2}\right)\left(-\frac{1}{\left[(\varrho')^2 + \left(z - \frac{L}{2}\right)^2\right]^{1/2}}\right) = +\frac{I}{2}$$

bearing in mind that $r^+ > 0$ always.

Exercise

Locate a point source of current I at a height h over a plane of separation between two homogeneous regions with electrical properties σ_1, ε_1 and σ_2, ε_2. The situation is illustrated in Fig. 2.11, where the current point source is located on the z axis at $z = h$ in the upper region, and the plane of separation is at $z = 0$.

Show that the potential at $P(x, y, z)$ for $z > 0$ image is given by

$$\Psi = \frac{I}{4\pi(\sigma_1 + j\omega\varepsilon_1)}\left(\frac{1}{R} + \frac{K}{R'}\right)$$

Quasi-static fields 35

where

$$R = [x^2 + y^2 + (z - h)^2]^{1/2}$$

$$R' = [x^2 + y^2 + (z + h)^2]^{1/2}$$

$$K = [(\sigma_1 + j\omega\varepsilon_1) - (\sigma_2 + j\omega\varepsilon_2)]/[(\sigma_1 + j\omega\varepsilon_1) + (\sigma_2 + j\omega\varepsilon_2)]$$

Show also that the potential at P for $z < 0$ is given by

$$\Psi = \frac{I}{4\pi(\sigma_1 + j\omega\varepsilon_1)} T \frac{1}{r}$$

where

$$T = 1 + K = 2(\sigma_1 + j\omega\varepsilon_1)/[(\sigma_1 + j\omega\varepsilon_1) + (\sigma_2 + j\omega\varepsilon_2)]$$

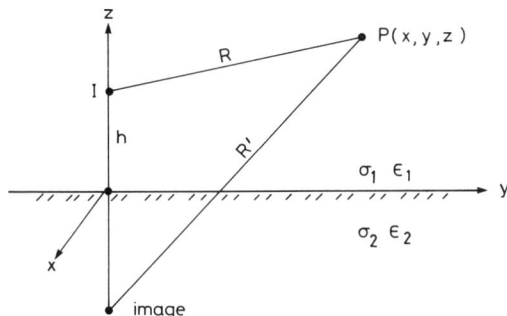

Fig. 2.11 Current point source at height h over the interface and the 'image' location at distance h below the interface

Solution It will suffice to verify the forms given for the potential in the two regions: (a) $\nabla^2 \Psi = 0$ everywhere except at $(0, 0, h)$; (b) Ψ has the proper behaviour as $R \to 0$; (c) Ψ vanishes as R and $R' \to \infty$; (d) Ψ is continuous across the interface $z = 0$; and (e) J_z, the normal current density, is also continuous across $z = 0$. Note here that for $z > 0$

$$J_z = -(\sigma_1 + j\omega\varepsilon_1)\frac{\partial \Psi}{\partial z}$$

while for $z < 0$

$$J_z = -(\sigma_2 + j\omega\varepsilon_2)\frac{\partial \Psi}{\partial z}$$

Also

$$-\frac{\partial}{\partial z}\frac{1}{R} = \frac{z - h}{R^3}$$

and

$$-\frac{\partial}{\partial z}\frac{1}{R'} = \frac{z + h}{R'^3}$$

are useful identities.

2.7. Bibliography

RICHARD BECKER, *Electromagnetic Fields and Interactions*, Part B, 'The electrostatic field', Dover Reprint, 1964.

G. WENDT, Statische Felder und Stationare Strome, *Handbuch der Physik*, Band XVI, Springer Verlag, 1958.

R. PLONSEY and R. E. COLLIN, *Principles and Applications of Electromagnetic Fields*, Chap 2–6, McGraw Hill, 1961.

J. R. WAIT, Chapter 24 in *Antenna Theory* (ed. R. E. Collin and F. J. Zucker), McGraw Hill, 1969.

J. R. WAIT, Chapter 1 in *Geo-Electromagnetism*, Academic Press, 1982.

J. R. WAIT, Chapter 2 in *Electromagnetic Wave Theory*, Harper and Row, 1985 (Revised Printing).

W. R. Smythe, *Static and Dynamic Electricity*, 3rd edn, McGraw Hill, 1968.

Chapter 3

Dynamic fields

3.1. Ampère's and Faraday's laws: Maxwell's equations

In the general case, Ampère's law can be stated as follows:

$$\oint \mathbf{H} \cdot d\mathbf{l} = \iint_A \mathbf{J} \cdot \mathbf{i}_n \, da \tag{3.1}$$

where the integral on the left-hand side is the line integral of the magnetic field around a closed contour or path; $d\mathbf{l}$ is the vector of length dl at a point on the path. The surface integral on the right-hand side is tha total current that flows normally throught the capping surface; $\mathbf{i}_n$ is the outward-pointing unit vector. The situation is illustrated in Fig. 3.1. The direction of the closed contour and the positive side of the surface are in accordance with the right-hand screw rule. Equivalently, one can use the right-hand rule where the fingers are in the direction of $d\mathbf{l}$ and the thumb is in the direction of $\mathbf{i}_n$.

Now we again invoke Ohm's law in the complex form

$$\mathbf{J} = (\sigma + j\varepsilon\omega)\mathbf{E} \tag{3.2}$$

and insert this into the right-hand side of eqn. 3.1. The other key step is to replace the left-hand side of eqn. 3.1 by a surface integral with the help of Stokes's theorem as stated by eqn. 1.76. Thus eqn. 3.1 is equivalent to

$$\iint_A (\nabla \times \mathbf{H}) \cdot \mathbf{i}_n \, da = \iint_A (\sigma + j\varepsilon\omega)\mathbf{E} \cdot \mathbf{i}_n \, da \tag{3.3}$$

If this result is to be true for all surfaces A, we can legitimately equate the integrands. Thus

$$\nabla \times \mathbf{H} = (\sigma + j\varepsilon\omega)\mathbf{E} \tag{3.4}$$

which is one of Maxwell's equations in point form. Here it has been deduced from the integral form of Ampère's law (eqn. 3.1) without any special restrictions on σ and ε (i.e. the medium can be inhomogeneous). However, it is important that the sources of the electromagnetic field are outside the region being considered.

Sometimes eqn. 3.4 is called the *Ampère-Maxwell equation* because it evolved from Ampère's law. It was thanks to the genius of James Clerk Maxwell that the displacement term $j\varepsilon\omega\mathbf{E}$ was included on the right-hand side of eqn. 3.4.

Another important and basic law of electricity was proposed by Michael Faraday. In our context it would read

$$\oint \mathbf{E} \cdot d\mathbf{l} = -\iint_A j\omega \mathbf{B} \cdot \mathbf{i}_n \, da \tag{3.5}$$

Fig. 3.1 gives a geometrical interpretation. The quantity on the left-hand side of eqn. 3.5 is the line integral of the electric field around a closed circuit or path. The quantity on the right-hand side is the net flux of $j\omega \boldsymbol{B} \cdot \boldsymbol{i}_n$ through the capping surface.

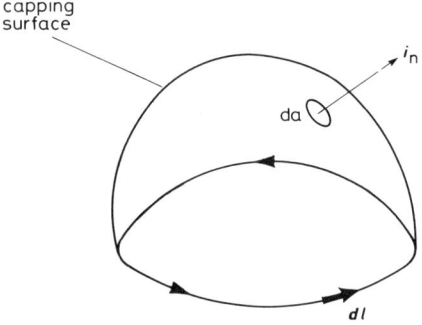

Fig. 3.1 Closed circuit with capping surface A

We now note that $\boldsymbol{B} = \mu \boldsymbol{H}$ and invoke Stokes's law to permit us to write eqn. 3.5 in the form

$$\iint_A (\nabla \times \boldsymbol{E}) \cdot \boldsymbol{i}_n \, da = - \iint_A j\mu\omega \boldsymbol{H} \cdot \boldsymbol{i}_n \, da \tag{3.6}$$

when we equate integrands, eqn. 3.6 leads to the second of Maxwell's equations:

$$\nabla \times \boldsymbol{E} = -j\mu\omega \, \boldsymbol{H} \tag{3.7}$$

where μ, the magnetic permeability, is not restricted to be independent of position. Equation 3.7 is often called the *Faraday-Maxwell equation*.

It is useful to note that, if $\omega \to 0$, eqn. 3.7 reduces to

$$\nabla \times \boldsymbol{E} = 0 \tag{3.8}$$

in which case $\boldsymbol{E} = -\nabla \Phi$, where Φ is a scalar potential. But unless ω is sufficiently small the electric field is not derivable from a scalar potential, as we shall indicate below.

Before considering specific applications we will consider one formal extension of Maxwell's equations. That is, we allow for the presence of a source current in an explicit fashion. Thus we recognise that the current density $\boldsymbol{J}$, as it appears in a general statement of Ampère's law, is decomposed as follows:

$$\boldsymbol{J} = (\sigma + j\varepsilon\omega)\boldsymbol{E} + \boldsymbol{J}_s \tag{3.9}$$

where $\boldsymbol{J}_s$ represents the driving or source current density such as may be produced by a transmitting antenna. In regions where $\boldsymbol{J}_s = 0$, our previous development is valid. Thus the so-called 'source-free form' of Maxwell's equation refers to eqns 3.4 and 3.7. Now, to allow explicitly for the source current density, the two Maxwell equations in complex vector form are written

$$\nabla \times \boldsymbol{H} = (\sigma + j\varepsilon\omega)\boldsymbol{E} + \boldsymbol{J}_s \tag{3.10}$$

$$\nabla \times \boldsymbol{E} = -j\mu\omega \, \boldsymbol{H} \tag{3.11}$$

The corresponding time domain vector forms are written

$$\nabla \times \boldsymbol{h}(t) = \left(\sigma + \varepsilon \frac{\partial}{\partial t}\right) \boldsymbol{e}(t) + \boldsymbol{j}_s(t) \tag{3.12}$$

$$\nabla \times \boldsymbol{e}(t) = -\mu \frac{\partial}{\partial t} \boldsymbol{h}(t) \tag{3.13}$$

which follows immediately from eqns. 3.10 and 3.11. Here $\boldsymbol{j}_s(t)$ is the time-dependent source vector current density.

3.2 Plane wave transmission

Maxwell's equations are the bases of most quantitative investigations of electromagnetic wave transmission. We begin our analysis by considering plane wave transmission. Essentially this means that the surface of constant amplitude and the surface of constant phase are parallel planes. The meaning of this statement will become evident in what follows.

We choose our co-ordinate system (x, y, z) such that the derivatives in the x and y direction are zero. In regions devoid of sources, eqns. 3.7 and 3.4 are then written

$$\begin{vmatrix} \boldsymbol{i}_x & \boldsymbol{i}_y & \boldsymbol{i}_z \\ 0 & 0 & \partial/\partial z \\ E_x & E_y & E_z \end{vmatrix} = -j\mu\omega(\boldsymbol{i}_x H_x + \boldsymbol{i}_y H_y + \boldsymbol{i}_z H_z) \tag{3.14}$$

and

$$\begin{vmatrix} \boldsymbol{i}_x & \boldsymbol{i}_y & \boldsymbol{i}_z \\ 0 & 0 & \partial/\partial z \\ H_x & H_y & H_z \end{vmatrix} = (\sigma + j\varepsilon\omega)(\boldsymbol{i}_x E_x + \boldsymbol{i}_y E_y + \boldsymbol{i}_z E_z) \tag{3.15}$$

We now equate, in turn, the x, y and z components of each side of the preceding two equations. This process yields the following set of first-order differential equations:

$$\frac{dE_y}{dz} = j\mu\omega H_x \tag{3.16}$$

$$\frac{dE_x}{dz} = -j\mu\omega H_y \tag{3.17}$$

$$0 = j\mu\omega H_z \tag{3.18}$$

$$\frac{dH_y}{dz} = -(\sigma + j\varepsilon\omega) E_x \tag{3.19}$$

$$\frac{dH_x}{dz} = (\sigma + j\varepsilon\omega) E_y \tag{3.20}$$

$$0 = (\sigma + j\varepsilon\omega) E_z \tag{3.21}$$

The vast simplification is the result of our rather severe physical assumptions: namely, that variations in the x and y directions are set equal to zero. As a consequence, the z components of the fields are zero (i.e. $E_z = H_z = 0$). Also note that we have replaced $\partial/\partial z$ by d/dz because z is now the only variable.

An important observation is that the pair of equations 3.17 and 3.19 for E_x and H_y are coupled. Similarly, the pair of equations 3.16 and 3.20 are coupled. But these two pairs are independent of each other. It is also worthy of note that, up to this point, the intrinsic properties σ, ε and μ are allowed to vary with z. However, in what follows we will assume that for the region under consideration σ, ε and μ do not vary with z.

The next steps is to insert eqn. 3.17 for H_y into eqn. 3.19. Then we obtain the second-order equation

$$\frac{d^2 E_x}{dz^2} - \gamma^2 E_x = 0 \tag{3.22}$$

where

$$\gamma^2 = j\mu\omega(\sigma + j\varepsilon\omega) \tag{3.23}$$

In a similar fashion we insert eqn. 3.19 for E_x into eqn. 3.17, leading to

$$\frac{d^2 H_y}{dz^2} - \gamma^2 H_y = 0 \tag{3.24}$$

Similar results are obtained by working with the pair of equations 3.16 and 3.20. Thus we conclude that all components of the field satisfy

$$\left(\frac{d^2}{dz^2} - \gamma^2\right)\Psi = 0 \tag{3.25}$$

where Ψ represents any one of E_x, H_x, E_y, H_y (remember of course that E_z and H_z are zero).

The complex quantity γ is termed the *propagation constant*, even though it is a function of frequency. It is defined here as

$$\gamma = [j\mu\omega(\sigma + j\varepsilon\omega)]^{1/2} \tag{3.26}$$

where the sign of the radical is always chosen such that $\mathrm{Re}\,\gamma > 0$.

Exercise

Show explicitly, for the time-dependent form of Maxwell's equations, in a homogeneous source-free region, with invariance in the x and y directions, that $e_x(t)$, $e_y(t)$, $h_x(t)$ and $h_y(t)$ satisfy

$$\left(\frac{\partial^2}{\partial z^2} - \sigma\mu\frac{\partial}{\partial t} - \mu\varepsilon\frac{\partial^2}{\partial t^2}\right)\Psi(t) = 0 \tag{3.27}$$

where $\Psi(t)$ is any one of the forementioned field components.

3.3 Plane wave solutions

We now wish to solve eqn. 3.25. By simple inspection, we see that both $\exp(+\gamma z)$ and $\exp(-\gamma z)$ will satisfy eqn. 3.25. Thus a general solution is

$$\Psi = A\,e^{-\gamma z} + B\,e^{\gamma z} \tag{3.28}$$

where A and B, at this stage, can be any complex constants. Now we interpret this result. Set

$$\gamma = \alpha + j\beta \tag{3.29}$$

where α and β are the real and imaginary parts. Now $\alpha > 0$ because we have already specified that Re $\gamma > 0$.

Both sides of eqn. 3.28 are now multiplied by exp $(j\omega t)$, in accord with our convention, and the resulting equation is written

$$\Psi \, e^{j\omega t} = A \, e^{-\alpha z} \, e^{-j(\beta z - \omega t)} + B \, e^{\alpha z} \, e^{j(\beta z + \omega t)} \quad (3.30)$$

Of course Re $\Psi \exp(j\omega t)$ is the actual field component $\psi(t)$, but we will deal only with the phasor form.

We note that the phase terms in eqn. 3.30 can be written

$$\exp\left[\mp j\beta(z \mp vt)\right]$$

if $v = \omega/\beta$. The phase (i.e. the exponent) does not change if

$$z \mp t = \text{constant} \quad (3.31)$$

Thus, in the first case,

$$\partial z/\partial t = +v \quad (3.32)$$

is by definition the *phase velocity* for waves propagating in the positive z direction. In the second case, the phase is constant when

$$\partial z/\partial t = -v \quad (3.33)$$

which corresponds to waves propagating in the negative z direction. We further note that the phase changes by 2π radians when z changes by $2\pi/\beta$. Thus we write $2\pi/\beta = \lambda$ or $\beta = 2\pi/\lambda$ where, by definition, λ is the *wavelength*.

In eqn. 3.30 it is clear that the factor $\exp(-\alpha z)$ becomes smaller as z increases in the positive direction, whereas $\exp(+\alpha z)$ becomes smaller as z increases in the negative direction. In fact

$$\lim_{(z \to +\infty)} \exp(-\alpha z) = 0$$

and

$$\lim_{(z \to -\infty)} \exp(+\alpha z) = 0$$

By convention we call α the *attenuation constant* or *attenuation coefficient*.

The above statements are summarized by writing eqn. 3.30 in the equivalent form

$$\Psi \, e^{j\omega t} = A \, e^{-\alpha z} \exp[-j(2\pi/\lambda)(z - vt)] \quad (3.34)$$
$$\text{(propagates to } z = +\infty)$$

$$+ B \, e^{+\alpha z} \exp[+j(2\pi/\lambda)(z + vt)]$$
$$\text{(propagates to } z = -\infty)$$

where $v = \omega/\beta = \lambda$ where f is the frequency in hertz. At this point we have not specified A or B; they will depend on the nature of the source of excitation, and we will return to this question later.

The first special case to consider is propagation in free space. Then $\sigma = 0$. $\varepsilon = \varepsilon_0 = 8 \cdot 854 \times 10^{-12}$ F/m and $\mu = \mu_0 = 4\pi \times 10^{-7}$ H/m. Clearly in this case

$$\gamma = jk = j(\varepsilon_0 \mu_0)^{1/2} \omega \quad \text{and} \quad \alpha = 0$$

Here $k = \omega/c$, where $c = (\varepsilon_0\mu_0)^{-1/2}$; k is the free space wave number. The value of the phase velocity is given by $c = 2.998 \times 10^8 \simeq 3.0 \times 10^8$ m/s. We can also write $c = \lambda_0 f$ where λ_0 is the free space wavelength.

Exercise

Naval communications use VLF (f is from 3 to 30 kHz) so that the signals can be received at periscope depth on submarines. What is the range of free space wavelengths involved?

Solution

100 to 10 km.

Exercise

The so-called *X band* in microwave technology is centred at a wavelength of 3 cm. What frequency does this correspond to?

Solution

10^{10} Hz or 10 GHz (gigahertz).

Exercise

The earth's circumference is approximately 40 000 km. What frequency corresponds to having this circumference equal to 4/3 of the free space wavelength?

Solution

10 Hz or 10 c/s. This is the approximate resonant frequency of the earth–ionosphere cavity; that is, the earth and the ionosphere act as the boundary walls of a cavity resonator.

Exercise

In radio-oceanography the ground wave has a phase velocity of approximately c. Strong backscatter signals occur when the distance between ocean wave crests is one half the radio wavelength. If the frequency of maximum echo is 20 MHz, what is the estimated distance between crests?

Solution

The radio wavelength is 15 m, so the distance between crests is 7·5 m.

Exercise

In dealing with the source-free region $z \geqslant z_0$, we specify that the wave function $\Psi(z)$ is Ψ_0 at $z = z_0$. Then, assuming that the form of eqn. 3.30 is valid, show that

$$\Psi(z) = \Psi_0 \exp[-\alpha(z - z_0)] \exp[-j\beta(z - z_0)]$$

which is good for all $z > z_0$.

Note that in eqn. 3.30 B must be zero if Ψ is to vanish as $z \to +\infty$. Then $\Psi(z_0) = \Psi_0$ gives $A = \Psi_0 \exp[+(\alpha + j\beta)z_0]$.

3.4 Transmission in material media

To quantify the transmission of plane waves in material media with homogeneous properties, we need to solve explicitly for α and β. To this end we square both sides of eqn. 3.29 to obtain

$$j\mu\omega\sigma - \varepsilon\mu\omega^2 = \alpha^2 + 2j\alpha\beta - \beta^2$$

Equation real and imaginary parts yields

$$\beta = \mu\omega\sigma/(2\alpha) \tag{3.35}$$

$$\alpha^2 - \beta^2 = -\varepsilon\mu\omega^2 \tag{3.36}$$

Eliminating β from this pair leads to the quadratic equation in α^2 of the form

$$(\alpha^2) + \varepsilon\mu\omega^2(\alpha^2) - (\mu\omega\sigma/2)^2 = 0 \quad (3.37)$$

The solutions are

$$\alpha^2 = \{-\varepsilon\mu\omega^2 \pm [(\varepsilon\mu\omega^2)^2 + (\mu\omega\sigma)^2]^{1/2}\}/2 \quad (3.38)$$

Now we have stipulated that α is real and positive; thus the $+$ sign before the radical must be chosen. Therefore

$$\alpha = \left\{\frac{\mu\omega}{2}[(\sigma^2 + \varepsilon^2\omega^2)^{1/2} - \varepsilon\omega]\right\}^{1/2} \quad (3.39)$$

Using eqn. 3.35, we find that

$$\beta = \left\{\frac{\mu\omega}{2}[(\sigma^2 + \varepsilon^2\omega^2)^{1/2} + \varepsilon\omega]\right\}^{1/2} \quad (3.40)$$

In the limiting case of nondissipative media (i.e. $\sigma = 0$), the attenuation is zero ($\alpha = 0$) and the phase factor simplifies to

$$\beta = (\varepsilon\mu)^{1/2}\omega \quad (3.41)$$

which can be written

$$\beta = \omega/v \quad (3.42)$$

where $v = (\varepsilon\mu)^{-1/2}$ is the phase velocity and λ is the wavelength.

Exercise

Consider plane wave transmission in distilled water, where we assume the following properties: $\mu = \mu_0$, $\varepsilon/\varepsilon_0 = 81$ and $\sigma = 0$. Show that $v = c/9$ and $\lambda = \lambda_0/9$; c and λ_0 are the corresponding phase velocity and wavelength, respectively, at the same frequency for free space.

Another important limiting case is when the conduction currents are much greater than the displacement currents (i.e. $\sigma \gg \varepsilon\omega$). Then from eqns. 3.39 and 3.40 we see that

$$\alpha \simeq (\mu\omega\sigma/2)^{1/2} \simeq \beta \quad (3.43)$$

Exercise

Show for the case of copper, where $\sigma \simeq 5\cdot8 \times 10^7$ and $\mu = \mu_0 = 4\pi \times 10^{-7}$, that $\alpha \simeq 15\cdot1 f^{1/2}$, where f is the frequency in hertz. What are the limits here on f assuming, for example, that $\varepsilon/\varepsilon_0 = 10$?

Solution

$2\pi f \ll 5\cdot8 \times 10^7/(8\cdot85 \times 10^{-11})$ or $f \ll 10^{17}$ Hz.

Exercise

Again for the copper of the previous exercise what is the *skin depth* $\delta = 1/\alpha$ when $f = 10$ MHz ?

Solution

$$\delta \simeq \left(\frac{2}{\sigma\mu_0\omega}\right)^{1/2} = \frac{1}{47\cdot8}\left(\frac{10}{f}\right)^{1/2} = \frac{10^{-3}}{47\cdot8} = 0\cdot02\,\text{mm}$$

Exercise

Again for copper, what is the phase velocity?

Solution

$$v = \omega/\beta = \left(\frac{2\omega}{\sigma\mu_0}\right)^{1/2} = \omega\delta = 2\pi f\delta = \frac{2\pi}{47\cdot 8}$$

$$\times 10^4 \text{ m/s} \simeq 1\cdot 31 \text{ km/s}$$

Yet another limiting case is when the dissipation is small and the frequency is high. Then, for example,

$$(\sigma^2 + \varepsilon^2\omega^2)^{1/2} - \varepsilon\omega = \varepsilon\omega\left[\left(1 + \frac{\sigma^2}{\varepsilon^2\omega^2}\right)^{1/2} - 1\right]$$

$$\simeq \varepsilon\omega\left[1 + \frac{\sigma^2}{2\varepsilon^2\omega^2} - 1\right] \simeq \frac{\sigma^2}{2\varepsilon\omega} \quad (3.44)$$

Then eqn. 3.39 reduces to

$$\alpha \simeq \left(\frac{\mu\omega}{2}\frac{\sigma^2}{2\varepsilon\omega}\right)^{1/2} \simeq \tfrac{1}{2}\eta\sigma$$

where $\eta = (\mu/\varepsilon)^{1/2}$ in ohms. This simple and useful formula for the attenuation rate α is valid when $\sigma \ll \varepsilon\omega$. To within the same approximation $\beta \simeq (\varepsilon\mu)^{1/2}\omega$ and the phase velocity $v \simeq (\varepsilon\mu)^{-1/2}$, which, to this order, does not depend on σ.

Exercise

Consider the propagation of microwaves in a freshwater lake where the conductivity $\sigma = 10^{-4}$ s/m and the relative permittivity $\varepsilon/\varepsilon_0 = 81$. What is the attenuation rate in dB/m?

Solution

Now

$$\eta = (\mu/\varepsilon)^{1/2} = (1/9)(\mu_0/\varepsilon_0)^{1/2} = (1/9)(120\pi) = 40\pi/3$$

then

$$\alpha = (1/2)(40\pi/3) \times 10^{-4} = (2\pi/3) \, 10^{-3} \simeq 2\cdot 09 \text{ Np/km}$$

or

$$\alpha = (2\pi/3) \times 8\cdot 686 \simeq 18\cdot 1 \text{ dB/km}$$

Exercise

What are the restrictions on frequency in order for the preceding result to be reasonable?

Solution

$2\pi f \gg \sigma/\varepsilon$ or $f \gg 23$ kHz. In other words, it should be in the megahertz range or higher.

3.5 Excitation of plane waves

To properly discuss the propagation of plane waves, it is important to consider the source. Our source model is a sheet current with a uniform density j_x A/m contained in the plane $z = 0$. This can be envisaged as an array of many fine parallel wires, each carrying a current j_x/N, were N is the number of wires per meter. A uniform current sheet corresponds to the limit $N \to \infty$ while j_x is finite.

Dynamic fields

The situation is shown in Fig. 3.2, where the surrounding homogeneous medium has electrical properties σ, ε and μ. Only a square meter of the infinite current sheet is shown. We now wish to deduce the plane waves that are generated by this configuration.

The first step is to apply Ampère's law to the slender rectangular circuit shown in Fig. 3.2. This circuit is drawn such that the long sides are in the y direction, and they pass along the positive z side and return along the negative z side of the current sheet. Thus the total current

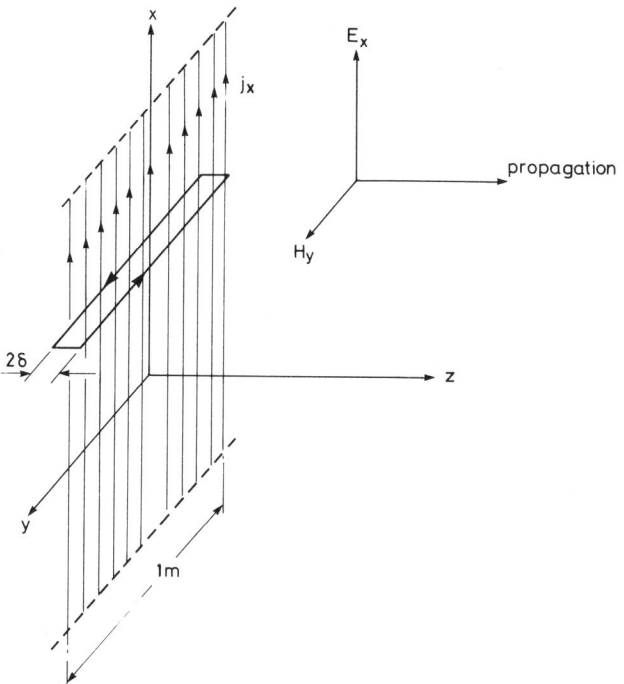

Fig. 3.2 Idealised current sheet source of density j_x A/m in the x direction

enclosed is j_x amperes if the length of the rectangle is one metre and its width is 2δ. The contributions or effects of the small ends of the rectangle vanish in the limit as $\delta \to 0$.

Ampère's law states that the total integral of H around the circuit is equal to the current passing through it. Thus we can write

$$\lim_{\delta \to 0} [H_y(z = -\delta) - H_y(z = +\delta)] = j_x \qquad (3.45)$$

A more concise statement is

$$H_y(z = -0) - H_y(z = +0) = j_x \qquad (3.46)$$

which says that the discontinuity in the tangential magnetic field at the sheet is equal to the orthogonal surface current density.

We now write down appropriate expressions for the magnetic field component H_y on the two sides of the sheet. Referring to the general form given by eqn. 3.28, we now assert that

$$H_y = B e^{-\gamma z} \quad \text{for } z > 0 \qquad (3.47)$$

$$H_y = A e^{+\gamma z} \quad \text{for } z < 0 \tag{3.48}$$

where B and A are yet to be determined. The magnetic field must satisfy the physical requirement $|H_y| \to 0$ as $|z| \to \infty$. In other words the electromagnetic field is attenuated as we move away from either side of the sheet. Also, of course, the waves generated are outgoing.

Now, according to eqn. 3.19,

$$E_x = -[1/(\sigma + j\varepsilon\omega)]\partial H_y/\partial z \tag{3.49}$$

which holds for both $z > 0$ and $z < 0$. Thus

$$E_x = \eta H_y \quad \text{for } z > 0 \tag{3.50}$$

$$E_x = -\eta H_y \quad \text{for } z < 0 \tag{3.51}$$

where

$$\eta = \frac{\gamma}{\sigma + j\varepsilon\omega} = \frac{j\mu\omega}{\gamma} = \left(\frac{j\mu\omega}{\sigma + j\varepsilon\omega}\right)^{1/2} \tag{3.52}$$

is called the *characteristic impedance*. Because of symmetry, E_x is the same on both sides of the current sheet. That is,

$$E_x(z) = E_x(-z) \tag{3.53}$$

From eqn. 3.50 and 3.51 it is evident that

$$H_y(z) = -H_y(-z) \tag{3.54}$$

Consequently $B = -A$, and therefore

$$H_y = -A e^{-\gamma z} \quad \text{for } z > 0$$

$$= +A e^{+\gamma z} \quad \text{for } z < 0 \tag{3.55}$$

When this expression for H_y is inserted in eqn. 3.45, we immediately deduce that $A = +j_x/2$. Thus, in summary, we have predicted that the current sheet source j_x at $z = 0$ generates the following nonzero field components:

$$\left.\begin{array}{l} E_x = -j_x \eta\, e^{-\gamma z}/2 \\ H_y = -j_x\, e^{-\gamma z}/2 \end{array}\right\} \quad \text{for } z > 0 \quad \begin{array}{l}(3.56)\\(3.57)\end{array}$$

$$\left.\begin{array}{l} E_x = -j_x \eta\, e^{+\gamma z}/2 \\ H_y = +j_x\, e^{+\gamma z}/2 \end{array}\right\} \quad \text{for } z < 0 \quad \begin{array}{l}(3.58)\\(3.59)\end{array}$$

Exercise

Consider the same problem as discussed but with the source current sheet having only a j_y component. Show that the following fields are generated:

$$\left.\begin{array}{l} E_y = -j_y \eta\, e^{-\gamma z}/2 \\ H_x = +j_y\, e^{-\gamma z}/2 \end{array}\right\} \quad \text{for } z > 0 \quad \begin{array}{l}(3.60)\\(3.61)\end{array}$$

$$\left.\begin{array}{l} E_y = -j_y \eta\, e^{+\gamma z}/2 \\ H_x = -j_y\, e^{+\gamma z}/2 \end{array}\right\} \quad \text{for } z < 0 \quad \begin{array}{l}(3.62)\\(3.63)\end{array}$$

Dynamic fields

Exercise

We wish to generate a circularly polarised electromagnetic plane wave for the region $z > 0$. At and n the positive side of the current sheet the electric field vector is given by

$$e(t) = E_0(i_x \cos \omega t + i_y \sin \omega t)$$

Show, by superimposing two orthogonal source currents at $z = 0$ with densities $\mathrm{Re}(j_x\, e^{j\omega t})$ and $\mathrm{Re}(j_y\, e^{j\omega t})$, that this objective can be met. Also determine the values of the phasors j_x and j_y.

Solution

According to eqn. 1.62, the complex vector associated with $e(t)$ is $E_0(i_x - ji_y)$. This must be equated to $-(j_x\eta/2)i_x - (j_y\eta/2)i_y$. Then the required excitation is

$$j_x = -(2/\eta)E_0 \tag{3.64}$$

and

$$j_y = +j(2/\eta)E_0 = (2/\eta)E_0\, e^{j\pi/2} \tag{3.65}$$

3.6 Oblique excitation of plane waves

In our preceding analysis we have considered excitation by a current sheet with constant amplitude and phase. It is instructive to consider the case where the phase has a progressive change in, say, the y direction. Thus the source current density now has the form

$$j_x(y) = j_0\, e^{-jpy} \tag{3.66}$$

where p is a constant (with dimensions of radians per metre). The source condition is now

$$H_y(z = -0) - H_y(z = +0) = j_0\, e^{-jpy} \tag{3.67}$$

Clearly, H_y for $|z| > 0$ must be a function of both y and z, but invariance with x is maintained. From the general form of Maxwell's equations, as given by eqns. 3.10 and 3.11, we are led to write

$$-j\mu\omega H_y = \frac{\partial E_x}{\partial z} \tag{3.68}$$

$$j\mu\omega H_z = \frac{\partial E_x}{\partial y} \tag{3.69}$$

and

$$(\sigma + j\varepsilon\omega)E_x = \frac{\partial H_z}{\partial y} - \frac{\partial H_y}{\partial z} \tag{3.70}$$

We now insert expressions for H_y and H_z, as given by eqns. 3.68 and 3.69, into eqn. 3.70. Assuming constancy of μ, this leads to

$$\left(\frac{\partial^2}{\partial y^2} + \frac{\partial^2}{\partial z^2} - \gamma^2\right)E_x = 0 \tag{3.71}$$

where

$$\gamma^2 = j\mu\omega(\sigma + j\varepsilon\omega) \tag{3.72}$$

We try a solution of the form

$$E_x = f(z)\, e^{-jpy} \tag{3.73}$$

which is suggested by the nature of the source condition. Also, on physical grounds, we anticipate that the y variation of the resultant solutions should contain $\exp(-jpy)$. If such were not the case, we would not be able to match the boundary conditions as indicated below.

Equation 3.73 for E_x is now inserted into eqn. 3.71, which tells us that

$$\left(\frac{d^2}{dz^2} - u^2\right)E_x = 0 \tag{3.74}$$

where

$$u^2 = \gamma^2 + p^2 \tag{3.75}$$

The general solution of eqn. 3.74 is

$$E_x = (C\,e^{uz} + D\,e^{-uz})\,e^{-jpy} \tag{3.76}$$

where C and D are coefficients yet to be determined, and

$$u = (\gamma^2 + p^2)^{1/2} \tag{3.77}$$

with $\mathrm{Re}\,u > 0$.

According to eqns. 3.68 and 3.69,

$$H_y = -\frac{e^{-jpy}}{j\mu\omega}(Cu\,e^{uz} - Du\,e^{-uz}) \tag{3.78}$$

$$H_z = \frac{e^{-jpy}}{j\mu\omega}p(C\,e^{uz} + D\,e^{uz}) \tag{3.79}$$

Equivalent forms of eqns. 3.78 and 3.79 are

$$H_y = (-MC\,e^{uz} + MD\,e^{-uz})\,e^{-jpy} \tag{3.80}$$

$$H_z = (-PC\,e^{uz} - PD\,e^{-uz})\,e^{-jpy} \tag{3.81}$$

where

$$M = \frac{u}{j\mu\omega} = \frac{(\gamma^2 + p^2)^{1/2}}{j\mu\omega} \tag{3.82}$$

$$P = p/(\mu\omega) \tag{3.83}$$

The parameters M and P, as defined above, have dimensions of siemens and they may be interpreted as *wave admittances*.

In the context of our specific problem, we again may assert that

$$E_x(-z) = E_x(+z) \tag{3.84}$$

because of the obvious even symmetry of the electric field on both sides of the current sheet. Thus we are led to write

$$E_x = E_0\,e^{uz}\,e^{-jpy} \qquad z < 0 \tag{3.85}$$

$$E_x = E_0\,e^{-uz}\,e^{-jpy} \qquad z > 0 \tag{3.86}$$

These forms are consistent with eqn. 3.76 if we invoke the even property of E_x and bear in mind that $|E_x| \to 0$ as $|z| \to \infty$. Now it follows from eqns. 3.80 and 3.81, or directly from Maxwell's equations, that

$$H_y = -ME_0 \, e^{uz} \, e^{-jpy} \quad z < 0 \tag{3.87}$$

$$H_y = +ME_0 \, e^{-uz} \, e^{-jpy} \quad z > 0 \tag{3.88}$$

where M is given by eqn. 3.82. Furthermore

$$H_z = -PE_0 \, e^{uz} \, e^{-jpy} \quad z < 0 \tag{3.89}$$

$$H_z = -PE_0 \, e^{-uz} \, e^{-jpy} \quad z > 0 \tag{3.90}$$

where P is given by eqn. 3.83.

The single coefficient E_0 is yet to be determined. By invoking eqn. 3.67 together with eqns. 3.87 and 3.88, if follows that

$$-2ME_0 \, e^{-jpy} = j_0 \, e^{-jpy} \tag{3.91}$$

or

$$E_0 = -(j_0/2)/M \tag{3.92}$$

Thus the field components as given by eqns. 3.85 to 3.91 are now known in terms of the specified source current density $j_0 \exp(-jpy)$.

To discuss the above result it is desirable to let the conductivity vanish. Then

$$\gamma = j(\varepsilon\mu)^{1/2}\omega \tag{3.93}$$

Also we assume free space conditions, so that $\varepsilon = \varepsilon_0$ and $\mu = \mu_0$ on both sides of the current sheet. Thus we write $\gamma = jk$, where $k = (\varepsilon_0\mu_0)^{1/2}\omega = 2\pi/\lambda_0$. k is the so-called free space *wave number*, and λ_0 is the free space wavelength. Now we see that

$$u = (p^2 - k^2)^{1/2} \tag{3.94}$$

or

$$u = j(k^2 - p^2)^{1/2} \tag{3.95}$$

We are free here to set

$$(k^2 - p^2)^{1/2} = k \cos\theta$$

Clearly $\cos\theta$ is real if $k^2 > p^2$. Also note that θ is related to p by $p = k \sin\theta$. Now, for $z > 0$, using eqn. 3.86, it is evident that

$$E_x = E_0 \, e^{-jky\sin\theta} \, e^{-jkz\cos\theta} \tag{3.96}$$

This is a plane wave propagating at an angle θ from the normal or z direction. The planes of constant phase, as indicated in Fig. 3.3, are at right angles to the direction of propagation. Clearly, in the case where $p = 0$ (constant phase source), $\theta = 0$ and we have

$$E_x = E_0 \, e^{-jkz} \tag{3.97}$$

which is a plane wave propagating normal to the current sheet source. The other limiting case is when $p = k$. This case corresponds to having the phase shift of the source current density identical to that for plane waves in free space. Then $\sin\theta = 1$, and therefore

$$E_x = E_0 \, e^{-jky} \tag{3.98}$$

which is a plane wave propagating in the positive y direction. Obviously by choosing p via the relation $\sin\theta = p/k$, any desired angle θ can be achieved. Thus the direction of the radiated plane wave can be 'steered'.

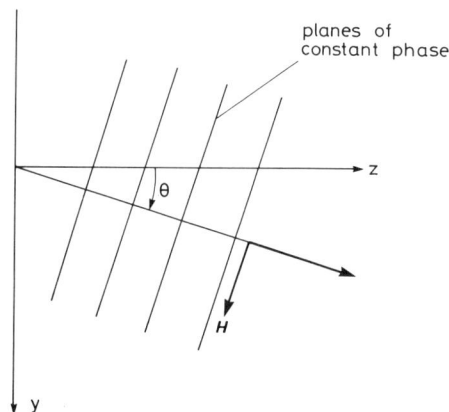

Fig. 3.3 Plane wave radiated from sheet current source at $z = 0$ with phase variation according to $\exp(-jpy)$. Note that $\boldsymbol{E} = \boldsymbol{i}_x E_x$ and $\boldsymbol{H} = \boldsymbol{i}_y H_y + \boldsymbol{i}_z H_z$

It follows from eqns. 3.88 or 3.89, or more directly from eqns. 3.68 and 3.69, that

$$H_y = -\frac{1}{j\mu\omega}\frac{\partial E_x}{\partial z} \tag{3.99}$$

$$H_z = \frac{1}{j\mu\omega}\frac{\partial E_x}{\partial y} \tag{3.100}$$

then, using eqn. 3.96, we find that

$$H_y = \frac{E_0}{\eta_0}\cos\theta\, e^{-jky\sin\theta}\, e^{-jkz\cos\theta} \tag{3.101}$$

and

$$H_z = -\frac{E_0}{\eta_0}\sin\theta\, e^{-jky\sin\theta}\, e^{-jkz\cos\theta} \tag{3.102}$$

where $\eta_0 = \mu_0\omega/k = (\mu_0/\varepsilon_0)^{1/2}$. As indicated in Fig. 3.3, the magnetic field is a vector $\boldsymbol{H}$ which is at right angles to $\boldsymbol{E}(=\boldsymbol{i}_x E_x)$ and the direction of propagation.

Exercise

Assume that the region $z > 0$ is homogeneous with properties σ, ε and μ. At the 'aperture' plane $z = 0$, assume that the magnetic field vector is given by

$$\boldsymbol{H} = \boldsymbol{i}_y H_0\, e^{-jqx} \tag{3.103}$$

where H_0 is known. Determine the magnetic field for $z > 0$.

Solution

Assume that a general form for $\boldsymbol{H}$, for $z > 0$, is

$$\boldsymbol{H} = \boldsymbol{i}_y F(z)\, e^{-jqx} \tag{3.104}$$

where

$$\left(\frac{d^2}{dz^2} - v^2\right) F(z) = 0 \tag{3.105}$$

with

$$v = (\gamma^2 + q^2)^{1/2} \tag{3.106}$$

$$\gamma^2 = j\mu\omega(\sigma + j\varepsilon\omega)$$

The solution of eqn. 3.105 is

$$F(z) = H_0 e^{-vz} \tag{3.107}$$

where Re $v > 0$; this satisfies the source condition at $z = 0$.

Exercise

In the preceding example, replace the source condition at $z = 0$ by

$$\mathbf{H} = \mathbf{i}_y H_0 \cos qx \tag{3.108}$$

Show that the solution for $z > 0$ is given by

$$\mathbf{H} = \mathbf{i}_y F(z) \cos qx \tag{3.109}$$

Solution

Simply note that $\cos qx = (e^{jqx} + e^{-jqx})/2$; thus the results for $\exp(\pm jqx)$ excitations are superimposed.

Exercise

Assume that the region $z < z_0$ is homogeneous. At $z = 0$ specify that

$$\mathbf{E} = \mathbf{i}_x E_0 e^{jsy} \tag{3.110}$$

Show that for all $z < z_0$.

$$\mathbf{E} = \mathbf{i}_x E_0 e^{jsy} e^{v(z-z_0)} \tag{3.111}$$

where $v = (\gamma^2 + s^2)^{1/2}$

Exercise

Show that, when $p > k$, for the current sheet in free space, the resultant fields are localized to the neighbourhood of the source.

Solution

Note that u is real and eqns. 3.85 and 3.86 have the form

$$E_x = E_0 e^{-jpy} e^{-u|z|} \tag{3.112}$$

which decays exponentially in the direction of $+z$ and $-z$ away from the sheet. Also note that $\sin\theta = p/k > 1$, which means that θ must be complex. (This question arises in connection with the propagation of surface waves, as discussed in Chapter 6.)

3.7 Further exercises

Exercise

Consider plane wave propagation in the z direction, with reference to a rectangular co-ordinate system (x, y, z), in a homogeneous medium with properties σ, ε and μ. The polarisation is such that the electric field has only a y component E_y given by $E_y = E_0 e^{-\gamma z}$ and, of course, the magnetic field is polarized in the x direction with $H_x = -\eta^{-1} E_0 e^{-\gamma z}$.

Now change the co-ordinate system to (x', y, z') such that the new z' axis makes an angle θ with the original z axis. The situation is depicted in Fig. 3.4.

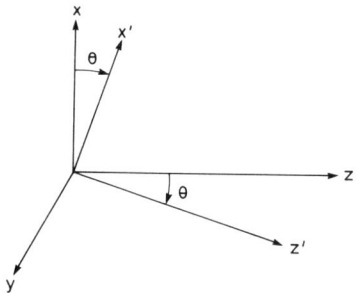

Fig. 3.4 The two co-ordinate systems (x, y, z) and (x', y, z')

In terms of the new system, show that

$$E_y = E_0 \, e^{-\gamma z' \cos\theta} \, e^{-\gamma x' \sin\theta} \qquad (3.113)$$

Also confirm that

$$\left(\frac{\partial^2}{(\partial x')^2} + \frac{\partial^2}{(\partial z')^2} - \gamma^2\right) E_y = 0 \qquad (3.114)$$

Furthermore, show that the magnetic field has components in the new system given by

$$H_{x'} = -\frac{\cos\theta}{\eta} E_0 \, e^{-\gamma z' \cos\theta} \, e^{-\gamma x' \sin\theta} \qquad (3.115)$$

$$H_{z'} = \frac{\sin\theta}{\eta} E_0 \, e^{-\gamma z' \cos\theta} \, e^{-\gamma x' \sin\theta} \qquad (3.116)$$

where $\eta = j\mu\omega/\gamma$.

Solution

Direct substitution of E_y into the two-dimensional wave equation above shows indeed it is satisfied, noting that $\cos^2\theta + \sin^2\theta = 1$. With reference to Fig. 3.5, a simple geometrical argument with a little trigonometry shows that

$$z = z' \cos\theta + x' \sin\theta$$

as required. In accordance with Maxwell's equations

$$j\mu\omega H_{x'} = \partial E_y/\partial z'$$

$$j\mu\omega H_{z'} = -\partial E_y/\partial x'$$

Thus we obtain eqns. 3.115 and 3.116 from eqn. 3.113.

Exercise

Reflection of plane waves is considered in some detail in the next chapter. As a preliminary exercise, consider the problem shown in Fig. 3.6. A plane wave polarised with the electric field in the x direction is normally incident on a plane interface, at $z = 0$, between two homogeneous regions. A subscript 1 designates the region to the left, and a subscript 2 designates the region to the right. Deduce the form of the reflected and transmitted waves for the situation indicated. Hint: use the fact that the tangential components of the electromagnetic fields are continuous at $z = 0$.

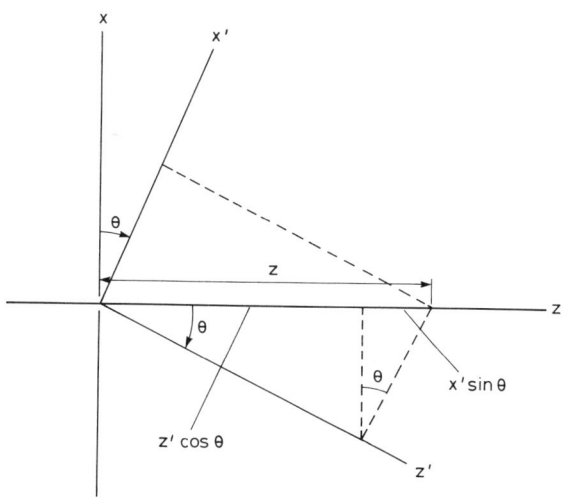

Fig. 3.5 Relevant geometry, giving z in terms of x' and z'

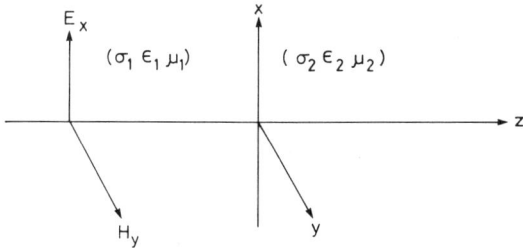

Fig. 3.6 Plane wave incident from the left normally onto the plane interface

Solution

For $z < 0$ it is evident that the electric field is the sum of an incident part and a reflected part, as follows:

$$E_x = E_x^{inc} + E_x^{refl}$$

$$= E_0(e^{-\gamma_1 z} + R\, e^{\gamma_1 z}) \qquad (3.117)$$

where E_0 is a specified constant, and R is a *reflection coefficient* to be determined. Similarly for the region $z > 0$ we see that

$$E_x = E_x^{trans} = E_0 T\, e^{-\gamma_2 z} \qquad (3.118)$$

where T is a *transmission coefficient* also to be determined. Now from Maxwell's equations we know that $-j\mu\omega H_y = \partial E_x/\partial z$. Therefore the corresponding expressions for the magnetic field are

$$H_y = E_0 \eta_1^{-1}(e^{-\gamma_1 z} - R e^{\lambda_1 z}) \quad \text{for } z < 0 \quad (3.119)$$

$$H_y = E_0 \eta_2^{-1} T e^{-\gamma_2 z} \quad \text{for } z > 0 \quad (3.120)$$

On matching the corresponding E_x and the corresponding H_y components at the interface $z = 0$, we obtain the two algebraic equations:

$$1 + R = T \quad (3.121)$$

$$\eta_1^{-1}(1 - R) = \eta_2^{-1} T \quad (3.122)$$

On solving for R and T, we get

$$R = (\eta_2 - \eta_1)/(\eta_2 + \eta_1) \quad (3.123)$$

$$T = 2\eta_2/(\eta_2 + \eta_1) \quad (3.124)$$

which are the desired results. We note that if σ_2 approaches ∞, corresponding to a metallic boundary, $R \simeq -1$ as expected; also, if η_2 approaches ∞, $R \simeq +1$, which is not so obvious. In the above we have defined

$$\eta_1 = \left(\frac{j\mu_1 \omega}{\sigma_1 + j\omega\varepsilon_1}\right)^{1/2}, \quad \eta_2 = \left(\frac{j\mu_2 \omega}{\sigma_2 + j\omega\varepsilon_2}\right)^{1/2}$$

and

$$\gamma_1 = [j\mu_1\omega(\sigma_1 + j\omega\varepsilon_1)]^{1/2}, \quad \gamma_2 = [j\mu_2\omega(\sigma_2 + j\omega\varepsilon_2)]^{1/2}$$

in accordance with our previous conventions. In the next chapter these results are generalized to oblique incidence

Exercise

Starting with Maxwell's equations for homogeneous media, derive the form of the second-order equation satisfied by the rectangular components of the electric and the magnetic fields. Perform the vector operations explicitly in rectangular co-ordinates.

Solution

Maxwell's equations in source-free regions are:

$$\nabla \times \mathbf{H} = (\sigma + j\omega\varepsilon)\mathbf{E}$$

$$\nabla \times \mathbf{E} = -j\omega\mu\mathbf{H}$$

These are easily combined to yield

$$\nabla \times \nabla \times \mathbf{E} = -\gamma^2 \mathbf{E}, \quad \gamma^2 = j\omega\mu(\sigma + j\omega\varepsilon)$$

which is also satisfied by $\mathbf{H}$. We now proceed in a straightforward manner:

$$\nabla \times \mathbf{E} = \begin{vmatrix} \mathbf{i}_x & \mathbf{i}_y & \mathbf{i}_z \\ \dfrac{\partial}{\partial x} & \dfrac{\partial}{\partial y} & \dfrac{\partial}{\partial z} \\ E_x & E_y & E_z \end{vmatrix}$$

$$= \mathbf{i}_x\left(\frac{\partial E_z}{\partial y} - \frac{\partial E_y}{\partial z}\right) + \mathbf{i}_y\left(\frac{\partial E_x}{\partial z} - \frac{\partial E_z}{\partial x}\right) + \mathbf{i}_z\left(\frac{\partial E_y}{\partial x} - \frac{\partial E_x}{\partial y}\right)$$

Dynamic fields

We again take the curl to get

$$\nabla \times (\nabla \times E) = \begin{vmatrix} i_x & i_y & i_z \\ \dfrac{\partial}{\partial x} & \dfrac{\partial}{\partial y} & \dfrac{\partial}{\partial z} \\ \left(\dfrac{\partial E_z}{\partial y} - \dfrac{\partial E_y}{\partial z}\right) & \left(\dfrac{\partial E_x}{\partial z} - \dfrac{\partial E_z}{\partial x}\right) & \left(\dfrac{\partial E_y}{\partial x} - \dfrac{\partial E_x}{\partial y}\right) \end{vmatrix}$$

$$= i_x \left(\frac{\partial^2 E_y}{\partial y \partial x} - \frac{\partial^2 E_x}{\partial y^2} - \frac{\partial^2 E_x}{\partial z^2} + \frac{\partial^2 E_z}{\partial z \partial x} \right) + i_y(\cdots) + i_z(\cdots)$$

$$= i_x \left[\frac{\partial}{\partial x}\left(\frac{\partial E_x}{\partial x} + \frac{\partial E_y}{\partial y} + \frac{\partial E_z}{\partial z} \right) - \frac{\partial^2 E_x}{\partial x^2} - \frac{\partial^2 E_x}{\partial y^2} - \frac{\partial^2 E_x}{\partial z^2} \right] + \cdots$$

$$= i_x \left(\frac{\partial}{\partial x} \nabla \cdot E - \nabla^2 E_x \right) + i_y \left(\frac{\partial}{\partial y} \nabla \cdot E - \nabla^2 E_y \right)$$

$$+ i_z \left(\frac{\partial}{\partial z} \nabla \cdot E - \nabla^2 E_z \right)$$

Now $\nabla \cdot E = 0$. Thus $(\nabla^2 - \gamma^2)\Psi = 0$, where $\Psi = E_x, E_y, E_z$.
Working with the magnetic field, we also get the same result.

Exercise

Using a separation of variables method, obtain a general solution of the *Helmholtz equation* $(\nabla^2 - \gamma^2)\Psi = 0$ in rectangular co-ordinates without any restriction on the x, y and z variations. Regard γ as a complex constant.

Solution

The partial differential equation (PDE) to be solved is

$$\left(\frac{\partial^2}{\partial x^2} + \frac{\partial^2}{\partial y^2} + \frac{\partial^2}{\partial z^2} - \gamma^2 \right)\Psi = 0$$

We try for a solution in the form $\Psi = X(x)\,Y(y)\,Z(z)$, which, as indicated, is a product of a function of x, a function of y and a function of z. Now we note that

$$\frac{\partial}{\partial x}\Psi = YZ\left(\frac{\partial}{\partial x}\right)X = YZ\left(\frac{dX}{dx}\right)$$

and

$$\frac{\partial^2}{\partial x^2}\Psi = YZ\left(\frac{d^2 X}{dx^2}\right)$$

Thus it follows that the above PDE is of the form

$$YZ\frac{d^2 X}{dx^2} + ZX\frac{d^2 Y}{dy^2} + XY\frac{d^2 Z}{dz^2} - \gamma^2 XYZ = 0$$

We divide through by XYZ to yield

$$\frac{1}{X}\frac{d^2 X}{dx^2} + \frac{1}{Y}\frac{d^2 Y}{dy^2} + \frac{1}{Z}\frac{d^2 Z}{dz^2} = \gamma^2$$

This tells us that we have a function of x, plus a function of y, plus a function of z, all equal to a constant γ^2. If this is to be true for all x, y and z, we must require that each of these functions are equal to constants. That is,

$$\frac{1}{X}\frac{d^2 X}{dx^2} = \gamma_1^2, \qquad \frac{1}{Y}\frac{d^2 Y}{dy^2} = \gamma_2^2, \qquad \frac{1}{Z}\frac{d^2 Z}{dz^2} = \gamma_3^2$$

where $\gamma_1^2 + \gamma_2^2 + \gamma_3^2 = \gamma^2$ is a constraint. Clearly individual solutions are of the form

$$X(x) = \exp(\pm \gamma_1 x), \quad Y(y) = \exp(\pm \gamma_2 y), \quad Z(z) = \exp(\pm \gamma_3 z)$$

The desired general solution is then

$$\Psi(x, y, z) = (A_1 e^{\gamma_1 x} + B_1 e^{-\gamma_1 x})(A_2 e^{\gamma_2 y} + B_2 e^{-\gamma_2 y})(A_2 e^{\gamma_3 z} + B_3 e^{-\gamma_3 z})$$

where A_1, B_1 etc. are disposable constants.

Exercise

Obtain a more convenient solution of the Helmholtz equation when it is written in the form $(\nabla^2 + k^2)\Psi = 0$, where k is real.

Solution

The same procedure is used, but now the individual solutions are chosen to be

$$X(x) = \exp(\pm jk_1 x), \quad Y(y) = \exp(\pm jk_2 y), \quad Z(z) = \exp(\pm jk_3 z)$$

where $k_1^2 + k_2^2 + k_3^2 = k^2$. Thus the general solution is best written

$$\Psi = (a_1 \cos k_1 x + b_1 \sin k_1 x)(a_2 \cos k_2 y + b_2 \sin k_2 y)$$

$$\times (a_3 \cos k_3 z + b_3 \sin k_3 z)$$

3.8 Review problems

(1) Show that $\nabla^2(1/r) = 0$, where $r = (x^2 + y^2 + z^2)^{1/2} = (\varrho^2 + z^2)^{1/2}$, by working out the results explicitly in both rectangular and cylindrical co-ordinates.
(2) Show that curl grad $(1/r) = 0$ by working in rectangular co-ordinates.
(3) Show that a symmetrical solution of $(\nabla^2 + k^2)V = 0$ is $\exp(\pm jkr)/r$.
(4) What is the general solution of the wave equation in problem (3) in rectangular co-ordinates?
(5) Repeat the exercise that leads to eqn. 2.25, but now take $j_x(t) = e^{-bt}u(t)$, $b > 0$.
(6) Do the exercise that leads to eqn. 2.35, but change the locations of source and sink to $x = +l/2$ and $-l/2$ on the x axis.
(7) Work out expressions for the E_x, E_y and E_z components in problem (6).
(8) Express the formula for the potential of the dipole given by eqn. 2.38 in terms of spherical co-ordinates. Then deduce E_r and E_ϕ directly using any of the formulas given in the appropriate part of Section 1.17. Also deduce J_r and J_θ.
(9) Sketch the current lines in the medium using previous expressions for $\mathbf{J}$.
(10) Verify the six vector calculus equations at the end of Section 1.17

Dynamic fields

by working out each side of each equation explicitly in rectangular co-ordinates.

(11) Let a travelling sound wave $u^{\text{inc}} = e^{-jkz}$ (of unit amplitude), coming from $z = -\infty$, be incident on a barrier at $z = z_0$ (see Fig. 3.7a). Deduce the form of the reflected wave u^{refl} assuming that the total field u^{tot} vanishes at $z = z_0$ (i.e. a 'soft' wall). As usual the time factor is $\exp(j\omega t)$, where $k = \omega/v$ and v is the velocity of sound. Sketch the intensity of the total field as a function of distance for several wavelengths from the barrier (to the left of course). Assume that intensity $= |u^{\text{tot}}|^2$.

(12) Repeat problem (11) for the case where the incident wave is given by $u^{\text{inc}} = e^{-jkz\cos\theta} e^{jkx\sin\theta}$, where θ is the angle of incidence (see Fig. 3.7b).

(13) Repeat both problems (11) and (12) for the case where the barrier is rigid such that $\partial u^{\text{tot}}/\partial z = 0$ at $z = z_0$.

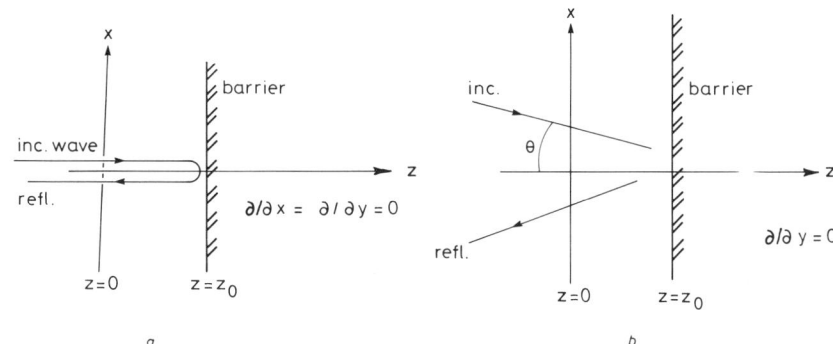

Fig. 3.7 *a* Normal incidence *b* Oblique incidence

Solutions

(1) Here we proceed by noting that

$$\frac{\partial}{\partial x}\frac{1}{r} = \left(\frac{\partial}{\partial r}\frac{1}{r}\right)\frac{\partial r}{\partial x} = -\frac{x}{r^3}$$

Now

$$\frac{\partial^2}{\partial x^2}\frac{1}{r} = -\frac{1}{r^3} + \frac{3x^2}{r^5}$$

Similarly

$$\frac{\partial^2}{\partial y^2}\frac{1}{r} = -\frac{1}{r^3} + \frac{3y^2}{r^5}$$

$$\frac{\partial^2}{\partial z^2}\frac{1}{r} = -\frac{1}{r^3} + \frac{3z^2}{r^5}$$

Thus

$$\nabla^2 \frac{1}{r} = \left(\frac{\partial^2}{\partial x^2} + \frac{\partial^2}{\partial y^3} + \frac{\partial^2}{\partial z^2}\right)\frac{1}{r} = -\frac{3}{r^3} + \frac{3(x^2+y^2+z^2)}{r^5} = 0$$

In cylindrical co-ordinates we only need show that

$$\nabla^2 \frac{1}{r} = \left[\frac{1}{\varrho}\frac{\partial}{\partial \varrho}\left(\varrho \frac{\partial}{\partial \varrho}\right) + \frac{\partial^2}{\partial z^2}\right](\varrho^2 + z^2)^{-1/2} = 0$$

details are omitted.

(2) curl grad $(1/r) = i_x\left(\dfrac{\partial^2}{\partial x \partial z}\dfrac{1}{r} - \dfrac{\partial^2}{\partial z \partial x}\dfrac{1}{r}\right) + i_y(\cdots) + i_z(\cdots) = 0$

(3) We proceed by noting that

$$\frac{\partial}{\partial x}\frac{e^{-\gamma r}}{r} = -x\left(\frac{1}{r^3} + \frac{\gamma}{r^2}\right)e^{-\gamma r}$$

and

$$\frac{\partial^2}{\partial x^2}\frac{e^{-\gamma r}}{r} = -\left(\frac{1}{r^3} + \frac{\gamma}{r^2}\right)e^{-\gamma r} + x\left(\frac{3x}{r^5} + \frac{2\gamma x}{r^4}\right)e^{-\gamma r}$$

$$+ x\left(\frac{1}{r^3} + \frac{\gamma}{r^2}\right)\gamma e^{-\gamma r}\frac{x}{r}$$

Similar forms apply for the y and z derivatives. Thus

$$(\nabla^2 - \gamma^2)\frac{e^{-\gamma r}}{r} = \left\{\left[-\frac{3}{r^3} + \frac{3(x^2 + y^2 + z^2)}{r^5}\right]r\right.$$

$$- \gamma\left[\frac{3}{r^2} - \frac{3(x^2 + y^2 + z^2)}{r^4}\right]r$$

$$\left. + \left[\frac{\gamma^2(x^2 + y^2 + z^2)}{r^2} - \gamma^2\right]\right\}\frac{e^{-\gamma r}}{r} = 0$$

(4) Possible solutions are

$$e^{\pm jk_1 x}, \quad e^{\pm jk_2 y}, \quad e^{\pm jk_3 z}$$

where $k_1^2 + k_2^2 + k_3^2 = k^2$.

The 'general solution' is the most general combination:

$$V = (a_1 e^{jk_1 x} + b_1 e^{-jk_1 x})(a_2 e^{jk_2 y} + b_2 e^{-jk_2 y})(a_3 e^{jk_3 z} + b_3 e^{-jk_3 z})$$

where a_1, b_1 etc. are constants. An equivalent form is

$$V = (a_1' \sin k_1 x + b_1' \cos k_1 x)(a_2' \cos k_2 y + b_2' \cos k_2 y)$$

$$(a_3' \sin k_3 z + b_3' \cos k_3 z)$$

where a_1', b_1' etc. are another set of constants.

(5) We follow the relevant solution method

$$\mathscr{L} j_x(t) = \mathscr{L} e^{-bt}u(t) = \frac{1}{s + b} = J_x(s)$$

Dynamic fields

Then
$$e_x(t) = \mathscr{L}^{-1} E_x(s) = \mathscr{L}^{-1} [(s + b)(\sigma + \varepsilon s)]^{-1}$$
$$= \mathscr{L}^{-1} \frac{1}{\sigma - b\varepsilon} \left(\frac{1}{s + b} - \frac{1}{s + \sigma/\varepsilon} \right)$$
$$= \frac{1}{\sigma - b\varepsilon} (e^{-bt} - e^{-\frac{\sigma t}{\varepsilon}}) u(t)$$

(6) In place of eqn. 2.35, we have
$$\psi = [I/4\pi(\sigma + j\varepsilon\omega)] (r_+^{-1} - r_-^{-1})$$
where
$$r_+ = \left[\left(x - \frac{l}{2}\right)^2 + y^2 + z^2 \right]^{1/2}$$
$$r_- = \left[\left(x + \frac{l}{2}\right)^2 + y^2 + z^2 \right]^{1/2}$$

(7) The limiting case of the dipole is treated in the same fashion. This amounts to replacing z by x, y by z, and x by y, in eqns. 2.38 to 2.42.

(8) Now eqn. 2.38 is written
$$\psi = Il \cos \theta / 4\pi(\sigma + j\omega\varepsilon) r^2$$

As usual $\boldsymbol{E} = -\operatorname{grad} \psi$, or $E_r = -\partial\psi/\partial r$ and $E_\theta = -(1/r)\partial\psi/\partial\theta$.
Carrying out the differentiation, we obtain eqns. 2.43 and 2.44.
Then
$$J_r = Il \cos \theta / 2\pi r^3$$
$$J_\theta = Il \sin \theta / 4\pi r^3$$

(9) To sketch the field lines, we may assign any value to $Il/2\pi$ and then merely calculate J_r and J_θ for various fixed points (r, θ). The vector $\boldsymbol{J} = \boldsymbol{i}_r J_r + \boldsymbol{i}_\theta J_\theta$ points in the direction of a field line. A more systematic procedure is to note the geometry in Fig. 3.8, where a

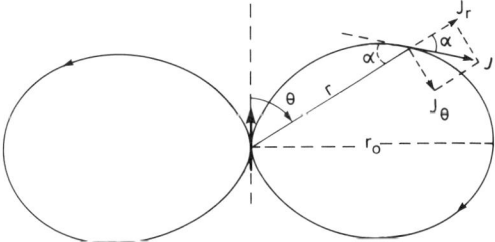

Fig. 3.8 Geometry of field or current line for electric dipole source

typical field line is shown. Now $\tan \alpha = J_\theta/J_r = \tan \theta/2$, where α is the angle subtended by the field line and the radius r drawn from the origin. But also $\tan \alpha = r d\theta/dr$. Thus
$$\frac{dr}{r} = \frac{2}{\tan \theta} d\theta \quad \text{or} \quad \ln r = \ln(\sin^2 \theta) + \text{const.}$$

Now if we say that $r = r_0$ at $\theta = \pi/2$, it is clear that

$$\ln r/r_0 = \ln(\sin^2\theta) \quad \text{or} \quad \frac{r}{r_0} = \sin^2\theta$$

This is the equation of the *field line*, for θ ranging from 0 to π, as shown in Fig. 3.8. Thus, at least for a dipole, all field lines have the same shape.

(10) (a) Now

$$\text{curl } A = i_x\left(\frac{\partial A_z}{\partial y} - \frac{\partial A_y}{\partial z}\right) + i_y(\cdots) + i_z(\cdots)$$

Then

$$\text{div curl } A = \frac{\partial}{\partial x}\left(\frac{\partial A_z}{\partial y} - \frac{\partial A_y}{\partial z}\right) + \frac{\partial}{\partial y}\left(\frac{\partial A_x}{\partial z} - \frac{\partial A_z}{\partial x}\right)$$

$$+ \frac{\partial}{\partial z}\left(\frac{\partial A_y}{\partial x} - \frac{\partial A_x}{\partial y}\right) = 0$$

(b) Now

$$\text{grad } V = i_x\frac{\partial V}{\partial x} + i_y\frac{\partial V}{\partial y} + i_z\frac{\partial V}{\partial z}$$

Then

$$\text{curl grad } V = i_x\left[\frac{\partial}{\partial y}\left(\frac{\partial V}{\partial z}\right) - \frac{\partial}{\partial z}\left(\frac{\partial V}{\partial y}\right)\right] + i_y(\cdots) + i_z(\cdots) = 0$$

In the following cases, proceed directly as indicated:

(c)

$$\text{div } V\boldsymbol{J} = \frac{\partial}{\partial x}(VJ_x) + \frac{\partial}{\partial y}(VJ_y) + \frac{\partial}{\partial z}(VJ_z)$$

$$= V\left(\frac{\partial J_x}{\partial x} + \frac{\partial J_y}{\partial y} + \frac{\partial J_z}{\partial z}\right) + J_x\frac{\partial V}{\partial x} + J_y\frac{\partial V}{\partial y} + J_z\frac{\partial V}{\partial z}$$

$$= V \text{ div } \boldsymbol{J} + \boldsymbol{J}\cdot\text{grad } V$$

(d)

$$\text{curl } V\boldsymbol{A} = i_x\left[\frac{\partial}{\partial y}(VA_z) - \frac{\partial}{\partial z}(VA_y)\right]$$

$$+ i_y\left[\frac{\partial}{\partial z}(VA_x) - \frac{\partial}{\partial x}(VA_z)\right]$$

$$+ i_z\left[\frac{\partial}{\partial x}(VA_y) - \frac{\partial}{\partial y}(VA_x)\right]$$

$$= i_x\left[V\left(\frac{\partial A_z}{\partial y} - \frac{\partial A_y}{\partial z}\right) - \left(A_y\frac{\partial V}{\partial z} - A_z\frac{\partial V}{\partial y}\right)\right]$$

Dynamic fields

$$+ i_y [(\cdots) - (\cdots)] + i_z[(\cdots) - (\cdots)]$$

$$= V \operatorname{curl} A - A \times \operatorname{grad} V$$

(e)

$$\operatorname{div}(A \times J) = \operatorname{div}[i_x(A_y J_z - A_z J_y) + i_y(\cdots) + i_z(\cdots)]$$

$$= \frac{\partial}{\partial x}(A_y J_z - A_z J_y) + \frac{\partial}{\partial y}(A_z J_x - A_x J_z) + \frac{\partial}{\partial z}(A_x J_y - A_y J_x)$$

$$= J_x\left(\frac{\partial A_z}{\partial y} - \frac{\partial A_y}{\partial z}\right) + J_y\left(\frac{\partial A_x}{\partial z} - \frac{\partial A_z}{\partial x}\right) + J_z\left(\frac{\partial A_y}{\partial x} - \frac{\partial A_x}{\partial y}\right)$$

$$- A_x\left(\frac{\partial J_z}{\partial y} - \frac{\partial J_y}{\partial z}\right) - A_y\left(\frac{\partial J_x}{\partial z} - \frac{\partial J_z}{\partial x}\right) - A_z\left(\frac{\partial J_y}{\partial x^2} - \frac{\partial J_x}{\partial y}\right)$$

$$= J \cdot \operatorname{curl} A - A \cdot \operatorname{curl} J$$

(f)

$$\operatorname{curl} A = \operatorname{curl} i_z A_z = i_x \frac{\partial A_z}{\partial y} - i_y \frac{\partial A_z}{\partial x}$$

$$\operatorname{curl} \operatorname{curl} A = i_x \frac{\partial^2 A_z}{\partial z \partial x} + i_y \frac{\partial^2 A_z}{\partial y \partial z} - i_z\left(\frac{\partial^2 A_z}{\partial x^2} + \frac{\partial^2 A_z}{\partial y^2}\right)$$

$$= i_x \frac{\partial}{\partial x}\left(\frac{\partial A_z}{\partial z}\right) + i_y \frac{\partial}{\partial y}\left(\frac{\partial A_z}{\partial z}\right) + i_z \frac{\partial}{\partial z}\left(\frac{\partial A_z}{\partial z}\right)$$

$$- i_z\left(\frac{\partial^2 A_z}{\partial z^2} + \frac{\partial^2 A_z}{\partial x^2} + \frac{\partial^2 A_z}{\partial y^2}\right)$$

$$= \operatorname{grad} \frac{\partial A_z}{\partial z} - i_z \nabla^2 A_z$$

(11) Choose $u^{\text{refl}} = D \, e^{jkz}$, where D is a constant to be found. Then the total field is:

$$u^{\text{tot}} = u^{\text{inc}} + u^{\text{refl}}$$

$$= e^{-jkz} + D \, e^{jkz}$$

where $k = 2\pi/\lambda$ and λ is wavelength.

The boundary condition at $z = z_0$ is now applied:

$$e^{-jkz_0} + D \, e^{jkz_0} = 0$$

Thus

$$D = -e^{-2jkz_0}$$

and

$$u^{\text{tot}} = e^{-jkz} - e^{-2jkz_0} \, e^{jkz}$$

The intensity is

$$|u^{tot}|^2 = 4 \sin^2 kl$$

where $l = z_0 - z$ and this is plotted in Fig. 3.9 as a function of the distance l from the barrier.

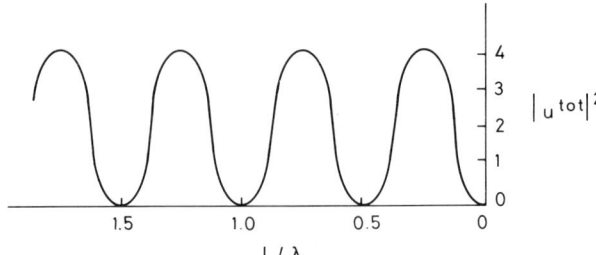

Fig. 3.9 Standing wave pattern: $\sin^2$ function

(12) Now we choose

$$u^{refl} = F\, e^{jkz \cos \theta}\, e^{jkx \sin \theta}$$

where F is to be determined by boundary condition at $z = z_0$. Then we get

$$F = -e^{-2jkz_0 \cos \theta}$$

The intensity is then found to be

$$|u^{tot}|^2 = 4 \sin^2(kl \cos \theta)$$

The standing wave pattern is precisely the same as in Fig. 3.9, but l is replaced by $l \cos \theta$.

(13) The expression for the total field has the same form as in problem (11). The condition that the normal derivative of the total field at the barrier vanish is

$$\frac{d}{dz}(e^{-jkz} + D\, e^{jkz})\big|_{z=z_0} = 0$$

Solving for D gives

$$D = +e^{-2jkz_0}$$

The desired expression is

$$u^{tot} = e^{-jkz} + e^{-2jkz_0}\, e^{jkz}$$

and the corresponding intensity is

$$|u^{tot}|^2 = 4 \cos^2 kl$$

This quantity is plotted in Fig. 3.10 as a function of l/λ.

In the case of oblique incidence, we have the same form for the reflected wave as in problem (12), but now find that

$$F = +e^{-2jkz_0 \cos \theta}$$

and the corresponding intensity is

$$|u^{tot}|^2 = 4\cos^2(kl\cos\theta)$$

This leads to the same standing wave pattern as in Fig. 3.10, but l is replaced by $l\cos\theta$.

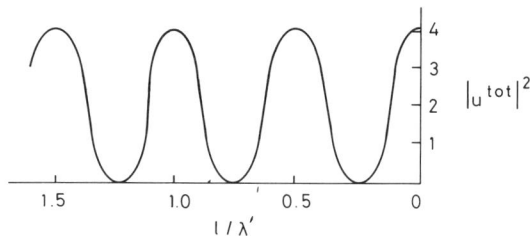

Fig. 3.10 Standing wave pattern: $\cos^2$ function

3.9 Bibliography

C. T. A. JOHNK, *Engineering Electromagnetics and Waves*, Chapter 3, Wiley, 1975.

D. S. JONES, *The Theory of Electromagnetism*, Chapter 1, Pergamon, 1964.

E. C. JORDAN and K. G. BALMAIN, *EM Waves and Radiating Systems*, Chapters 4 and 5 (highly recommended), Prentice Hall, 1968.

J. C. MAXWELL, *A Treatise on Electricity and Magnetism*, 3rd edn (as revised by J. J. Thompson), Clarendon Press, 1891 (reprint published by Dover, 1954). Note by J. J. (as he was affectionately called): 'I have attempted to verify the results which Maxwell gives without proof; I have not in all instances succeeded in arriving at the results given by him, and in such cases I have indicated the differences.' In particular note pp. 245–57, Vol. 2, for discussion of displacement current.

J. A. STRATTON, *Electromagnetic Theory*, Chapters 1 and 5, McGraw Hill, 1941.

W. L. WEEKS, *Electromagnetic Theory for Engineering Applications*, Wiley, 1964.

Chapter 4

Reflection and refraction of plane waves

4.1 Perfect reflector

To introduce the subject of reflection of electromagnetic waves, we consider a fairly simple example. A homogeneous half-space with free space parameters ε_0 and μ_0 occupies the region $z > 0$. The surface $z = 0$ is a perfect electrical conductor. A plane wave is now incident at an angle θ from the upper half-space, as indicated in Fig. 4.1. The incident wave in this example is polarised such the electric field vector $\boldsymbol{E}^{\text{inc}}$ is parallel to the conducting surface or ground plane. Then we orient our co-ordinate system such that $\boldsymbol{E}^{\text{inc}} = E_y^{\text{inc}} \boldsymbol{i}_y$. Thus, with reference to Fig. 4.1, we see that

$$\boldsymbol{E}^{\text{inc}} = E_0 \boldsymbol{i}_y \, e^{-jkx \sin \theta} \, e^{jkz \cos \theta} \tag{4.1}$$

where $k = (\varepsilon_0 \mu_0)^{1/2} \omega = 2\pi/\lambda_0$. Here E_0 is the complex value of the field E_y^{inc} at the origin ($x = 0, z = 0$), and is the same for all values of y. We also note that, at the ground plane ($z = 0$),

$$E_y^{\text{inc}}]_{z=0} = E_0 \, e^{-jkx \sin \theta} \tag{4.2}$$

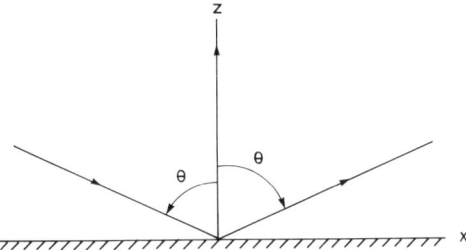

Fig. 4.1 Geometry for reflection of a plane wave from an interface at $z = 0$

This quantity has a constant amplitude, but there is a linear change of phase in the x direction.

Now in order that the tangential component E_y of the electric field vanishes at $z = 0$, the reflected wave $\boldsymbol{E}^{\text{refl}}$ must be chosen such that

$$\boldsymbol{E}^{\text{refl}} = E_y^{\text{refl}} \boldsymbol{i}_y \tag{4.3}$$

where

$$E_y^{\text{refl}}]_{z=0} = -E_0 \, e^{-jkx \sin \theta} \tag{4.4}$$

Now the resultant electric field for $z > 0$ is written
$$\mathbf{E} = \mathbf{i}_y(E_y^{\text{inc}} + E_y^{\text{refl}}) = \mathbf{i}_y E_y \tag{4.5}$$

Application of Maxwell's equations shows that
$$\mathbf{H} = \mathbf{i}_x(H_x^{\text{inc}} + H_x^{\text{refl}}) + \mathbf{i}_z(H_z^{\text{inc}} + H_z^{\text{refl}})$$
$$= \mathbf{i}_x H_x + \mathbf{i}_z H_z \tag{4.6}$$

where
$$H_x = \frac{1}{j\mu_0 \omega} \frac{\partial E_y}{\partial z} \tag{4.7}$$

$$H_z = -\frac{1}{j\mu_0 \omega} \frac{\partial E_y}{\partial x} \tag{4.8}$$

and
$$j\varepsilon_0 \omega E_y = \frac{\partial H_x}{\partial z} - \frac{\partial H_z}{\partial x} \tag{4.9}$$

On inserting eqns. 4.7 and 4.8 into 4.9, we obtain the *wave equation* for E_y:
$$\left(\frac{\partial^2}{\partial x^2} + \frac{\partial^2}{\partial z^2} + k^2\right) E_y = 0 \tag{4.10}$$

where
$$k = (\varepsilon_0 \mu_0)^{1/2} \omega$$

Now we see that because both E_y and E_y^{inc} satisfy eqn. 4.10, E_y^{refl} also satisfies eqn. 4.10. Obviously the desired solution is
$$E_y^{\text{refl}} = -E_0 \, e^{-jkx\sin\theta} \, e^{-jkz\cos\theta} \tag{4.11}$$

This is a wave propagating in the positive z direction; it reduces to eqn. 4.4 at $z = 0$.

The resultant field is given by
$$E_y = E_y^{\text{inc}} + E_y^{\text{refl}}$$
$$= E_0 \, e^{-jkx\sin\theta} (e^{jkz\cos\theta} - e^{-jkz\cos\theta})$$
$$= E_0 2j \, e^{-jkx\sin\theta} \sin(kz\cos\theta) \tag{4.12}$$

Clearly this expression for the electric field vanishes at the perfectly conducting ground plane. Also we note that $E_y = 0$ for $kz\cos\theta = n\pi$, where $n = 1, 2, 3, \ldots$. Furthermore $|E_y|$ has a maximum value of $2|E_0|$ whenever $kz\cos\theta = (2n-1)\pi/2$ where $n = 1, 2, 3 \ldots$ The function $|E_y/E_0|^2$, which is a relative intensity, is plotted in Fig. 4.2 as a function of $kz\cos\theta$. This is sometimes called a *standing wave* pattern. It arises whenever there is a strong interference between the incident and the reflected wave.

Exercise

With reference to Fig. 4.1, consider the case where the incident wave is polarized such that

$$\boldsymbol{H}^{\mathrm{inc}} = \boldsymbol{i}_y H_0 \, e^{-jkx\sin\theta} \, e^{jkz\cos\theta} \qquad (4.13)$$

Assuming that the ground plane is perfectly conducting, show that the relation

$$\boldsymbol{H}^{\mathrm{refl}} = \boldsymbol{i}_y H_0 \, e^{-jkx\sin\theta} \, e^{-jkz\cos\theta} \qquad (4.14)$$

is the appropriate expression for the reflected wave.
Hint: note that the expression

$$E_x = -\frac{1}{j\varepsilon_0 \omega} \frac{\partial H_y}{\partial z} \qquad (4.15)$$

must vanish at $z = 0$.

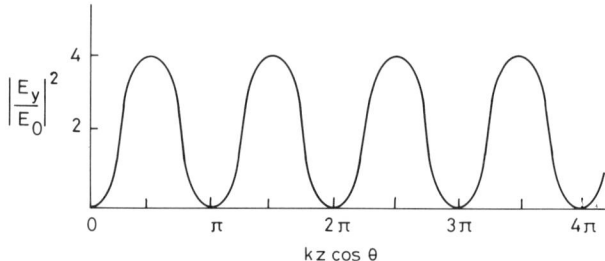

Fig. 4.2 Standing wave pattern for the intensity above a perfect reflector

4.2 Reflection from a dielectric half-space

We must now deal with the important case where a plane wave, from free space, is incident on a dielectric half-space (for $z < 0$) (see Fig. 4.3). Again we define the electric field to be polarised in accordance with eqn. 4.1. We also assume that the resultant electric field for $z > 0$ has the form given by eqn. 4.5. But now we modify eqn. 4.12 to be of the form

$$\begin{aligned} E_y &= E_y^{\mathrm{inc}} + E_y^{\mathrm{refl}} \\ &= E_0 \, e^{-jkx\sin\theta} \left(e^{jkz\cos\theta} + R_\perp \, e^{-jkz\cos\theta} \right) \end{aligned} \qquad (4.16)$$

where $R_\perp$ is yet to be determined. The subscript $\perp$ here is to signify that the electric field (parallel to the y axis) is perpendicular to the $y = 0$ plane, which is called the 'plane of incidence'.

From the Faraday-Maxwell equation, the resultant magnetic field components are first

$$H_x = \frac{1}{j\mu_0\omega} \frac{\partial E_y}{\partial z} = E_0 M_0 \, e^{-jkx\sin\theta} \left(e^{jkz\cos\theta} - R_\perp \, e^{-jkz\cos\theta} \right) \qquad (4.17)$$

where

$$M_0 = (k/\mu_0\omega) \cos\theta = \eta_0^{-1} \cos\theta \qquad (4.18)$$

Reflection and refraction of plane waves

and second

$$H_z = -\frac{1}{j\mu_0\omega}\frac{\partial E_y}{\partial x} = E_0 P_0 \, e^{-jkx\sin\theta}(e^{jkz\cos\theta} + R_\perp \, e^{-jkz\cos\theta}) \quad (4.19)$$

where

$$P_0 = (k/\mu_0\omega)\sin\theta = \eta_0^{-1}\sin\theta \quad (4.20)$$

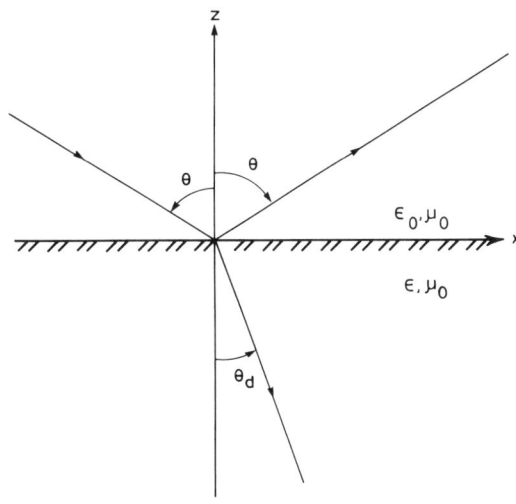

Fig. 4.3 Geometry for a plane wave incident at an angle θ on a dielectric half-space. The plane of incidence is $y = 0$ or the x–z plane

We may readily verify that eqns. 4.16, 4.17 and 4.19 satisfy eqn. 4.10.

We now need to set up expressions for the fields in the lower dielectric half-space. The permittivity is ε, the conductivity is zero, and the permeability is assumed to be μ_0 (as for free space). The key step is to note that the factor $\exp(-jkx\sin\theta)$ that is present in the incident electromagnetic field must carry through everywhere. Otherwise the boundary conditions could not be met for all values of x along the interface $z = 0$. Thus the suggested form of the solution for the electric field, for $z < 0$, is

$$\mathbf{E}_d = \mathbf{i}_y E_{dy} \quad (4.21)$$

where

$$E_{dy} = f(z) \, e^{-jkx\sin\theta} \quad (4.22)$$

where the function $f(z)$ is yet to be determined.

In direct analogy to eqn. 4.10, we see that E_{dy} must satisfy

$$\left(\frac{\partial^2}{\partial x^2} + \frac{\partial^2}{\partial z^2} + k_d^2\right)E_{dy} = 0 \quad (4.23)$$

where

$$k_d = (\varepsilon\mu_0)^{1/2}\omega$$

then inserting eqn. 4.22 into eqn. 4.23 we see that

$$\left[\frac{d^2}{dz^2} + (k_d^2 - k^2 \sin^2 \theta)\right] f(z) = 0 \tag{4.24}$$

The desired form of the solution is written

$$f(z) = E_0 T_\perp \exp\left[+j(k_d^2 - k^2 \sin^2 \theta)^{1/2} z\right] \tag{4.25}$$

where $T_\perp$ is a constant yet to be determined (the factor E_0 is included for later convenience). The $+j$, rather than $-j$, is chosen in the exponent so that only outwardly propagating waves are present when $z \to -\infty$.

At this point we may define an angle θ_d via the substitution

$$(k_d^2 - k^2 \sin^2 \theta)^{1/2} = k_d \cos \theta_d \tag{4.26}$$

Now the desired form of the solution, for $z < 0$, reads

$$E_{dy} = E_0 T_\perp e^{-jkx \sin \theta} e^{jk_d z \cos \theta_d} \tag{4.27}$$

But according to eqn. 4.26,

$$k \sin \theta = k_d \sin \theta_d \tag{4.28}$$

so that

$$E_{dy} = E_0 T_\perp e^{-jk_d x \sin \theta_d} e^{jk_d z \cos \theta_d} \tag{4.29}$$

This is clearly identified as a plane wave propagating at angle θ_d with the negative z axis. Equation 4.28 is *Snell's law*, which has been known in optics for more than a century. As indicated, when $N_d = k_d/k$, the refractive index is greater than one and the angle θ_d is less than θ, and hence the wave normal is bent toward the surface normal.

The components of the magnetic field in the dielectric ($z < 0$) are easily found to be, first

$$H_{dx} = \frac{1}{j\mu_0 \omega} \frac{\partial E_{dy}}{\partial z} = E_0 M_d e^{-jkx \sin \theta} T_\perp e^{jk_d z \cos \theta_d} \tag{4.30}$$

where

$$M_d = (k_d/\mu_0 \omega) \cos \theta_d \tag{4.31}$$

and second

$$H_{dz} = \frac{-1}{j\mu_0 \omega} \frac{\partial E_{dy}}{\partial x} = E_0 P_d e^{-jkx \sin \theta} T_\perp e^{jk_d z \cos \theta_d} \tag{4.32}$$

where

$$P_d = (k_d/\mu_0 \omega) \sin \theta_d \tag{4.33}$$

From the foregoing it is clear that $R_\perp$ and $T_\perp$ can be interpreted as a reflection and a transmission coefficient, respectively, at the interface $z = 0$. We shall determine $R_\perp$ and $T_\perp$ by applying appropriate boundary conditions at $z = 0$. Specifically it will be noted that *the tangential field components E_y and H_x are continuous at $z = 0$*. Such continuity is clearly evident when we consider hypothetical circuits as shown in Fig. 4.4. In the upper diagram, the line integral of H around the slender

Reflection and refraction of plane waves

rectangular path vanishes in the limit when 2δ tends to zero. Similarily, in the lower diagram, the line integral of E around a slender rectangular path vanishes in the limit when 2δ tends to zero.

The preceding statements are written in mathematical form as follows:

$$\lim_{\delta \to 0} [H_{dx}(z = -\delta) - H_x(z = +\delta)] = 0 \tag{4.34}$$

$$\lim_{\delta \to 0} [E_{dy}(z = -\delta) - E_y(z = +\delta)] = 0 \tag{4.35}$$

These *boundary conditions*, which are a direct consequence of Maxwell's equations, must hold for all values of x and y. Using the field expressions given by eqns. 4.16, 4.17, 4.27 and 4.30, the boundary conditions 4.34 and 4.35 tell us that

$$M_d T_\perp = M_0(1 - R_\perp) \tag{4.36}$$

and

$$T_\perp = 1 + R_\perp \tag{4.37}$$

Then clearly

$$\frac{1 - R_\perp}{1 + R_\perp} = \frac{M_d}{M_0} \tag{4.38}$$

or

$$R_\perp = \frac{M_0 - M_d}{M_0 + M_d} = \frac{k \cos \theta - k_d \cos \theta_d}{k \cos \theta + k_d \cos \theta_d} \tag{4.39}$$

Furthermore

$$T_\perp = \frac{2M_0}{M_0 + M_d} = \frac{2 k \cos \theta}{k \cos \theta + k_d \cos \theta_d} \tag{4.40}$$

Using eqn. 4.26 the reflection coefficient can also be written in equivalent form as follows:

$$R_\perp = \frac{k \cos \theta - (k_d^2 - k^2 \sin^2 \theta)^{1/2}}{k \cos \theta + (k_d^2 - k^2 \sin^2 \theta)^{1/2}} \tag{4.41}$$

or

$$R_\perp = \frac{\cos \theta - (N_d^2 - \sin^2 \theta)^{1/2}}{\cos \theta + (N_d^2 - \sin^2 \theta)^{1/2}} \tag{4.42}$$

where

$$N_d = k_d/k = (\varepsilon/\varepsilon_0)^{1/2}$$

This form for $R_\perp$ is the *Fresnel reflection coefficient* for a half-space of *refractive index* N_d when the electric field is perpendicular to the plane of incidence. Clearly, if $N_d \gg 1$, $R_\perp$ tend to -1 and the interface appears to act as a perfect conductor for all values of θ from 0 to $\pi/2$. Also we note that if $\theta = 0$, corresponding to normal incidence,

$$R_\perp = (1 - N_d)/(1 + N_d) \tag{4.43}$$

which has a magnitude $|R_\perp| \leq 1$ for $N_d \geq 1$.

It is now instructive to repeat the preceding derivation for the case where the magnetic field vector is perpendicular to the plane of incidence. In this case we may choose the form given by eqn. 4.13. Thus both the incident and the reflected waves have only y components of the magnetic field. In place of eqn. 4.16, we now write, for $z > 0$,

$$H_y = H_y^{\text{inc}} + H_y^{\text{refl}} \qquad (4.44)$$

$$= H_0 e^{-jkx\sin\theta} (e^{jkz\cos\theta} + R_\| e^{-jkz\cos\theta})$$

where $R_\|$ is the reflection coefficient. The subscript $\|$ signifies that the electric field is parallel to the plane of incidence.

From the Ampère-Maxwell equation, we now deduce, for $z > 0$, first that

$$E_x = -\frac{1}{j\varepsilon_0\omega}\frac{\partial H_y}{\partial z}$$

$$= -H_0 K_0 e^{-jkx\sin\theta} (e^{jkz\cos\theta} - R_\| e^{-jkz\cos\theta}) \qquad (4.45)$$

where

$$K_0 = \frac{k\cos\theta}{\varepsilon_0\omega} = \eta_0 \cos\theta \qquad (4.46)$$

and second that

$$E_z = \frac{1}{j\varepsilon_0\omega}\frac{\partial H_z}{\partial x} = -L_0 H_y \qquad (4.47)$$

where

$$L_0 = \frac{k\sin\theta}{\varepsilon_0\omega} = \eta_0 \sin\theta \qquad (4.48)$$

The expressions for the field components in the dielectric half-space $z < 0$ parallel those given by eqns. 4.27, 4.30 and 4.32. In fact, H_y satisfies the same equation 4.23. Thus we easily deduce that, for $z < 0$,

$$H_{dy} = H_0 T_\| e^{-jkx\sin\theta} e^{jk_d z\cos\theta_d} \qquad (4.49)$$

$$E_{dx} = -K_d H_{dy} \qquad (4.50)$$

where

$$K_d = \frac{k_d \cos\theta_d}{\varepsilon\omega} \qquad (4.51)$$

and

$$E_{dz} = -L_d H_{dy} \qquad (4.52)$$

where

$$L_d = \frac{k\sin\theta}{\varepsilon\omega} \qquad (4.53)$$

We now employ the same argument as before and match the tangential field components. On carrying out this process leads to

Reflection and refraction of plane waves

$$K_d T_\| = K_0(1 - R_\|) \tag{4.54}$$

and

$$T_\| = 1 + R_\| \tag{4.55}$$

These are solved as before for the reflection and transmission coefficients to yield

$$R_\| = (K_0 - K_d)/(K_0 + K_d) \tag{4.56}$$

$$T_\| = 2K_0/(K_0 + K_d) \tag{4.57}$$

An equivalent statement of eqn. 4.56 is

$$R_\| = \frac{N_d^2 \cos\theta - (N_d^2 - \sin^2\theta)^{1/2}}{N_d^2 \cos\theta + (N_d^2 - \sin^2\theta)^{1/2}} \tag{4.58}$$

which is the Fresnel reflection coefficient for a half-space of refractive index N_d when the electric field is parallel to the plane of incidence (i.e. E is parallel to the (x, y) plane in Fig. 4.4).

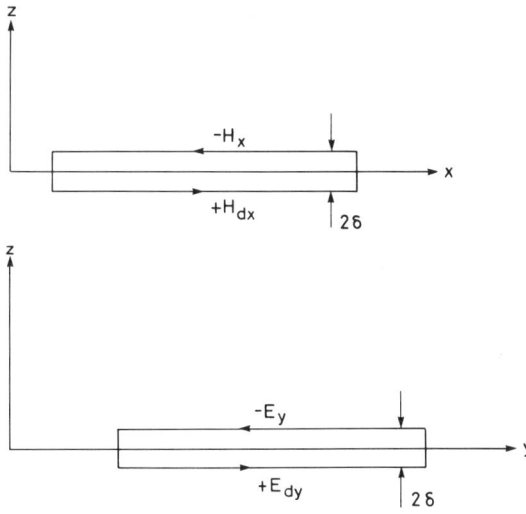

Fig. 4.4 Application of Ampère's and Faraday's laws to show that tangential magnetic and electric fields are continuous at an interface

In the limiting case where $N_d \gg 1$, $R_\|$ tends to $+1$, which has the same behaviour as a perfectly conducting ground plane (i.e. the reflected wave is given by eqn. 4.14). In the case of normal incidence, where $\theta = 0$, we see that

$$R_\| = (N_d - 1)/(N_d + 1) \tag{4.59}$$

which has a magnitude $|R_\|| \leqslant 1$ for $N_d \geqslant 1$. The sign difference between eqns. 4.43 and 4.59 is merely a result of our conventions in defining $R_\|$ and $R_\perp$.

The behaviour of the reflection coefficient magnitudes is illustrated in Fig. 4.5, where we plot $R_\|$ and $R_\perp$ as functions of θ from 0 to $\pi/2$. We note that in the limiting case of grazing incidence, both reflection coefficients have the same limit. That is

$$\lim_{\theta \to \pi/2} R_\perp \to -1$$

$$\lim_{\theta \to \pi/2} R_\| \to -1$$

The major difference between the curves shown in Fig. 4.5 is the null that occurs for $R_\|$. As can be seen clearly from eqn. 4.58, $R_\| = 0$ when

$$N_d^4 \cos^2 \theta = N_d^2 - \sin^2 \theta \qquad (4.60)$$

or when

$$\theta = \theta_b \qquad (4.61)$$

where

$$\tan \theta_b = N_d$$

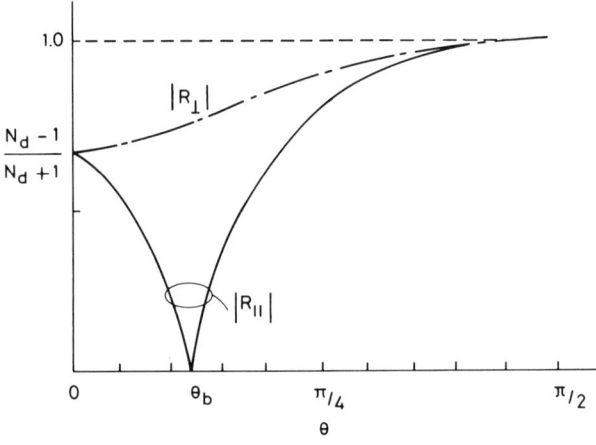

Fig. 4.5 The magnitude of the reflection coefficients for oblique incidence on a dielectric half-space of refractive index N_d

This value of θ_b is termed the *Brewster angle*. At this angle, *all the energy in the incident wave (for parallel type polarisation) is being transmitted into the dielectric*. In the notation of eqn. 4.56, we see that the condition of pure matching occurs if the effective wave impedances are identical; that is, $K_0 = K_d$.

Exercise

For the case of perpendicular incidence, i.e. $\boldsymbol{E}^{\text{inc}} = \boldsymbol{i}_y E_y^{\text{inc}}$, show, for the dielectric model, that the *surface admittance* is

$$[H_x/E_y]_{z=0} = M_d \qquad (4.62)$$

where M_d is given by eqn. 4.31.

Exercise

For the case of parallel incidence, i.e. $\boldsymbol{H}^{\text{inc}} = \boldsymbol{i}_y H_y^{\text{inc}}$, show that the *surface impedance* is

$$[E_x/H_y]_{z=0} = -K_d \qquad (4.63)$$

for the same dielectric half-space, where K_d is given by eqn. 4.51.

Exercise

Reconcile the seeming paradox that, according to eqn. 4.58,

$$\lim_{N_d \to \infty} R_\parallel \to +1$$

for *all* θ, whereas

$$\lim_{\theta \to \pi/2} R_\parallel = -1$$

Exercise

In considering the reflection of microwaves from atmospheric layers, it is often permissible to say that $N_d = 1 + \delta N$ where $|\delta N| \ll 1$. Then show that, for grazing angles $\theta \simeq (\pi/2) - \psi$ where $\psi \ll 1$, it is possible to approximate both reflection coefficients as

$$R_\perp \simeq R_\parallel \simeq -\delta N/(2\psi^2) \tag{4.64}$$

provided $\psi^2 \gg \delta N$ (i.e. not too near grazing angles).

Solution

Note that

$$(N_d^2 - \sin^2\theta)^{1/2} = [(1 + \delta N)^2 - 1 + \sin^2\psi]^{1/2}$$

$$= [2\delta N + (\delta N)^2 + \sin^2\psi]^{1/2} \simeq (2\delta N + \psi^2)^{1/2}$$

$$\simeq \psi\left(1 + \frac{2\delta N}{\psi^2}\right)^{1/2} \simeq \psi\left(1 + \frac{\delta N}{\psi^2}\right)$$

$$\simeq \psi + \frac{\delta N}{\psi} \tag{4.65}$$

Furthermore,

$$N^2 \cos\theta = (1 + \delta N)^2 \sin\psi \simeq \sin\psi \simeq \psi \tag{4.66}$$

and $\cos\theta = \sin\psi \simeq \psi$. Then both denominators in eqns. 4.42 and 4.58 are approximated by 2ψ. Such approximations are valid when $\delta N \ll \psi^2 \ll 1$. For example, $\delta N = 10^{-6}$, $\psi = 10^{-2}(0.57°)$ would satisfy the conditions, in which case

$$R_\perp \simeq R_\parallel \simeq -\delta N/(2\psi^2) = -0.005 \tag{4.67}$$

4.3 Critical reflection

There is an important class of reflection problems that arise when the relative refractive index N_d is less than one. A physical example, in the context of the geometry shown in Fig. 4.3, is when the lower region, for $z < 0$, is an ideal electron plasma. For present purposes, we can use the following simple formula for the effective refractive index:

$$N_d^2 = 1 - \frac{\omega_0^2}{\omega^2} \tag{4.68}$$

The plasma frequency ω_0 is defined by

$$\omega_0^2 = ne^2/\varepsilon_0 m \tag{4.69}$$

where e is the charge on the electron, m is the mass of the electron, ε_0 is the permittivity of free space, and n is the number of electrons per cubic metre. In the case where $\omega_0^2 < \omega^2$, it is clear that N_d is real and less than one.

Now, in general, the radical in the reflection coefficient formulas for $R_\perp$ and $R_\parallel$ can be written

$$(N_d^2 - \sin^2\theta)^{1/2} = -j(\sin^2\theta - N_d^2)^{1/2} \qquad (4.70)$$

Then eqns. 4.42 and 4.58 take the form

$$R_\perp = \frac{\cos\theta + j(\sin^2\theta - N_d^2)^{1/2}}{\cos\theta - j(\sin^2\theta - N_d^2)^{1/2}} \qquad (4.71)$$

$$R_\parallel = \frac{N_d^2\cos\theta + j(\sin^2\theta - N_d^2)^{1/2}}{N_d^2\cos\theta - j(\sin^2\theta - N_d^2)^{1/2}} \qquad (4.72)$$

If $\sin\theta > N_d$, the right-hand sides of eqns. 4.71 and 4.72 have a magnitude of one. The angle θ_c, as defined by $\sin\theta_c = N_d$, is termed the *critical angle*. Thus, when $\theta > \theta_c$, we see that $|R_\perp| = |R_\parallel| = 1$. In this case there is complete reflection, but a phase shift occurs that is easily computed from eqns. 4.71 or 4.72.

For both perpendicular and parallel polarisations, the transmitted wave in the lower medium is characterised by

$$\exp[jk(N_j^2 - \sin^2\theta)^{1/2}z]\exp(-jkx\sin\theta)$$

or equivalently by

$$\exp[k(\sin^2\theta - N_d^2)^{1/2}z]\exp(-jkx\sin\theta)$$

Clearly when $\theta > \theta_c$, corresponding to total reflection, the exponent is real in the z-dependent factor. In this case, the wave in the z direction is heavily damped. Such a wave is sometimes described as being *evanescent*.

The ray picture for reflection from such underdense media is shown in Fig. 4.6. The rays with the single arrows correspond to the noncritical

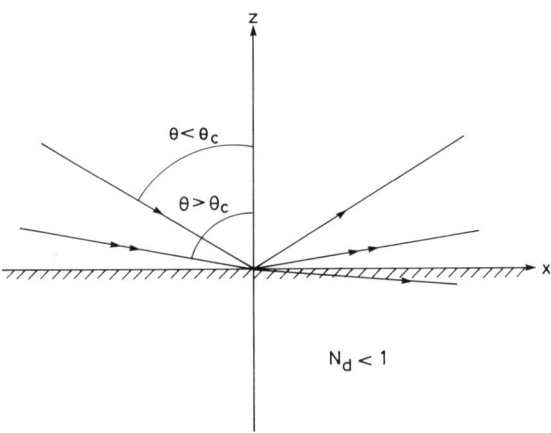

Fig. 4.6 Ray picture for reflection from an underdense medium in the case where $\theta < \theta_c$ (no critical reflection) and where $\theta > \theta_c$ (critical reflection)

case (i.e. $\theta < \theta_c$) when there is both a reflected and a transmitted ray. The rays with the double arrows correspond to the critical condition (i.e. $\theta > \theta_c$) when there is no transmitted propagating wave.

Exercise

In discussing the problem of critical reflection, we have regarded the underdense medium as a lossless plasma where indeed the refractive index may be less than one. But, in fact, critical reflection can occur in any situation where the incident plane wave, incident at say an angle θ_2, is coming from the relatively denser of the two dielectric half-spaces. Show that the condition for critical reflection is $\theta_2 > \theta_c$ if $\sin \theta_c = (\varepsilon_1/\varepsilon_2)^{1/2}$, where ε_1 and ε_2 are the permittivities of the two media (here $\varepsilon_1 < \varepsilon_2$). Thus, in the case of air (medium 1) and pure water (medium 2), where $\varepsilon_2/\varepsilon_1 = 81$, show that the critical angle is given by $\theta_c = \arcsin(1/9) \simeq 6\cdot4°$. Thus, for all angles of incidence θ_2 from $6\cdot4°$ to $90°$, the transmitted energy from the water, into the air, is zero.

Exercise

Show that the reflection coefficient formulas given by eqn. 4.42 can be obtained by matching only the normal and the tangential components of the magnetic field.

Solution

On matching H_z, given by eqn. 4.19, and H_{dz}, given by eqn. 4.32, at $z = 0$, we obtain eqn. 4.37.

4.4 Reflection from a lossy half-space

The solution for a plane wave incident from free space on a general dissipative half-space is a relatively minor extension of the foregoing development. At the same time, we allow the magnetic permeability in the half-space to have a value μ. The situation is illustrated in Fig. 4.7.

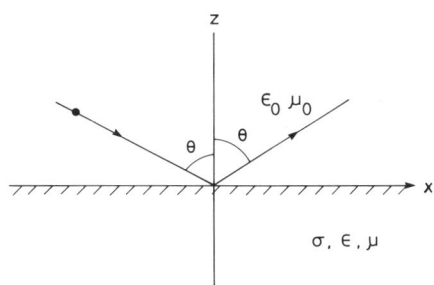

Fig. 4.7 Geometry for reflection of a plane wave from a lossy half-space occupying the region $z < 0$

As before, the region $z > 0$ is assumed to be free space with properties ε_0 and μ_0.

The incident wave is polarised such that the magnetic field vector $\mathbf{H}^{\text{inc}}$ is parallel to the y axis. This is sometimes called *transverse magnetic* (TM) to signify that the vector magnetic field is transverse or perpendicular to the plane of incidence (i.e. the x, z plane). Thus we may write

$$\mathbf{H}^{\text{inc}} = \mathbf{i}_y H_0 \, e^{u_0 z} \, e^{-jqx} \qquad (4.73a)$$

where H_0 is the specified amplitude, $u_0 = jk \cos \theta$ and $q = k \sin \theta$. As before, $k = (\varepsilon_0 \mu_0)^{1/2} \omega = \omega/c = 2\pi/\lambda_0$ is the free space wave number. This slight change of notation is convenient in what follows.

The reflected wave is now postulated to have the form

$$\mathbf{H}^{\text{refl}} = \mathbf{i}_y H_0 R_\parallel \, e^{-u_0 z} \, e^{-jqx} \qquad (4.73b)$$

Then the resultant magnetic field, for the region $z > 0$, must have the form

$$H_y = H_0(e^{u_0 z} + R_\| e^{-u_0 z}) e^{-jqx} \tag{4.74}$$

The corresponding electric field components are obtained from

$$E_x = -\frac{1}{j\varepsilon_0 \omega} \frac{\partial H_y}{\partial z} \tag{4.75}$$

$$E_z = \frac{1}{j\varepsilon_0 \omega} \frac{\partial H_y}{\partial x} \tag{4.76}$$

When dealing with the lower half-space, for $z < 0$, we observe that Maxwell's equations take the form

$$\frac{\partial E_x}{\partial z} - \frac{\partial E_z}{\partial x} = -j\mu\omega H_y \tag{4.77}$$

$$E_x = -\frac{1}{\sigma + j\varepsilon\omega} \frac{\partial H_y}{\partial z} \tag{4.78}$$

$$E_z = \frac{1}{\sigma + j\varepsilon\omega} \frac{\partial H_y}{\partial x} \tag{4.79}$$

These are combined in the usual manner to show that, for $z < 0$,

$$\left(\frac{\partial^2}{\partial x^2} + \frac{\partial^2}{\partial z^2} - \gamma^2\right) H_y = 0 \tag{4.80}$$

where

$$\gamma = [j\mu\omega(\sigma + j\varepsilon\omega)]^{1/2}$$

Then we seek a solution of eqn. 4.80 in the form

$$H_y = H_0 T_\| e^{uz} e^{-jqx} \tag{4.81}$$

where $T_\|$ is to be determined. Clearly, if eqn. 4.81 is to satisfy eqn. 4.80, we must have

$$u = (q^2 + \gamma^2)^{1/2} \tag{4.82}$$

where the radical is chosen such that $\operatorname{Re} u > 0$.

The boundary conditions are now applied at $z = 0$. Specifically, we match H_y as given by eqn. 4.74 with H_y as given by eqn. 4.81. Also we match E_x as obtained from eqn. 4.75 with E_x as obtained from eqn. 4.78. This process leads to the explicit formulas

$$R_\| = (K_0 - K)/(K_0 + K) \tag{4.83}$$

$$T_\| = 2K_0/(K_0 + K) \tag{4.84}$$

where

$$K_0 = u_0/j\varepsilon_0 \omega \tag{4.85}$$

$$K = u/(\sigma + j\varepsilon\omega) \tag{4.86}$$

In the limiting case where $\sigma = 0$ and $\mu = \mu_0$, the expressions for the reflection and transmission coefficients above become identical to those previously derived for the pure dielectric half-space (e.g. compare eqns. 4.83 and 4.84 with 4.56 and 4.57, respectively).

The solution for the problem of an incident plane wave polarised with the electric field in the y direction is carried out in the same fashion. This polarisation is referred to as *transverse electric* (TE) because the electric vector is transverse or perpendicular to the plane of incidence.

With reference to Fig. 4.7, we now specify that

$$\boldsymbol{E}^{\text{inc}} = \boldsymbol{i}_y E_0 \, e^{u_0 z} \, e^{-jqx} \tag{4.87}$$

where $u_0 = jk \cos \theta$ and $q = k \sin \theta$, as before. Again we postulate that the reflected wave is given by

$$\boldsymbol{E}^{\text{refl}} = \boldsymbol{i}_y E_0 R_\perp \, e^{-u_0 z} \, e^{-jqx} \tag{4.88}$$

The resultant electric field, which has only a y component, is then

$$E_y = E_0 (e^{u_0 z} + R_\perp \, e^{-u_0 z}) \, e^{-jqx} \qquad z > 0 \tag{4.89}$$

$$E_y = E_0 T_\perp \, e^{uz} \, e^{-jqx} \qquad z < 0 \tag{4.90}$$

By matching the tangential fields E_y and H_x at $z = 0$, we deduce that

$$R_\perp = (M_0 - M)/(M_0 + M) \tag{4.91}$$

$$T_\perp = 2M_0/(M_0 + M) \tag{4.92}$$

where

$$M_0 = u_0/j\mu_0 \omega \tag{4.93}$$

$$M = u/j\mu\omega \tag{4.94}$$

$$u = (q^2 + \gamma^2)^{1/2}$$

Again we note that in the limiting case where $\sigma = 0$ and $\mu = \mu_0$, the expressions for the reflection and transmission coefficients become identical to their pure dielectric counterparts (e.g. compare eqns. 4.91 and 4.92 with 4.39 and 4.40, respectively).

To deal with a concrete numerical example, we select parameters that would correspond quite well to reflection of radio waves from a moderately conducting soil: $\sigma = 10^{-2}$ s/m, $\varepsilon/\varepsilon_0 = 4$ and $\mu = \mu_0$. Values of $|R_\parallel|$ and arg $R_\parallel$ are plotted in Fig. 4.8 as a function of θ for three different frequencies of 1·5, 15 and 150 MHz. Similar plots are shown in Fig. 4.9 for values of $|R_\perp|$ and arg $R_\perp$. These show very clearly that the complex values of the reflection coefficients, for both wave polarisations, approach -1 as the grazing angle $(90° - \theta)$ tends to zero. Also, most importantly, we see that the magnitude $|R_\parallel|$ has a minimum in the region between 60° and 80°; this is the same phenomenon as the zero of $R_\parallel$ at the Brewster angle θ_b for the dielectric half-space. Because of the nonzero value of the conductivity for the plots in Fig. 4.8 there is no real Brewster angle where $|R_\parallel|$ vanishes, but a relatively sharp minimum may occur at the higher frequencies if $\sigma/\varepsilon\omega \ll 1$.

The Brewster absorption phenomenon, for TM waves, can be exploited to reduce the effect of glare at optical wavelengths if the TE polarised wave is filtered out by suitable means (such as employed in anti-glare sunglasses).

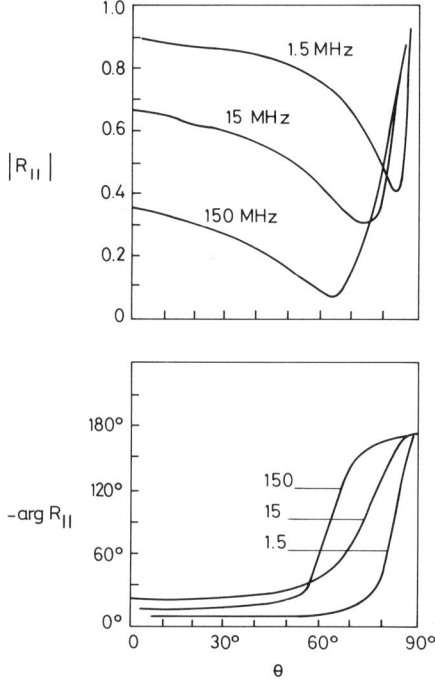

Fig. 4.8 Amplitude and phase (argument) of the reflection coefficient R, for TM polarisation, for ground conductivity $\sigma = 10^{-2}$ s/m and relative permittivity $\varepsilon/\varepsilon_0 = 4$

4.5 Further exercises

Exercise

In the limiting case of a highly conducting half-space model where $\sigma \gg \varepsilon\omega$, show that the reflection coefficients, as defined by eqns. 4.83 and 4.91, can be approximated by the simpler formulas

$$R_\parallel \simeq \frac{\eta_0 \cos\theta - Z_g}{\eta_0 \cos\theta + Z_g} \tag{4.95}$$

where

$$Z_g \simeq (j\mu\omega/\sigma)^{1/2}, \qquad \eta_0 = (\mu_0/\varepsilon_0)^{1/2} \tag{4.96}$$

and

$$R_\perp \simeq \frac{\eta_0^{-1}\cos\theta - Z_g^{-1}}{\eta_0^{-1}\cos\theta + Z_g^{-1}} \tag{4.97}$$

Exercise

For the conditions indicated in the previous exercise, show that the transmission coefficients, as defined by eqns. 4.84 and 4.92, can be approximated by

$$T_\| \simeq 2 \tag{4.98}$$

$$T_\perp \simeq 2(Z_g/\eta_0) \cos \theta \tag{4.99}$$

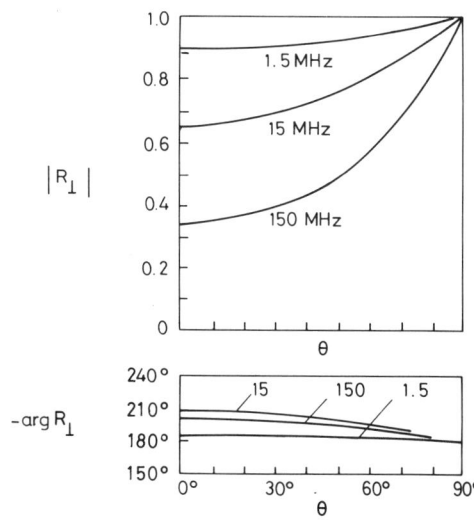

Fig. 4.9 Amplitude and phase (argument) of the reflection coefficient, for TE polarisation, for ground conductivity $\sigma = 10^{-2}$ s/m and relative permittivity $\varepsilon/\varepsilon_0 = 4$

Exercise Again for the same conditions as above, show that, for $z < 0$ within the conductor, for both wave polarisations, we have

$$H_y(z) \simeq H_y(0) \exp(-\gamma|z|) \tag{4.100}$$

$$E_y(z) \simeq E_y(0) \exp(-\gamma|z|) \tag{4.101}$$

Exercise Show that $R_\|$ and $R_\perp$ as defined generally by eqns. 4.83 and 4.91 can be reduced to the respective forms 4.58 and 4.42 provided that $\mu = \mu_0$ and that we define N_d, a complex refractive index, by $N_d^2 = (\sigma + j\varepsilon\omega)/j\varepsilon_0\omega$.

Exercise Show, for the general case, that for the two wave polarisations,

$$E_x(z = 0) = -K H_y(z = 0) \tag{4.102}$$

$$H_x(z = 0) = +M E_y(z = 0) \tag{4.103}$$

where K and M are defined by eqns. 4.86 and 4.94, respectively. Then show that if $\sigma \gg \varepsilon\omega$, $K \simeq Z_g$ and $M \simeq Z_g^{-1}$.

Exercise Show for the geometry of Fig. 4.7 that B_z and J_z are continuous across the interface.
Hint: note that div $\mathbf{B} = 0$ and div $\mathbf{J} = 0$ in regions of no sources.

Exercise

Again with reference to Fig. 4.7, show that E_z and H_z are discontinuous at the interface $z = 0$. In particular, show that

$$j\varepsilon_0 \omega E_z(z = +0) = (\sigma + j\varepsilon\omega)E_z(z = -0) \qquad (4.104)$$

and

$$\mu_0 H_z(z = +0) = \mu H_z(z = -0) \qquad (4.105)$$

Confirm that these two conditions are satisfied by eqns. 4.76 and 4.79 for the E_z components, and by H_z from eqns. 4.89 and 4.90 for the E_y fields.

Hint: eqns. 4.104 and 4.105 are merely statements that normal current flow and normal magnetic flux are continuous at $z = 0$. Also note that for $z > 0$ we have

$$H_z = \frac{1}{j\mu_0 \omega} \frac{\partial E_y}{\partial x}$$

and for $z < 0$ we have

$$H_z = -\frac{1}{j\mu\omega} \frac{\partial E_y}{\partial x}$$

Exercise

Show that the 'pattern' of a small vertically oriented electric field probe, located at $z = h$ above the interface $z = 0$ in Fig. 4.7, is given by

$$F(\theta) = \tfrac{1}{2} \sin\theta \, (e^{jkh\cos\theta} + R_\| e^{-jkh\cos\theta}) \qquad (4.106)$$

Hint: note that

$$E_z(z = h) = \frac{1}{j\varepsilon_0 \omega} \frac{\partial H_y}{\partial x}\bigg|_{z=h} = -\eta_0 \sin\theta \, H_y(z = h)$$

$$= 2E_0 F(\theta) \qquad (4.107)$$

which is the definition of the pattern function $F(\theta)$.

Exercise

For a perfectly conducting ground plane show that the pattern function as defined above is given by

$$F(\theta) = \sin\theta \, \cos(kh\cos\theta) \qquad (4.108)$$

Hint: note that right-hand side of eqn. 4.83 is equal to one when $90° - \theta > 0$.

Exercise

In general show that $F(\theta) \to 0$ as $\theta \to 90°$.
Hint: in this case, the right-hand side of eqn. 4.83 is equal to -1.

Exercise

Show that the wave tilt W as defined by the ratio

$$W = E_x(z = h)/E_z(z = h) \qquad (4.109)$$

is given by

$$W = \cot\theta \left(\frac{1 - R_\| e^{-j2kh\cos\theta}}{1 + R_\| e^{-j2kh\cos\theta}} \right) \qquad (4.110)$$

Hint: use eqns. 4.75 and 4.76 with 4.74, or note forms given by 4.45 and 4.47.

Reflection and refraction of plane waves

Exercise

In the previous exercise show, in the limit $h = +0$, that

$$W(h = +0) = K/(\eta_0 \sin \theta) \tag{4.111}$$

where K is given by eqn. 4.86. Then, in the limit $\theta \to 90°$, for $\mu = \mu_0$, show that

$$W \to W_0 = \frac{jk}{\gamma}\left(1 + \frac{k^2}{\gamma^2}\right)^{1/2} \tag{4.112}$$

Then evaluate W_0 for the case where $f = \omega/2\pi = 125 \text{ kHz}$, $\sigma = 10^{-3}$ s/m, $\varepsilon/\varepsilon_0 = 10$, $\varepsilon_0 = 8.85 \times 10^{-12}$, $\mu_0 = 4\pi \times 10^{-7}$.

Solution

$W_0 = 0.082 \angle 41.1°$. This example is relevant to low-frequency radio waves propagating along the earth's surface.

Exercise

(a) With reference to Fig. 4.7, we specify that the incident plane wave is defined by

$$\boldsymbol{H}^{inc} = i_y H_0 \, e^{jkz\cos\theta} \, e^{-jkx\sin\theta}$$

and we assume that $\sigma = 0$ and $\mu = \mu_0$ so that the lower half-space is a pure dielectric. Plot the field pattern for a vertical electric field probe at a height h in metres and for a frequency f in megahertz for a relative permittivity $\varepsilon/\varepsilon_0$, as follows:

Case	f	h	$\varepsilon/\varepsilon_0$	
(1)	10	15	3, 6, 9 and ∞	
(2)	5	30	3, 6, 9 and ∞	Figs. 4.10a, b, c, d
(3)	20	7.5	3, 6, 9 and ∞	
(4)	5	15	3, 6 and ∞	
(5)	2.5	30	3, 6 and ∞	Figs. 4.11a, b, c
(6)	10	7.5	3, 6 and ∞	
(7)	2	15	3 and ∞	
(8)	1	30	3 and ∞	Figs. 4.12a, b
(9)	4	7.5	3 and ∞	

Note that, when $\varepsilon/\varepsilon_0 = \infty$, the ground plane is behaving as a perfect reflector.

(b) Repeat the above tasks for the same incident wave but rotate the electric field probe to a horizontal position and orient it for maximum pickup.

(c) Finally, change the polarization of the incident wave such that

$$\boldsymbol{E}^{inc} = i_y E_0 \, e^{jkz\cos\theta} \, e^{-jkx\sin\theta}$$

(i.e. horizontally polarised incident wave). Now orient the electric field probe for maximum pick-up and again plot the field pattern.

Solution

(a) The desired pattern function here is $F(\theta)$ and it is defined precisely by eqn. 4.106 with

$$R_\parallel = \frac{(\varepsilon/\varepsilon_0)\cos\theta - (\varepsilon/\varepsilon_0 - \sin^2\theta)^{1/2}}{(\varepsilon/\varepsilon_0)\cos\theta + (\varepsilon/\varepsilon_0 - \sin^2\theta)^{1/2}}$$

which is equivalent to eqn. 4.58. Here we write

$$kh = \frac{\omega h}{c} = h\frac{2\pi f \times 10^6}{3 \times 10^8} = \frac{2\pi}{300} f(\text{mHz})h(\text{m})$$

So, clearly, results for the same product $f(\text{MHz})h(\text{m})$ will be identical for a fixed value of $\varepsilon/\varepsilon_0$. Note that these products are 150 for cases (1), (2) and (3), 75 for cases (4), (5) and (6), and 30 for cases (7), (8) and (9). Thus, as we have indicated in the above table, Figs. 4.10a, b, c and d cover cases (1), (2) and (3). In a similar fashion, Figs. 4.11a, b and c cover cases (4), (5) and (6), and Figs. 4.12 and b cover cases (7), (8) and (9). The quantities actually plotted are the magnitudes of $F(\theta)$ for θ ranging from 0° to 90° (i.e. normal to grazing incidence).

(b) The situation in (a) is now changed so that we are sensing the field

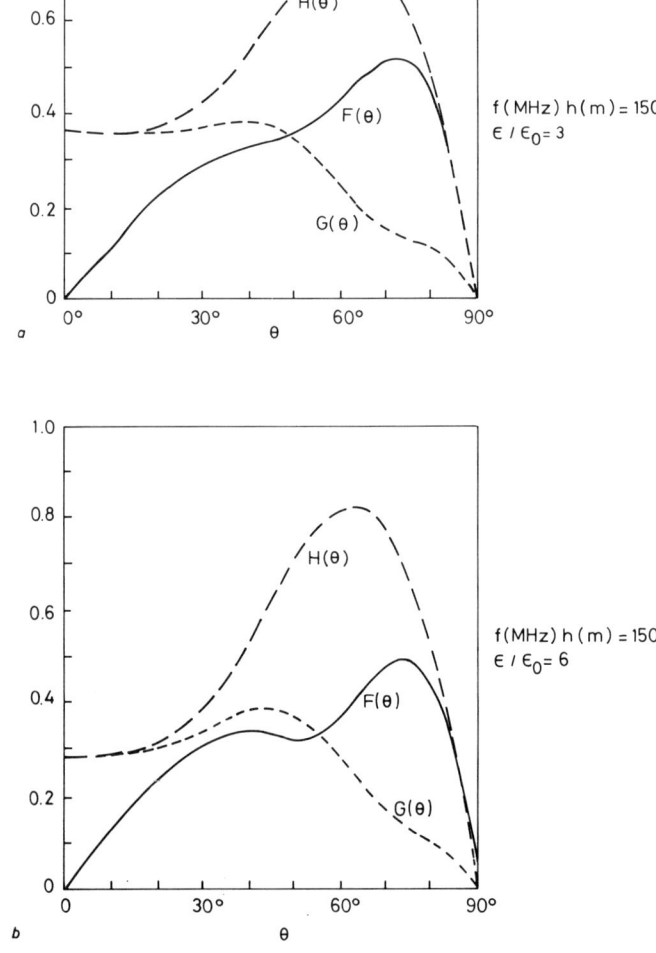

Fig. 4.10

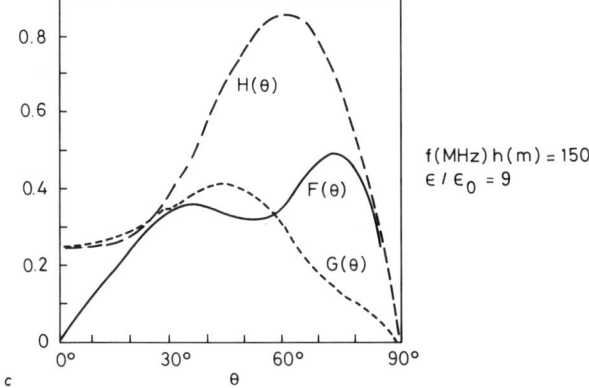

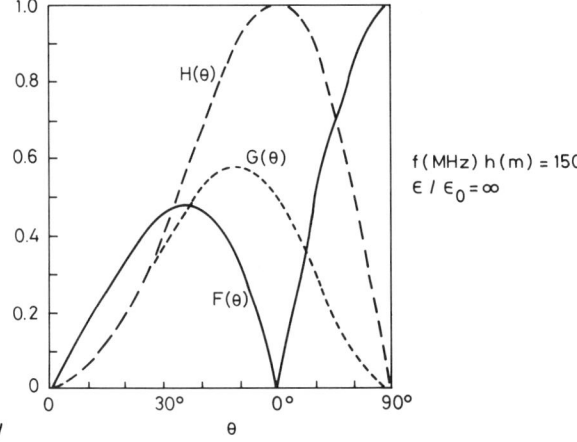

Fig. 4.10 continued

E_x at $z = h$ and $x = 0$. The dimensionless pattern function $G(\theta)$ is defined according to

$$G(\theta) = \frac{E_x(x = 0, z = h)}{2\eta_0 H_0}$$

where $\eta_0 = 120\pi$ and where E_x is given by eqn. 4.45. The magnitude of $G(\theta)$ is plotted in Figs. 4.10, 4.11 and 4.12.

(c) Now the electric field has only a y component so the preferred orientation of the probe is in the y direction. The pattern function $H(\theta)$ is now defined by

$$H(\theta) = \frac{E_y(x = 0, z = h)}{2E_0}$$

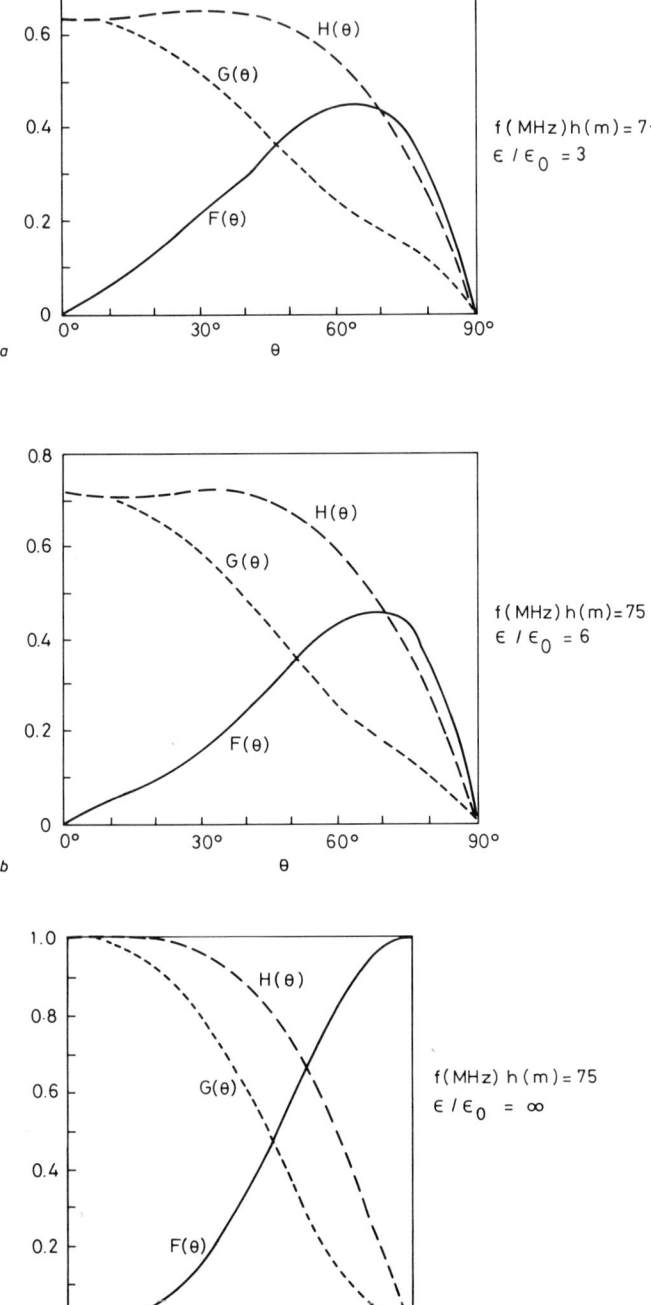

Fig. 4.11

Reflection and refraction of plane waves

where E_y is given by eqn. (4.16) with

$$R_\perp = \frac{\cos\theta - (\varepsilon/\varepsilon_0 - \sin^2\theta)^{1/2}}{\cos\theta + (\varepsilon/\varepsilon_0 - \sin^2\theta)^{1/2}}$$

The magnitude of $H(\theta)$ is plotted in Figs. 4.10, 4.11 and 4.12.

As we shall indicate in the next chapter, the functions $F(\theta)$, $G(\theta)$ and $H(\theta)$ also describe the radiation patterns of electric dipole transmitting antennas located over a dielectric half-space.

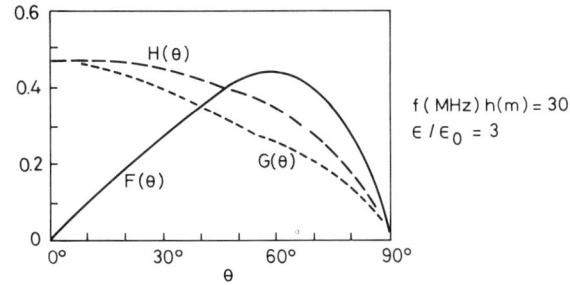

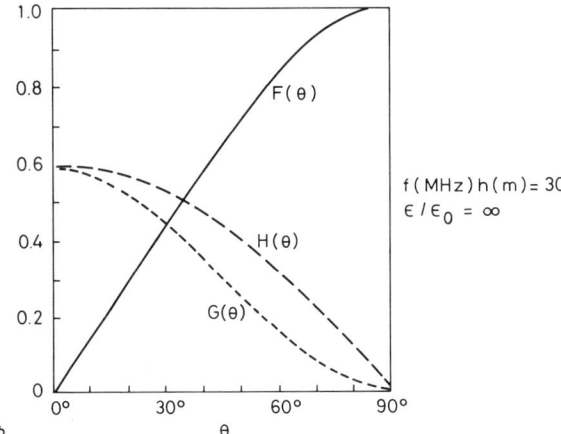

Fig. 4.12

4.6 Bibliography

C. C. JOHNSON, *Field and Wave Electrodynamics*, Chapter 3, McGraw Hill, 1965.
R. HARRINGTON, *Time Harmonic EM Fields*, Chapter 2, McGraw Hill, 1961.
M. KERKER, *The Scattering of Light*, Chapter 2, Academic Press, 1969.
M. BORN and E. WOLF, *Principles of Optics*, Section 1.5, Pergamon, 1959.
G. TYRAS, *Radiation and Propagation of EM Waves*, Section 2.4, Academic Press, 1969.
J. HEADING, *Ordinary Differential Equations*, Chapter 5 and 6, Elek Science, 1975.
J. R. WAIT, *EM Waves in Stratified Media*, Chapter 2, Pergamon, 1970.
S. A. SCHELKUNOFF, *Electromagnetic Waves*, Chapter 6, Van Nostrand, 1943.
L. C. SHEN and J. A. KONG, *Applied Electromagnetism*, Brookes Cole, 1982.
R. PLONSEY and R. E. COLLIN, *Principles and Applications of EM Fields*, McGraw Hill, 1961.
J. R. WAIT, *EM Wave Theory*, Chapter 4, Harper Row, 1985.

Chapter 5

Electromagnetic fields of current distributions

5.1 Vector potential

In Chapter 3 we dealt with the excitation of plane waves from an infinitely extended current sheet. In a sense, this idealised source could be regarded as a one-dimensional antenna. To deal with a confined source, we find it convenient to proceed in a different manner. In particular we shall consider a localised current distribution such as an electric dipole or linear current element. The electromagnetic fields of such a source in a homogeneous environment provide the framework to describe antenna performance.

As indicated, we assume initially that the medium is unbounded and homogeneous with conductivity σ, permittivity ε and permeability μ. For such a region, except at a source, the electromagnetic fields satisfy Maxwell's equations in the forms

$$-j\mu\omega H = \nabla \times E \tag{5.1}$$

$$(\sigma + j\varepsilon\omega)E = \nabla \times H \tag{5.2}$$

for the usual time factor $\exp(j\omega t)$. Now to facilitate our discussion, we introduce again an auxiliary function A that is called the magnetic vector potential. We will also require that rectangular components of A satisfy the Helmholtz equation everywhere except at the source. That is,

$$\nabla^2 A_p - \gamma^2 A_p = 0 \tag{5.3}$$

where $p = x, y, z$ and

$$\gamma^2 = j\mu\omega(\sigma + j\varepsilon\omega) \tag{5.4}$$

Another key point is that we select A so that the magnetic field vector H may be derived from it via the relation

$$H = \nabla \times A \tag{5.5}$$

Then, as a consequence of eqn. 5.2, we see that

$$E = \frac{1}{\sigma + j\varepsilon\omega} \nabla \times \nabla \times A \tag{5.6}$$

By making use of a known vector identity, eqn. 5.6 is equivalent to

$$E = [\nabla(\nabla \cdot A) - \tilde{\nabla}^2 A] \frac{1}{\sigma + j\varepsilon\omega} \tag{5.7}$$

where

$$\tilde{\nabla}^2 \mathbf{A} = \mathbf{i}_x \nabla^2 A_x + \mathbf{i}_y \nabla^2 A_y + \mathbf{i}_z \nabla^2 A_z \tag{5.8}$$

Now, according to eqn. 5.3

$$\tilde{\nabla}^2 \mathbf{A} = \gamma^2 \mathbf{A} \tag{5.9}$$

Then eqn. 5.7 is equivalent to

$$\mathbf{E} = \frac{1}{\sigma + j\varepsilon\omega} [-\gamma^2 \mathbf{A} + \nabla(\nabla \cdot \mathbf{A})] \tag{5.10}$$

The introduction of the vector potential in this manner may seem artificial at this stage. But its utility will be evident in the following example. As indicated in Fig. 5.1., we locate an electric dipole source or current element Il at the origin of a rectangular co-ordinate system (x, y, z). The dipole is oriented in the z direction.

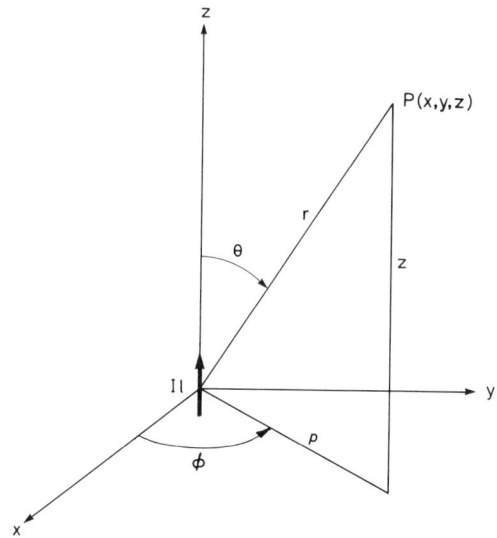

Fig. 5.1 Electric dipole or current element Il located at origin and oriented in z direction

A clue to the form of the vector potential $\mathbf{A}$ is found in the static solution. For example, we found in Chapter 2 (see eqn. 2.75) that if

$$\mathbf{A} = \mathbf{i}_z Il/4\pi r \tag{5.11}$$

where $r = (x^2 + y^2 + z^2)^{1/2}$, the correct form of the static magnetic field was obtained. Now, of course $1/r$ satisfies Laplace's equation i.e.

$$\left(\frac{\partial^2}{\partial x^2} + \frac{\partial^2}{\partial y^2} + \frac{\partial^2}{\partial z^2}\right)\frac{1}{r} = 0 \tag{5.12}$$

But here we need a solution of the Helmholtz equation that reduces to $1/r$ when $\gamma r \to 0$. Such a form is $r^{-1} \exp(\pm \gamma r)$, as one can verify by noting that

$$\left(\frac{\partial^2}{\partial x^2} + \frac{\partial^2}{\partial y^2} + \frac{\partial^2}{\partial z^2} - \gamma^2\right)\frac{e^{\pm \gamma r}}{r} = 0 \tag{5.13}$$

This step is left as an exercise for the reader.
The desired form for our solution is

$$\mathbf{A} = \mathbf{i}_z \frac{Il}{4\pi} \frac{e^{-\gamma r}}{r} \tag{5.14}$$

where we have rejected the non-physical solution that behaves as $r^{-1}\exp(\gamma r)$ because it becomes exponentially large (i.e. infinite) as $r \to \infty$.

As a check on the consistency of our results, we note that eqn. 5.10 in the quasi-static limit reduces to

$$\mathbf{E} = -\nabla \psi \tag{5.15}$$

if

$$\psi = -(\sigma + j\varepsilon\omega)^{-1}\nabla \cdot \mathbf{A} \tag{5.16}$$

where ψ is a scalar potential. In the case of the electric dipole or current element, we see that

$$\psi = -\frac{Il}{4\pi(\sigma + j\varepsilon\omega)} \frac{\partial}{\partial z} \frac{1}{r} \tag{5.17}$$

or

$$\psi = \frac{Ilz}{4\pi(\sigma + j\varepsilon\omega)r^3} \tag{5.18}$$

which is in agreement with eqn. 2.38.

It is now clear that the quasi-static solutions are generally valid if $|\gamma r| \ll 1$ so that terms like $\exp(-\gamma r)/r$ can be replaced by $1/r$. This is not as stringent as proceeding to the static limit where $\omega \to 0$. Thus in the quasi-static solutions discussed in Chapter 2, it is appropriate to retain the form $\sigma + j\varepsilon\omega$ which, of course, reduces to the real σ in the pure static limit.

5.2 Complete dipole field expressions

To obtain the complete expressions for the electromagnetic fields of the current element shown in Fig. 5.1, we need to perform the operations indicated by eqns. 5.5 and 5.10. Thus, for example,

$$H_x = \frac{\partial A_z}{\partial y} \tag{5.19}$$

$$H_y = -\frac{\partial A_z}{\partial x} \tag{5.20}$$

$$H_z = 0$$

$$(\sigma + j\varepsilon\omega)E_x = \frac{\partial^2 A_z}{\partial x \partial z} \tag{5.21}$$

Electromagnetic fields of current distributions

$$(\sigma + j\varepsilon\omega)E_y = \frac{\partial^2 A_z}{\partial y \partial z} \tag{5.22}$$

$$(\sigma + j\varepsilon\omega)E_z = -\gamma^2 A_z + \frac{\partial^2 A_z}{\partial z^2} \tag{5.23}$$

where

$$A_z = \frac{Il}{4\pi} \frac{e^{-\gamma r}}{r} \tag{5.24}$$

To obtain the corresponding field components in cylindrical components (ϱ, ϕ, z), we note that

$$H_\phi = -H_x \sin\phi + H_y \cos\phi \tag{5.25}$$

$$H_\varrho = H_x \cos\phi + H_y \sin\phi = 0 \tag{5.26}$$

where

$$\varrho = (x^2 + y^2)^{1/2}, \quad \tan\phi = y/x$$

Similarly we see that

$$E_\varrho = E_x \cos\phi + E_y \sin\phi \tag{5.27}$$

$$E_\phi = -E_x \sin\phi + E_y \cos\phi = 0 \tag{5.28}$$

Then to convert to spherical co-ordinates in the (r, θ, ϕ) system we use eqns. 2.46 and 2.47 together with eqns. 5.23, 5.25 and 5.27. Then we find that

$$H_\phi = \frac{Il}{4\pi r^2} (1 + \gamma r) e^{-\gamma r} \sin\theta \tag{5.29}$$

$$(\sigma + j\varepsilon\omega)E_r = \frac{Il}{2\pi r^3} (1 + \gamma r) e^{-\gamma r} \cos\theta \tag{5.30}$$

$$(\sigma + j\varepsilon\omega)E_\theta = \frac{Il}{4\pi r^3} (1 + \gamma r + \gamma^2 r^2) e^{-\gamma r} \sin\theta \tag{5.31}$$

When $|\gamma r| \ll 1$, it is evident that eqns. 5.29, 5.30 and 5.31 reduce to the quasi-static forms given by eqns. 2.66, 2.43 and 2.44, respectively.

A quantity of some interest is the *radial wave impedance* denoted Z_r. It was defined earlier by

$$Z_r = E_\theta / H_\phi \tag{5.32}$$

and thus, on using eqns. 5.29 and 4.31,

$$Z_r = \eta \left(1 + \frac{1}{\gamma r} + \frac{1}{\gamma^2 r^2}\right) \left(1 + \frac{1}{\gamma r}\right)^{-1} \tag{5.33}$$

where

$$\eta = \gamma/(\sigma + j\varepsilon\omega) = [j\mu\omega/(\sigma + j\varepsilon\omega)]^{1/2}$$

is the intrinsic or characteristic impedance of the medium, as defined earlier.

Another parameter of interest is the *wave ratio* V, which is defined by

$$V = E_r/E_\theta \tag{5.34}$$

Using eqns. 5.30 and 5.31, it is found that

$$V = \frac{2}{\gamma r} \frac{\left(1 + \frac{1}{\gamma r}\right)}{\left(1 + \frac{1}{\gamma r} + \frac{1}{\gamma^2 r^2}\right)} \cot\theta \tag{5.35}$$

Now clearly in the case where $|\gamma r| \gg 1$, $Z_r \simeq \eta$. Then also $V \ll 1$, except near $\theta = 0°$ where $\cot\theta$ becomes large. Thus we see, for the far field (i.e. $|\gamma r| \gg 1$), that the wave is behaving locally as a plane wave. That is, $|E_r| \ll |E_\theta|$ and $E_\theta \simeq \eta H_\phi$.

5.3 Free space fields and radiation patterns

To facilitate the discussion in what follows, we take the special case of free space conditions. Then $\sigma = 0$, $\varepsilon = \varepsilon_0$ and $\mu = \mu_0$; hence $\gamma = jk$ where $k = \omega/c$.

Now eqns. 5.29, 5.30 and 5.31 reduce to

$$H_\phi = \frac{Il}{4\pi r^2}(1 + jkr)e^{-jkr}\sin\theta \tag{5.36}$$

$$j\varepsilon_0\omega E_r = \frac{Il}{2\pi r^3}(1 + jkr)e^{-jkr}\cos\theta \tag{5.37}$$

$$j\varepsilon_0\omega E_\theta = \frac{Il}{4\pi r^3}(1 + jkr - k^2 r^2)e^{-jkr}\sin\theta \tag{5.38}$$

In the far field, where $kr \gg 1$ (i.e. $r \gg \lambda_0$), we obtain the simpler forms

$$H_\phi \simeq jk\frac{Il}{4\pi r}e^{-jkr}\sin\theta \tag{5.39}$$

$$E_r \simeq \frac{2}{jkr}j\mu_0\omega\frac{Il}{4\pi r}e^{-jkr}\cos\theta \tag{5.40}$$

$$E_\theta \simeq j\mu_0\omega\frac{Il}{4\pi r}e^{-jkr}\sin\theta \tag{5.41}$$

As indicated, both $|E_\theta|$ and $|H_\phi|$ vary as $1/r$, whereas $|E_r|$ varies as $1/r^2$. Then, if we retain only the inverse distance forms, the radiation fields of the current element are given by

$$E_\theta \simeq \eta_0 H_\phi \tag{5.42}$$

$$\simeq j\mu_0\omega\frac{Il}{4\pi r}e^{-jkr}\sin\theta \tag{5.43}$$

The angular dependence of $|E_\theta|$ and $|H_\phi|$, in the range $0 < \theta < \pi$, is proportional to $\sin\theta$. This factor, by definition, is the radiation field pattern. As illustrated in Fig. 5.2, the pattern can be visualised as a doughnut form when rotated in the ϕ direction about the polar axis. This plot is often called the *E-plane pattern* because the electric field vector lies in the plane of the figure. Of course, there is no ϕ dependence of the radiation fields. This fact is illustrated in Fig. 5.3 where we have simply noted that, over the range $0 < \phi < 2\pi$, $|E_\theta|$ and $|H_\phi|$ have an omnidirectional radiation pattern for any fixed value of θ. Such a pattern is often called an *H-plane pattern* because H lies in the plane of the figure. Obviously there are many applications, such as in broadcasting, where such an omnidirectional pattern is desirable.

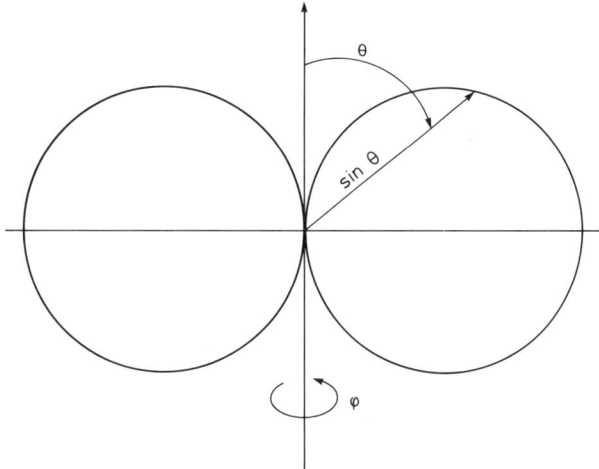

Fig. 5.2 Radiation field pattern (*E*-plane) for an electric dipole or small current element located in free space (i.e. the holeless doughnut)

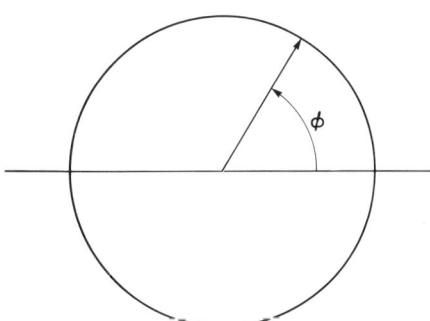

Fig. 5.3 Radiation field pattern (*H*-plane) for an electric dipole or small current element located in free space (i.e. top view of holeless doughnut)

Exercise

With reference to Fig. 5.1, locate another electric dipole of the same strength Il at $z = z_0$ on the axis $\theta = 0$ and oriented in the z direction. Show that the expression for the vector potential anywhere in the surrounding medium is given by

$$A_z = \frac{Il}{4\pi}\left(\frac{e^{-\gamma r}}{r} + \frac{e^{-\gamma \hat{r}}}{\hat{r}}\right) \tag{5.44}$$

where

$$r = (x^2 + y^2 + z^2)^{1/2} = (\varrho^2 + z^2)^{1/2}$$

$$\hat{r} = [x^2 + y^2 + (z - z_0)^2]^{1/2} = [\varrho^2 + (z - z_0)^2]^{1/2}$$

Also show that

$$H_\phi = \frac{Il}{4\pi}\left[\frac{1}{r^2}(1 + \gamma r)\,e^{-\gamma r}\sin\theta + \frac{1}{\hat{r}^2}(1 + \gamma\hat{r})\,e^{-\gamma\hat{r}}\sin\hat{\theta}\right] \quad (5.45)$$

where

$$\sin\theta = \varrho/r$$

$$\sin\hat{\theta} = \varrho/\hat{r}$$

Solution Perform the operation $H_\phi = -\partial A_z/\partial \varrho$ to yield eqn. 5.45.

Exercise In the previous exercise, assume free space conditions so that $\gamma = jk$. Then also assume that $kr \gg 1$. Now show that

$$A_z \simeq \frac{Il}{4\pi r}\,e^{-jkr}(1 + e^{jkz_0\cos\theta}) \quad (5.46)$$

where we may also assume that $r \gg z_0$.

Solution Note that

$$\frac{e^{-jk\hat{r}}}{\hat{r}} \simeq \frac{e^{-jk\hat{r}}}{r} \quad (5.47)$$

while, in the exponent,

$$\hat{r} = (r^2 - 2zz_0 + z_0^2)^{1/2}$$

$$\simeq r[1 - (2z_0/r)\cos\theta]^{1/2}$$

$$\simeq r - z_0\cos\theta \quad (5.48)$$

Exercise With the same assumptions, show that the radiation fields are given by

$$E_\theta \simeq \eta_0 H_\phi$$

$$\simeq j\mu_0\omega\,\frac{Il}{4\pi r}\,e^{-jkr}\sin\theta\,(1 + e^{jkz_0\cos\theta}) \quad (5.49)$$

Solution Here we simply note that $H_\phi \simeq jk\sin\theta\,A_z$.

5.4 Fields of a linear antenna

A useful generic model is a linear antenna of finite length that has a specified current $I(z')$ throughout its length. The situation is illustrated in Fig. 5.4, where we show the accompanying cylindrical (ϱ, ϕ, z) and spherical (r, θ, ϕ) co-ordinate systems. The primed co-ordinate z' measures the distance along the antenna from the origin. The co-ordinates of the end points of the antenna are $z' = z_1$ and $z' = z_2$. The observer is at the point P(ϱ, z) where the fields are independent of the azimuthal angle ϕ.

The medium surrounding the antenna is homogeneous with properties (σ, ε, μ). Thus we can deduce the fields at P as the superposition of

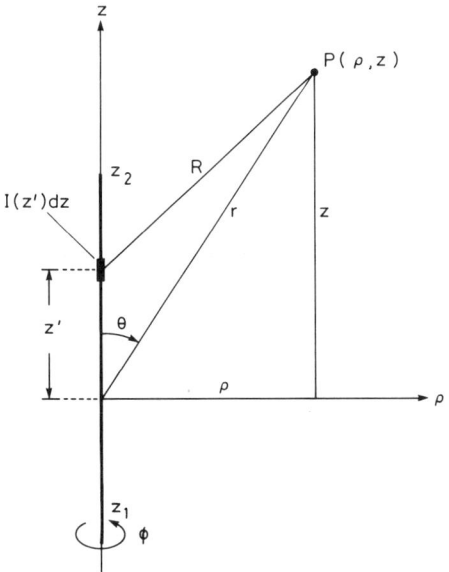

Fig. 5.4 Geometry for deducing fields at P for a linear current distribution on z axis

the elemental dipole fields for the configuration shown in Fig. 5.1. For example, the vector potential dA at P for the dipole $I(z')dz'$ at $z = z'$ is given by an adaptation of eqn. 5.14:

$$d\boldsymbol{A} = \boldsymbol{i}_z dA_z \tag{5.50}$$

where

$$dA_z = \frac{I(z') e^{-\gamma R}}{4\pi R} dz' \tag{5.51}$$

where

$$R = [\varrho^2 + (z - z')^2]^{1/2}$$

$$\gamma = [j\mu\omega(\sigma + j\varepsilon\omega)]^{1/2}$$

Then clearly the resultant vector $\boldsymbol{A}$ at P for the totality of the current elements is given by

$$\boldsymbol{A} = \boldsymbol{i}_z A_z \tag{5.52}$$

where

$$A_z = \frac{1}{4\pi} \int_{z_1}^{z_2} \frac{e^{-\gamma R}}{R} I(z') \, dz' \tag{5.53}$$

The corresponding field components are deduced from eqns. 5.19 to

5.23. In cylindrical co-ordinates these would read

$$, H_\phi = -\frac{\partial A_z}{\partial \varrho} \tag{5.54}$$

$$(\sigma + j\varepsilon\omega)E_\varrho = \frac{\partial^2 A_z}{\partial\varrho\partial z} \tag{5.55}$$

$$(\sigma + j\varepsilon\omega)E_z = \left(-\gamma^2 + \frac{\partial^2}{\partial z^2}\right)A_z \tag{5.56}$$

$$E_\phi = H_\varrho = H_z = 0$$

Apart from the initial filamental current assumption, the field expressions so derived are exact. But as a practical matter we need to specify the current distribution $I(z')$ and to perform the integration over z'.

Exercise

Consider the quasi-static limit where $|\gamma R| \ll 1$. Show that eqn. 5.53 reduces to eqn. 2.71 if $z_1 = -L/2$, $z_2 = +L/2$ and $I(z') = I$.

Solution

Merely set $\exp(-\gamma R) = 1$ and $I(z') = I$ in eqn. 5.53. Then apply eqn. 5.54 to arrive at eqn. 2.70.

Exercise

Consider the dipole limit where $|\gamma(z_2 - z_1)| \ll 1$ and $r \gg (z_2 - z_1)$. Show that eqn. 5.53 reduces to

$$A_z \simeq \frac{(Il)_{\text{eff}}}{4\pi\bar{R}} e^{-\gamma\bar{R}} \tag{5.57}$$

where

$$\bar{R} = \{\varrho^2 + [z - (z_1 + z_2)/2]^2\}^{1/2}$$

is the mean and where the effective moment is given by

$$(Il)_{\text{eff}} = \int_{z_1}^{z_2} I(z')\, dz' \tag{5.58}$$

Solution

Use the mean value theorem and bring $e^{-\gamma R}/R$ outside the integral in eqn. 5.53.

5.5 Radiation of linear antenna in free space

An important special case is when the linear antenna is located in free space. Then, as usual, $\sigma = 0$, $\varepsilon = \varepsilon_0$ and $\mu = \mu_0$. Thus in eqn. 5.5 $\gamma = jk$, where $k = \omega/c = (\varepsilon_0\mu_0)^{1/2}\omega$. The fields can be deduced from eqns. 5.54, 5.55 and 5.56, but $I(z')$ needs to be specified. A common assumption here is to say that the current is sinusoidally distributed; thus for the present situation

$$I(z') = A \sin(kz' + \alpha) \tag{5.59}$$

where A and α are constants to be determined by the location of the generator and the conditions at the end points of the antenna. Such an approximation does have some validity for a realisable metal wire of restricted length, *if the radius of the wire is sufficiently small*. Then the integration over z' can be carried out without further restrictions, but the results are cumbersome if we allow the field point to be arbitrarily located.

If we are primarily interested in the radiation fields, there is a vast simplification when distance R to the observer is much greater than the maximum dimension $(z_2 - z_1)$ of the antenna. This case will be considered below.

To facilitate our discussion we set $z_1 = -L/2$ and $z_2 = L/2$, and the current distribution is assumed to have the form

$$I(z') = I_m \sin\left[k\left(\frac{L}{2} - |z'|\right)\right] \tag{5.60}$$

for $-L/2 < z' < L/2$. This function is plotted in Fig. 5.5 where, as indicated, we have shown the linear antenna being fed by a generator at the centre. Here I_m is the maximum current and, of course, $I(0)$ is the generator current. We are now faced with evaluation of the following integral:

$$A_z = \frac{I_m}{4\pi} \int_{-L/2}^{L/2} \frac{e^{-jkR}}{R} \sin\left[k\left(\frac{L}{2} - |z'|\right)\right] dz' \tag{5.61}$$

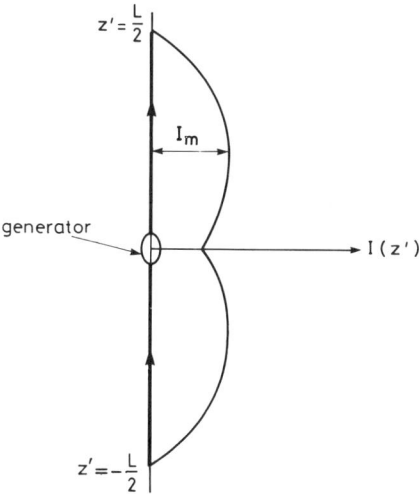

Fig. 5.5 Linear antenna of length L showing assumed sinusoidal current distribution

Actually eqn. 5.61 can be expressed in terms of cosine and sine integrals, but we simplify matters by invoking the far field approximations. Thus, for example,

$$\frac{e^{-jkR}}{R} \simeq \frac{1}{r} e^{-jkr} e^{jkz'\cos\theta} \tag{5.62}$$

which is the same simplification as depicted by eqns. 5.47 and 5.48. The key point is that, for $kr \gg 1$ and $r \gg L$, the amplitude factor $1/R$ is essentially the same as $1/r$ but the phase difference $k(R - r)$ may be significant (see Fig. 5.6).

Equations 5.61 and 5.62 may now be combined to yield the manageable integral

$$A_z = \frac{I_m e^{-jkr}}{4\pi r} \int_{-L/2}^{L/2} \sin\left[k\left(\frac{L}{2} - |z'|\right)\right] e^{jkz'\cos\theta} \, dz' \quad (5.63)$$

This form is readily evaluated to give

$$A_z = \frac{I_m e^{-jkr}}{2\pi r} \left[\frac{\cos\left(k\frac{L}{2}\cos\theta\right) - \cos\left(k\frac{L}{2}\right)}{k \sin^2\theta}\right] \quad (5.64)$$

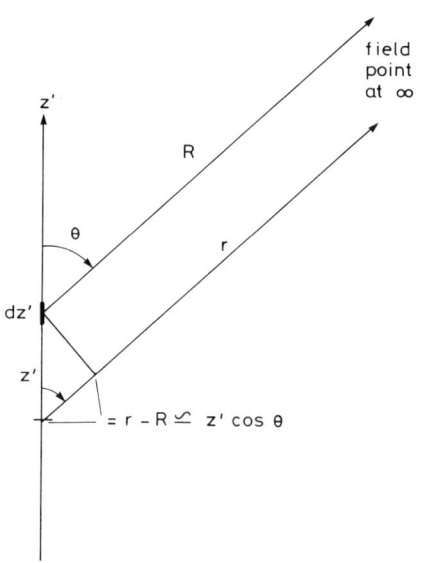

Fig. 5.6 Geometry for deducing phase differences in far field from different parts of the antenna.

To obtain the magnetic field with the same far field approximation we note that

$$H_\phi = -\frac{\partial A_z}{\partial \varrho} \simeq jk \sin\theta \, A_z$$

$$\simeq \frac{jI_m e^{-jkr}}{2\pi r} \left[\frac{\cos\left(k\frac{L}{2}\cos\theta\right) - \cos\left(k\frac{L}{2}\right)}{\sin\theta}\right] \quad (5.65)$$

In obtaining this result we have made use of the approximation

$$\frac{\partial}{\partial \varrho} \frac{e^{-jkr}}{r} = -\frac{1}{r^2}(1 + jkr) e^{-jkr} \frac{\partial r}{\partial \varrho}$$

$$\simeq -\frac{jk}{r} e^{-jkr} \sin\theta \quad (5.66)$$

which is certainly valid when $kr \gg 1$. In the far field we also note that $E_\theta \simeq \eta_0 H_\phi$ while the E_r component is out of contention.

The radiation field pattern is essentially the square bracket factor in eqn. 5.65. This E-plane pattern is plotted in Fig. 5.7, where θ ranges from 0° to 180° for various antenna lengths from less than $\lambda_0/4$ to $\lambda_0/2$.

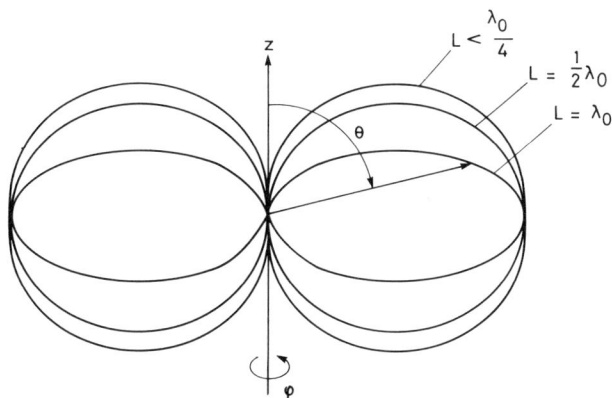

Fig. 5.7 Radiation field pattern for linear antenna of length L

The results are normalised at $\theta = 90°$, which is the maximum in each case. These results show clearly that as L increases the broadside maximum is sharpened somewhat. However, when L exceeds λ_0 the pattern begins to break up into secondary lobes that are usually undesirable. An example is shown in Fig. 5.8 where $L = 1 \cdot 5 \lambda_0$. Also, for long antennas, the initial sinusoidal current distribution is not a valid assumption even for thin antennas. This particular topic is outside the scope of our treatment.

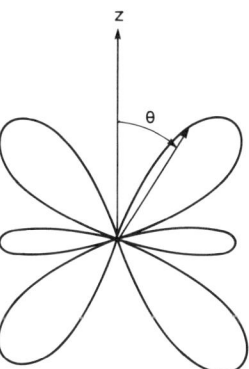

Fig. 5.8 Radiation field pattern for linear antenna of length $L = 1 \cdot 5 \lambda_0$

5.6 Antenna arrays

To obtain a measure of directivity, it is common practice to superimpose the fields of individual antennas that are separated in space. This is a vast subject with a wide literature. We will consider only the case of an equispaced array of parallel elements. The situation is illustrated in Fig. 5.9, where we have located N identical antenna elements with spacing d along the y axis. Each antenna has length L and carries a current $I_n(z')$. The field point is at P at distance r which is much greater than the free space wavelength. Also the distances $r, r_1, r_2, \ldots, r_{N-1}$ are all assumed to be much greater than the element spacing d.

Now the element $n = 0$ at the origin produces the electric field at P given by

$$\mathbf{E}_0 = \mathbf{i}_\theta E_{\theta 0} \tag{5.67}$$

where

$$E_{\theta 0} = j\mu_0 \omega \frac{e^{-jkr}}{4\pi r} F_0(\theta) \tag{5.68}$$

where

$$F_0(\theta) = \sin\theta \int_{-L/2}^{L/2} I_0(z') e^{jkz'\cos\theta} \, dz' \tag{5.69}$$

The element $n = 1$, displaced along the y axis by a distance d, produces the electric field at P given by

$$\mathbf{E}_1 = \mathbf{i}_\theta E_{\theta 1} \tag{5.70}$$

where

$$E_{\theta 1} = j\mu_0 \omega \frac{e^{-jkr}}{4\pi r} F_1(\theta) e^{+j\Psi} \tag{5.71}$$

where

$$F_1(\theta) = \sin\theta \int_{-L/2}^{L/2} I_1(z') e^{jkz'\cos\theta} \, dz' \tag{5.72}$$

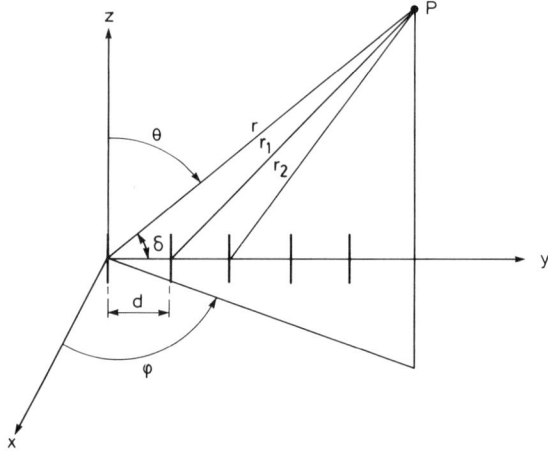

Fig. 5.9 An array of N vertical linear antennas with equal spacing d. The field point P is allowed to recede to 'infinity' so that $r, r_1, r_2, \ldots, r_{N-1}$ are effectively parallel

The phase angle Ψ is well approximated in the far field by

$$\Psi = k(r - r_1) \simeq kd \cos\delta \tag{5.73}$$

where, as indicated in Fig. 5.9, δ is the angle subtended by r and the y axis. Now $\cos\delta = y/r = \sin\theta \sin\phi$, so that

$$\Psi = kd \sin\theta \sin\phi \tag{5.74}$$

Clearly the total radiation field at P produced by the N elements is

$$E = i_\theta E_\theta \tag{5.75}$$

where

$$E_\theta = j\mu_0\omega \frac{e^{-jkr}}{4\pi r} \sum_{n=0}^{N-1} F_n(\theta) \, e^{jn\Psi} \tag{5.76}$$

where

$$F_n(\theta) = \sin\theta \int_{-L/2}^{L/2} I_n(z') \, e^{jkz'\cos\theta} \, dz' \tag{5.77}$$

To facilitate our discussion we will now assume that the currents $I_n(z)$ are identical except for a constant phase shift between elements. Thus, in effect,

$$I_n(z') = I_0(z') \, e^{-jn\psi} \tag{5.78}$$

where ψ is the phase shift between two adjacent elements and where $n = 0, 1, 2, \ldots, N - 1$. Now eqn. 5.76 can be written in the form

$$E_\theta = j\mu_0\omega \frac{e^{-jkr}}{4\pi r} F_0(\theta) \sum_{n=0}^{N-1} e^{jn(\Psi-\psi)} \tag{5.79}$$

Then, on using the standard summation formula

$$\sum_{n=0}^{N-1} Z^n = \frac{1 - Z^N}{1 - Z} \tag{5.80}$$

we obtain the following useful expression for the radiation field of the array:

$$E_\theta = j\mu_0\omega \frac{e^{-jkr}}{4\pi r} F_0(\theta) G(\theta, \phi) \tag{5.81}$$

where

$$G(\theta, \phi) = \frac{1 - \exp[jN(kd\sin\theta\sin\phi - \psi)]}{1 - \exp[j(kd\sin\theta\sin\phi - \psi)]} \tag{5.82}$$

Thus the resultant pattern is the product of the element pattern $F_0(\theta)$ and the array pattern $G(\theta, \phi)$.

To illustrate the directive properties of such an array, we consider two special cases. In the first case we take $\psi = 0$ so that the elements are being fed in phase. We choose $N = 8$ and $d = \lambda_0/4$ or $kd = \pi/2$. The field pattern in the plane $\theta = \pi/2$ is proportional to

$$\left| G\left(\frac{\pi}{2}, \phi\right) \right| = \left| \frac{\sin(2\pi \sin\phi)}{\sin\left(\frac{\pi}{4}\sin\phi\right)} \right| \tag{5.83}$$

This quantity has a maximum value of 8 at $\phi = 0°$ and $180°$. There are also subsidiary maxima that correspond to the minor lobes in the radiation pattern. The pattern function is also zero when $\phi = 90°$ and $270°$ in addition to zeros between the minor lobes. This H-plane pattern is plotted in Fig. 5.10 as a function of ϕ from 0 to $360°$. Such broadside radiation patterns have a main lobe in both the front and rear directions.

In the second special case we choose the phase shift ψ to be equal to kd. Again we select $N = 8$ so that phase shift ψ between elements is $\pi/2$. Now the radiation field pattern in the plane $\theta = \pi/2$ is proportional to

$$\left|G\left(\frac{\pi}{2}, \phi\right)\right| = \left|\frac{\sin\,[2\pi(1 - \sin\,\phi)]}{\sin\,[(\pi/4)\,(1 - \sin\,\phi)]}\right| \tag{5.84}$$

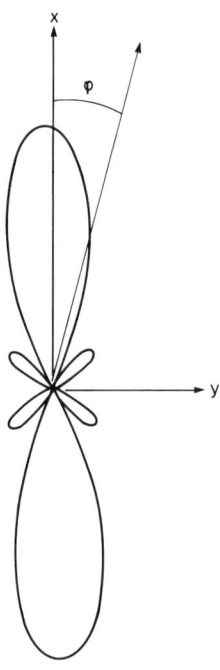

Fig. 5.10 Broadside radiation field pattern in H-plane, at $z = 0$, of an eight-element array with uniform excitation and equal spacing $d = \lambda_0/4$ ($\psi = 0$)

This function has a major maximum in the 'end-fire' direction $\phi = \pi/2$ which is parallel to the y axis. As shown in Fig. 5.11, there are minor lobes in the range from 0 to 360° beyond the major lobe. We also note that the width of the major lobe is somewhat greater than the major lobes for the broadside array of the same number of elements and spacing (e.g. compare Figs. 5.10 and 5.11).

5.7 Power considerations

An important aspect of antenna radiation is the energy flow in the medium and the power needed to maintain the current on the structure. This brings us to a discussion of *Poynting's theorem*.

We consider a homogeneous region with properties $(\sigma, \varepsilon, \mu)$. Maxwell's equations for the time-harmonic case are

$$\nabla \times \mathbf{E} = -j\mu\omega\mathbf{H} \tag{5.85}$$

$$\nabla \times \mathbf{H} = (\sigma + j\varepsilon\omega)\mathbf{E} + \mathbf{J}_\mathrm{i} \tag{5.86}$$

where $\mathbf{J}_\mathrm{i}$ is the impressed current density (e.g. as produced by the antenna structure). We now multiply the first Maxwell equation by the complex conjugate $\mathbf{H}^*$, and we multiply the complex conjugate of the second Maxwell equation by E; then we subtract one of the resulting equations from the other to yield

$$H^* \cdot \nabla \times E - E \cdot \nabla \times H^* = -E \cdot J_i^* - \sigma E \cdot E^*$$
$$+ j\omega(\varepsilon E \cdot E^* - \mu H \cdot H^*) \qquad (5.87)$$

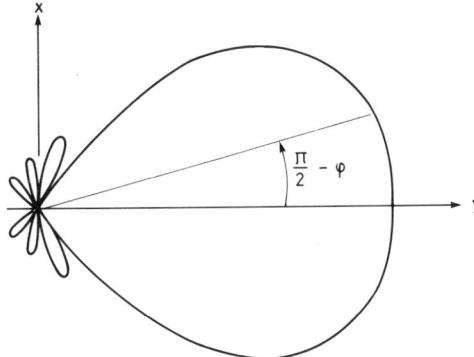

Fig. 5.11 End-fire radiation field pattern in H-plane, at $z = 0$, of an eight-element array with uniform excitation and equal spacing $d = \lambda_0/4$ ($\psi = \pi/2$). In this case there is a progressive phase shift in the excitation of the elements Fig. 5.12 Radiation resistance R_m in ohms of a linear antenna of length L

Now, using a known vector identity, the left-hand side of eqn. 5.87 is $\nabla \cdot (E \times H^*)$. This quantity is now integrated over a volume V with a subsequent application of Gauss' theorem to give

$$\int_V (H^* \cdot \nabla \times E - E \cdot \nabla \times H^*) \, dv = \iint_S (E \times H^*)_n \, da$$

where S is a surface enclosing V, and $(E \times H^*)_n$ is the outward normal component of the vector $E \times H^*$ at the surface element da.

On rearranging terms, eqns. 5.87 and 5.88 can be cast into the form

$$-\tfrac{1}{2} \int_V E \cdot J_i^* \, dv = \tfrac{1}{2} \int_V \sigma E \cdot E^* \, dv + \tfrac{1}{2} \int_S (E \times H^*)_n \, da$$
$$+ \frac{j\omega}{2} \int_V (\mu H \cdot H^* - \varepsilon E \cdot E^*) \, dv \qquad (5.89)$$

where we have divided throughout by 2. The real part of the left-hand side of eqn. 5.89 is the average power spent by the impressed forces to sustain the field. The first term on the right of eqn. 5.89 is the rate of energy dissipation in the volume V. The real part of the second term must then be the total power crossing the surface S, because the third term is purely imaginary. The vector $\tfrac{1}{2}(E \times H^*)$ is called the *complex Poynting vector*. According to *Poynting's theorem*, the real part of the power P crossing the closed surface is given by

$$P = \tfrac{1}{2} \text{Re} \int_S (E \times H^*)_n \, da \text{ W/m}^2 \qquad (5.90)$$

The real part of the complex Poynting vector can be expressed in the form

$$\tfrac{1}{2} \operatorname{Re}(\mathbf{E} \times \mathbf{H}^*) = \langle \mathbf{e}(t) \times \mathbf{h}(t) \rangle$$

$$= \frac{1}{2\pi} \int_0^{2\pi} [\mathbf{e}(t) \times \mathbf{h}(t)] \, \mathrm{d}(\omega t) \quad (5.91)$$

where $\mathbf{e}(t)$ and $\mathbf{h}(t)$ are the actual time-dependent field vectors for the sinusoidally varying source current density $j_i(t)$. Equation 5.91 follows directly from the vector identity given by eqn. 1.60. The time-dependent real vector $\mathbf{e}(t) \times \mathbf{h}(t)$ can be interpreted as the instantaneous power flow in the direction perpendicular to the plane containing $\mathbf{e}(t)$ and $\mathbf{h}(t)$. But there are still some fundamental questions on how one should identify the Poynting vector at a point in space with the local power flow at that point. Strictly speaking one should only regard the Poynting flux integrated over a *closed* surface, such as indicated by eqn. 5.90, as the power flow across the surface.

In the case of an antenna located in free space, the power supplied to the terminals must all be transmitted across an enclosing surface S if ohmic losses in the antenna can be neglected. Thus, in such cases, we can define a radiation resistance R by

$$R = 2P/|I_0|^2 = 2P/I_0 I_0^* \quad (5.92)$$

where I_0 is the (complex) current at the terminals. Therefore we can say that the radiation resistance R of the antenna located in an insulating medium is given by

$$R = \frac{1}{|I_0|^2} \operatorname{Re} \int_S (\mathbf{E} \times \mathbf{H}^*)_n \, \mathrm{d}a \quad (5.93)$$

where S can be any surface that encloses the antenna structure. Also we may assert, for such a loss-free antenna, that R is the input resistance (i.e. real part of the input impedance) at the antenna terminals.

A useful related concept is the gain of an antenna in a given direction. Such a gain g, in the present context, is defined by

$$g(\theta, \phi) = \frac{4\pi r^2}{P} p_r(\theta, \phi) \quad (5.94)$$

$$p_r(\theta, \phi) = \tfrac{1}{2} \operatorname{Re}(\mathbf{E} \times \mathbf{H}^*)_r \quad (5.95)$$

is the radial component of the Poynting vector in the direction θ, ϕ at a radial distance r. Here P is the total power radiated from the antenna.

An example of the power calculation is now given for the linear antenna in free space. The total power P is to be obtained from

$$P = \int_0^{2\pi} \int_0^{\pi} p_r r^2 \sin\theta \, \mathrm{d}\theta \, \mathrm{d}\phi \quad (5.96)$$

where

$$p_r = (1/2) \operatorname{Re} E_\theta H_\phi^* \quad (5.97)$$

We are now free to let r be sufficiently large that the far field approximations for the fields are valid. Then clearly

$$p_r \simeq (1/2) \operatorname{Re} \eta_0 H_\phi H_\phi^* = (\eta_0/2)|H_\phi|^2 \quad (5.98)$$

where

$$\eta_0 = (\mu_0/\varepsilon_0)^{1/2} = 120\pi$$

Now the ϕ integration in eqn. 5.96 is trivial. Then, using eqn. 5.65, we see that

$$2P = \frac{\eta_0}{2\pi}|I_m|^2 \int_0^\pi \frac{\left[\cos\left(k\frac{L}{2}\cos\theta\right) - \cos\left(k\frac{L}{2}\right)\right]^2}{\sin\theta} d\theta \quad (5.99)$$

Unless kL is large compared with 1, the integral in eqn. 5.99 can be evaluated by numerical or graphical methods very easily. Using such results we can deduce the radiation resistance R_m, which is defined by

$$R_m = 2P/|I_m|^2 \; \Omega \quad (5.100)$$

An example of how R_m varies with the antenna length L is shown in Fig. 5.12. In this particular definition of radiation resistance, the reference current is I_m as indicated in eqn. 5.60. As shown in Fig. 5.12, the value of R_m increases from zero for $L = 0$ in a manner proportional to L^2 when L/λ_0 is small. At $L = \lambda_0/2$ or $kL = \pi$, the computed value of R_m is 73·1 Ω. When L exceeds about λ_0, the value of R_m decreases somewhat because of partial cancellation of the fields radiated from different portions of the structure.

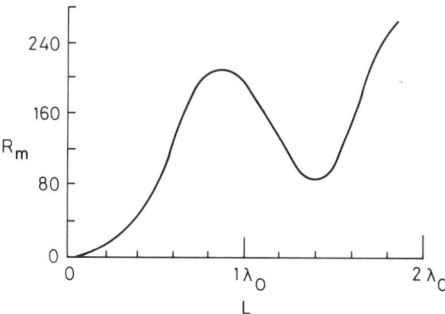

Fig. 5.12

When $L \leqslant \lambda_0$ the maximum field occurs at $\theta = \pi/2$. We can then use eqns. 5.65, 5.94 and 5.98 to show that the maximum gain g_m is given by

$$g_m = g\left(\theta = \frac{\pi}{2}\right) = \frac{\eta_0}{\pi R_m}\left(1 - \cos\frac{kL}{2}\right)^2 \quad (5.101)$$

which of course is independent of ϕ. In the dipole limit we obtain $g_m = 1\cdot 50$; for a half-wave antenna (i.e. $L = \lambda_0/2$) we obtain $g_m = 1\cdot 64$; and for a full-wave antenna (i.e. $L = \lambda_0$) we obtain $g_m = 2\cdot 42$.

5.8 Further exercises

Exercise

Show that the electromagnetic fields of the electric dipole given by eqns. 5.29, 5.30 and 5.31 satisfy Maxwell's equations when the latter are written in spherical co-ordinates (for no variation in the ϕ direction).

Solution

We merely substitute these three equations into

$$\frac{\partial}{\partial \theta}(\sin \theta \, H_\phi) = (\sigma + j\varepsilon\omega) r \sin \theta \, E_r \qquad (5.102)$$

$$\frac{\partial}{\partial r}(rH_\phi) = -(\sigma + j\varepsilon\omega) r E_\theta \qquad (5.103)$$

$$\frac{\partial}{\partial r}(rE_\theta) - \frac{\partial E_r}{\partial \theta} = -j\mu\omega r H_\phi \qquad (5.104)$$

Exercise

Show that there is a set of equations analogous to eqns. 5.29, 5.30 and 5.31 that correspond to a magnetic dipole source located at the origin.

Solution

In place of eqn. 5.5 we define a vector potential F via

$$E = -\nabla \times F \qquad (5.105)$$

Then, in analogy to eqn. 5.10, we deduce that

$$H = \frac{1}{j\mu\omega}[-\gamma^2 F + \nabla(\nabla \cdot F)] \qquad (5.106)$$

Furthermore, in analogy to eqn. 5.14 we can write

$$F = i_z \frac{p_m}{4\pi} \frac{e^{-\gamma r}}{r} \qquad (5.107)$$

where p_m is the magnetic dipole moment. In free space, the nonzero field components are found to be

$$E_\phi = -\frac{p_m}{4\pi r^2}(1 + jkr) e^{-jkr} \sin \theta \qquad (5.108)$$

$$j\mu_0 \omega H_r = \frac{p_m}{2\pi r^3}(1 + jkr) e^{-jkr} \cos \theta \qquad (5.109)$$

$$j\mu_0 \omega H_\theta = \frac{p_m}{4\pi r^3}(1 + jkr - k^2 r^2) e^{-jkr} \sin \theta \qquad (5.110)$$

which are analogous to eqns. 5.36, 5.37 and 5.38. In this case $p_m = j\omega q_m l$ in terms of the effective magnetic charges $+q_m$ and $-q_m$ separated by an infinitesimal distance l (e.g. see eqn. 2.61). An equivalent source model is a small loop of current I with its axis in the z direction. Then $p_m = j\mu_0 \omega I dA$, where dA is the infinitesimal loop area (see next exercise).

Exercise

Show that the mutual impedance Z_m between a short wire element of length l and a small loop of area dA is given by

$$Z_m = -j\mu_0 \omega \frac{dAl}{4\pi r^2}(1 + \gamma r) e^{-\gamma r} \qquad (5.111)$$

when oriented for maximum coupling in a homogeneous medium.

Solution

With reference to eqn. 5.29 we see that at $\theta = \pi/2$, H_ϕ is at maximum.

Electromagnetic fields of current distributions

Thus for maximum coupling the loop is contained in the plane $\phi = 0$ and located at $\theta = \pi/2$. By Faraday's law of induction the voltage v induced in the loop is $-j\mu\omega H_\phi dA$ for a current I in the wire element located at the origin. The mutual impedance, by definition, is given by $Z_m = v/I$. Now this result can also be obtained by treating the small loop as a source and regarding the wire element as a receiving antenna. For example, we would then work with eqn. 5.108 and deduce that the voltage v induced in the wire element, at $\theta = \pi/2$, was $E_\phi l$ (provided that the wire element was oriented in the ϕ direction) for a current I in the loop. Reciprocity ensures that both procedures give identical results for the mutual impedance. In fact, this is one way to demonstrate that the effective magnetic dipole moment of a small current-carrying loop is $p_m = j\mu_0 \omega I dA$.

Exercise

Show that the radial wave admittance for a magnetic dipole source, as defined by

$$Y_r = -H_\theta/E_\phi \tag{5.112}$$

is, for a homogeneous medium, given by

$$Y_r = \frac{1}{\eta}\left(1 + \frac{1}{\gamma r} + \frac{1}{\gamma^2 r^2}\right)\left(1 + \frac{1}{\gamma r}\right)^{-1} \tag{5.113}$$

Hint: this result is fully analogous to eqn. 5.33.

Exercise

Show that the radiation fields of a small current-carrying loop in free space are given by

$$H_\theta \simeq -\frac{1}{\eta_0} E_\phi \simeq (j\varepsilon_0 \omega)\frac{j\mu_0 \omega I \, dA}{4\pi r} e^{-jkr} \sin\theta \tag{5.114}$$

Hint: this result is fully analogous to eqn. 5.43.

Exercise

Consider a perfectly conducting ground plane at $z = 0$ and locate a vertical electric dipole Il at height $z = h$ in the homogeneous medium for $z > 0$. Show that the vector potential for $z > 0$ is given by

$$\mathbf{A} = \mathbf{i}_z A_z \tag{5.115}$$

where

$$A_z = \frac{Il}{4\pi}\left(\frac{e^{-\gamma r_1}}{r_1} + \frac{e^{-\gamma r_2}}{r_2}\right) \tag{5.116}$$

where

$$r_1 = [\varrho^2 + (z-h)^2]^{1/2} \tag{5.117}$$

$$r_2 = [\varrho^2 + (z+h)^2]^{1/2} \tag{5.118}$$

Solution

We verify this result by noting the following:
(a) For $z > 0$,

$$(\nabla^2 - \gamma^2)A_z = 0$$

except at the source.

(b) A_z has the proper behaviour as $r_1 \to 0$ (i.e. it behaves as $1/r_1$).
(c) A_z vanishes as both r_1 and $r_2 \to \infty$.
(d)
$$E_\varrho = \frac{1}{\sigma + j\varepsilon\omega} \frac{\partial^2 A_z}{\partial\varrho\partial z}$$
vanishes at $z = 0$.

Exercise

In terms of spherical co-ordinates (r, θ, ϕ), show that the radiation fields of the electric dipole in the previous exercise, for free space conditions, are given by

$$E_\theta \simeq \eta_0 H_\phi \simeq j\mu_0\omega \frac{Il}{2\pi r} e^{-jkr} F(\theta) \qquad (5.121)$$

where

$$F(\theta) = \tfrac{1}{2}(e^{jkh\cos\theta} + e^{-jkh\cos\theta})\sin\theta$$

$$= \cos(kh\cos\theta)\sin\theta \qquad (5.122)$$

Solution

We make the usual far field assumption and deduce that $r_1 = r - h\cos\theta$ and $r_1 = r + h\cos\theta$. Then eqn. 5.65 is applied to determine H_ϕ.

Exercise

Show how to modify the pattern function $F(\theta)$ in the preceding exercise to account for an imperfectly conducting half-space that occupies the region $z < 0$.

Solution

We merely replace $F(\theta)$ by the form given by eqn. 4.106, where $R_\parallel$ is the appropriate Fresnel reflection coefficient as given, for example, by eqn. 4.83 in terms of the properties of the lower half-space. It is important to note that this procedure has its roots in geometrical optics; it is only valid when we are dealing with the radiation fields where, strictly speaking, $r \to \infty$. Note also that according to eqn. 4.83, for a finitely conducting half-space, $F(\theta)$ always vanishes as $\theta \to \pi/2$, corresponding to grazing angles. One must use other more sophisticated methods to deduce the fields along the surface of the imperfectly conducting ground plane.

Exercise

Show how to deduce the fields of a horizontal electric dipole or current element Il located at height h over a perfectly conducting ground plane at $z = 0$.

Solution

In terms of rectangular co-ordinates (x, y, z) the dipole is located at $z = h$ on the z axis and is oriented in the direction of the x axis. Clearly the primary component of A will have only an x component (e.g. compare with eqn. 5.14). This suggests that we write the resultant vector potential as the sum of two terms such that $A = i_x A_x$, where

$$A_x = \frac{Il}{4\pi}\left(\frac{e^{-\gamma r_1}}{r_1} + R_h \frac{e^{-\gamma r_2}}{r_2}\right) \qquad (5.123)$$

and where r_1 and r_2 are as given by eqns. 5.117 and 5.118, respectively, but R_h is not known. Now we see from eqn. 5.10 that

Electromagnetic fields of current distributions

$$E_x = \frac{1}{\sigma + j\varepsilon\omega}\left(-\gamma^2 A_x + \frac{\partial^2 A_x}{\partial x^2}\right) \tag{5.124}$$

$$E_y = \frac{1}{\sigma + j\varepsilon\omega}\frac{\partial^2 A_x}{\partial x \partial y} \tag{5.125}$$

Clearly $E_x = E_y = 0$ at $z = 0$ if R_h is chosen to be -1. To deduce the complete fields for the horizontal electric dipole over the imperfectly conducting half-space is not a trivial extension of the result given above.

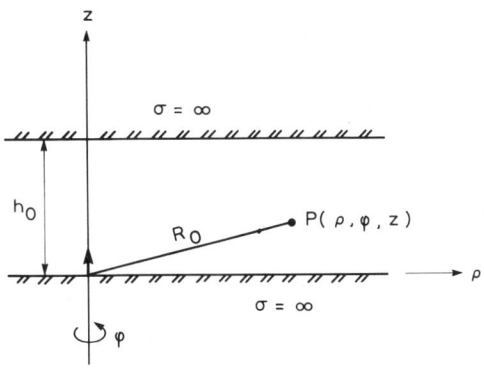

Fig. 5.13 Vertical electric dipole on lower surface of parallel plate region

Exercise

Consider a vertical electric dipole Il on the surface of a perfectly conducting plane at $z = 0$, as indicated in Fig. 5.13. Now locate another perfectly surface at $z = h_0$ and assume that the region $0 < z < h_0$ is free space and unbounded in the ϱ direction. Show that the magnetic field H_ϕ at $P(\varrho, \phi, z)$, within this parallel plate region, is given by

$$H_\phi = \frac{Il\varrho}{2\pi} \sum_{n=-\infty}^{+\infty} \frac{1}{R_n^3}(1 + jkR_n)\, e^{-jkR_n} \tag{5.126}$$

where the summation extends over all integers from $-\infty$ to $+\infty$, including zero. Here

$$R_n = [\varrho^2 + (2nh_0 - z)^2]^{1/2} \tag{5.127}$$

Solution

We postulate a series of images, as indicated in Fig. 5.14. For example, the contribution of the first image at $z = 2h_0$ cancels the field E_ϱ at $z = h_0$ from the source, so that the perfectly conducting boundary condition is satisfied. In a similar fashion we find that the contributions from the image at $z = -2h_0$ and the image at $z = 2h_0$ yield a zero field for E_ϱ at $z = 0$. Contributions from other image pairs add in a similar fashion to yield a null (i.e. zero) value of the resultant E_ϱ at $z = 0$ and at $z = h_0$. The process is summarised in a concise fashion by noting that, for the observer at $P(\varrho, \phi, z)$, we have the summation:

$$A_z = \frac{Il}{2\pi} \sum_{n=-\infty}^{+\infty} \frac{e^{-jkR_n}}{R_n} \tag{5.128}$$

where R_n, as defined by eqn. 5.127, is the distance from the image at $z = 2nh_0$ to the observer point P. For example R_1, as indicated in Fig. 5.14, is the distance from $z = 2h_0$ to the point P. Finally we note that the magnetic field at P, which has only a ϕ component, is obtained from

$$H_\phi = -\frac{\partial A_z}{\partial \varrho} \tag{5.129}$$

which yields eqn. 5.126. One may also verify that

$$j\varepsilon\omega E_\varrho = \left.\frac{\partial^2 A_z}{\partial \varrho \partial z}\right|_{\substack{z=0\\z=h_0}} = 0 \tag{5.130}$$

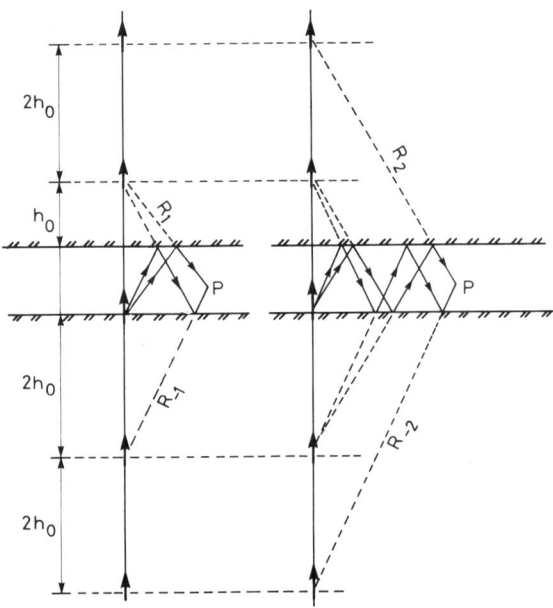

Fig. 5.14 Image locations of the source dipole on the vertical axis. Single- and double-bounce rays are shown separately

Exercise

Show that an alternative representation of the fields within the parallel plate region can be developed in terms of waveguide modes.

Solution

Here we accept the concept of the infinite line of images, but we examine the condition for resonance. For example, as indicated in Fig. 5.15, the rays making an angle θ_m with the vertical axis will be self-reinforcing if $2kh_0 \cos \theta_m = 2m\pi$, where $m = 0, \pm 1, \pm 2, \pm 3 \ldots$. Thus the field components for a given mode of order m will vary according to $\exp(-jk\varrho \sin \theta_m) \cos(kz \cos \theta_m)$. The sum of these discrete contributions will give a representation in terms of waveguide modes. Such a representation complements the ray picture and is most useful at large ranges (i.e. ϱ is large). The waveguide concept is discussed in detail in the next chapter.

Electromagnetic fields of current distributions

Exercise Consider a short electric antenna of length l carrying a current I located at height h over a perfectly conducting ground plane, as shown in Fig. 5.16. As indicated, the antenna or dipole is tilted by an angle θ_0 with respect to the vertical (i.e. the z axis in Fig. 5.16). Without loss of generality we may choose the rectangular system such that the dipole lies in the plane $x = 0$ and the ground plane itself is the surface $z = 0$. The first objective is to obtain expressions for the complete electromagnetic field valid everywhere in the upper half-space (i.e. the air).

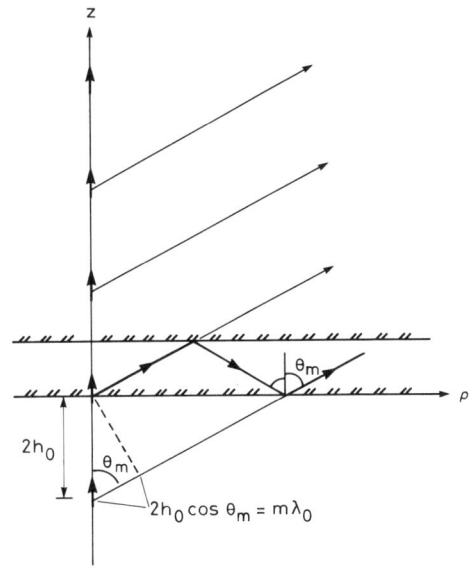

Fig. 5.15 Concept of waveguide mode as resonance of a bundle of rays

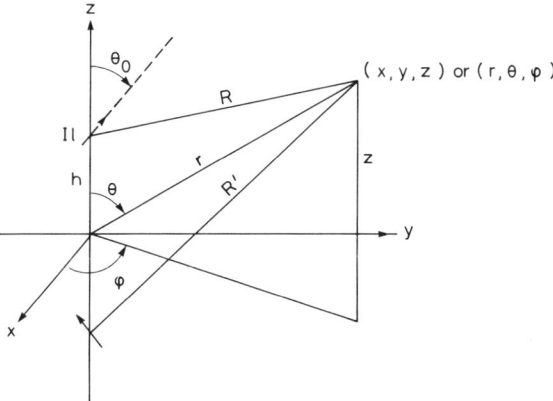

Fig. 5.16 Tilted electric dipole located over a ground plane, showing location of image dipole

Solution The key to the solution is to recognise that a tilted dipole of current moment Il can be resolved into a component IlC in the z direction and a component IlS in the y direction, where $C = \cos\theta_0$ and $S = \sin\theta_0$.

Now we can use image theory, where we note that the image of the vertical dipole component has the same polarity as the source but the image of the horizontal dipole has the opposite sign. Thus we are led to write the following expression for the vector potential A, for the region $z > 0$:

$$A = i_y A_y + i_z A_z \tag{5.131}$$

where

$$A_y = \frac{IlS}{4\pi}\left(\frac{e^{-jkR}}{R} - \frac{e^{-jkR'}}{R'}\right) \tag{5.132}$$

$$A_z = \frac{IlC}{4\pi}\left(\frac{e^{-jkR}}{R} + \frac{e^{-jkR'}}{R'}\right) \tag{5.133}$$

where

$$k = (\varepsilon_0 \mu_0)^{1/2}\omega = \omega/c = 2\pi/\text{wavelength}$$

$$R = [x^2 + y^2 + (z-h)^2]^{1/2}$$

$$R' = [x^2 + y^2 + (z+h)^2]^{1/2}$$

In the usual fashion, the fields are obtained from

$$H = \text{curl } A \tag{5.134}$$

and

$$j\varepsilon_0 \omega E = (k^2 + \text{grad div})A \tag{5.135}$$

In particular, the magnetic field components are deduced from

$$H_x = \frac{\partial}{\partial y}A_z - \frac{\partial}{\partial z}A_y \tag{5.136}$$

$$H_y = -\frac{\partial}{\partial x}A_z \tag{5.137}$$

$$H_z = \frac{\partial}{\partial x}A_y \tag{5.138}$$

Here we note that

$$\frac{\partial}{\partial x}\frac{e^{-jkR}}{R} = -jk\frac{e^{-jkR}}{R}\left(1 + \frac{1}{jkR}\right)\frac{x}{R} \tag{5.139}$$

$$\frac{\partial}{\partial y}\frac{e^{-jkR}}{R} = -jk\frac{e^{-jkR}}{R}\left(1 + \frac{1}{jkR}\right)\frac{y}{R} \tag{5.140}$$

$$\frac{\partial}{\partial z}\frac{e^{-jkR}}{R} = -jk\frac{e^{-jkR}}{R}\left(1 + \frac{1}{jkR}\right)\frac{z-h}{R} \tag{5.141}$$

Thus we have the following explicit expressions for the magnetic field components:

$$H_x = -\frac{Il}{4\pi}\left[\frac{jk}{R}e^{-jkR}\left(1+\frac{1}{jkR}\right)\left(\frac{yC}{R}-\frac{(z-h)S}{R}\right)\right.$$

$$\left.+\frac{jk}{R'}e^{-jkR'}\left(1+\frac{1}{jkR'}\right)\left(\frac{yC}{R'}+\frac{(z+h)S}{R'}\right)\right]$$

(5.142)

$$H_y = +\frac{Il}{4\pi}\left[\frac{jk}{R}e^{-jkR}\left(1+\frac{1}{jkR}\right)\frac{xC}{R}+\frac{jk}{R'}e^{-jkR'}\left(1+\frac{1}{jkR'}\right)\frac{xC}{R'}\right]$$

(5.143)

$$H_z = -\frac{Il}{4\pi}\left[\frac{jk}{R}e^{-jkR}\left(1+\frac{1}{jkR}\right)\frac{xS}{R}-\frac{jk}{R'}e^{-jkR'}\left(1+\frac{1}{jkR'}\right)\frac{xS}{R'}\right]$$

(5.144)

We may immediately verify that H_z vanishes at $z=0$, as required for a perfectly conducting surface. Also we can see that H_x and H_y, at $z=0$, are doubled from the corresponding values if the ground plane were not present. Explicit expressions for the electric field components can be obtained by working with eqn. 5.135, but we do show the results here.

Exercise

For the problem in the preceding exercise, deduce the radiation patterns of the tilted dipole with reference to spherical co-ordinates, as shown in Fig. 5.16. Also plot these patterns in the principal planes. Choose $h = \lambda_0/4$ ($kh = \pi/2$).

Solution

We are now talking about the far field, where the following conditions hold:

$$kr = k(x^2+y^2+z^2)^{1/2} \gg 1$$

$$r \gg h$$

Thus

$$1/R \simeq 1/R' \simeq 1/r$$

Also we see that

$$x/R \simeq x/R' \simeq x/r = \sin\theta\cos\phi \quad (5.145)$$

$$y/R \simeq y/R' \simeq y/r = \sin\theta\sin\phi \quad (5.146)$$

$$(z-h)/R \simeq (z+h)/R' \simeq z/r = \cos\theta \quad (5.147)$$

But in the phase terms, we keep first-order corrections; thus

$$kR = k(x^2+y^2+z^2-2zh+h^2)^{1/2} = k(r^2-2zh+h^2)^{1/2}$$

$$\simeq k(r^2-2zh)^{1/2} = kr[1-(2zh/r^2)]^{1/2} \simeq kr[1-(zh/r^2)]$$

$$\simeq kr - kh(z/r) = kr - kh\cos\theta \quad (5.148)$$

Similarily
$$kR' \simeq kr + kh\cos\theta \qquad (5.149)$$

Thus we have
$$e^{-jkR} \simeq e^{-jkr} e^{jkh\cos\theta} \qquad (5.150)$$

$$e^{-jkR'} \simeq e^{-jkr} e^{-jkh\cos\theta} \qquad (5.151)$$

Now the farfield forms are easily deduced from eqns. 5.142, 5.143 and 5.144:

$$H_x \simeq -\frac{Il}{4\pi}\frac{jk}{r}e^{-jkr}[C\sin\theta\sin\phi\,(e^{jkh\cos\theta} + e^{-jkh\cos\theta})$$
$$- S\cos\theta\,(e^{jkh\cos\theta} - e^{-jkh\cos\theta})] \quad (5.152)$$

$$H_y \simeq \frac{Il}{4\pi}\frac{jk}{r}e^{-jkr}[C\sin\theta\cos\phi\,(e^{jkh\cos\theta} + e^{-kkh\cos\theta})] \quad (5.153)$$

$$H_z \simeq -\frac{Il}{4\pi}\frac{jk}{r}e^{-jkr}[S\sin\theta\cos\phi\,(e^{jkh\cos\theta} - e^{-jkh\cos\theta})] \quad (5.154)$$

To convert to spherical co-ordinates, we merely note that
$$H_r = H_x\sin\theta\cos\phi + H_y\sin\theta\sin\phi + H_z\cos\theta \quad (5.155)$$

$$H_\phi = -H_x\sin\phi + H_y\cos\phi \qquad (5.156)$$

$$H_\theta = H_x\cos\theta\cos\phi + H_y\cos\theta\sin\phi - H_z\sin\theta \quad (5.157)$$

Then, on using eqns. 5.152, 5.153 and 5.154, we see that
$$H_r \simeq 0 \qquad (5.158)$$

$$H_\phi \simeq \frac{jkIl}{2\pi}\frac{e^{-jkr}}{r}[C\sin\theta\cos(kh\cos\theta) - jS\sin\phi\cos\theta\sin(kh\cos\theta)]$$
$$(5.159)$$

$$H_\theta \simeq -\frac{kIl}{2\pi}\frac{e^{-jkr}}{r}[S\sin(kh\cos\theta)\cos\phi] \qquad (5.160)$$

Furthermore, in the far field,
$$E_\theta \simeq \eta_0 H_\phi \quad \text{and} \quad E_\phi \simeq -\eta_0 H_\theta \qquad (5.161)$$

where $\eta_0 = 120\pi$. Also $E_r \simeq 0$.

The square bracket terms in eqns. 5.159 and 5.160 are, by definition, the radiation pattern functions. To illustrate the salient forms we set $kh = \pi/2$, corresponding to a height $h = \lambda_0/4$. Thus in Fig. 5.17 we show the so-called E-plane pattern for the case where the dipole is vertically oriented (i.e. $\theta_0 = 0°$). In Fig. 5.18 we show the E-plane pattern for the horizontally oriented dipole (i.e. $\theta_0 = 90°$), and in Fig. 5.19 we show the corresponding H-plane pattern. In Fig. 5.20 we show the E-plane pattern for the case where the dipole is tilted at 45°; the corresponding pattern of the total field intensity for the $\phi = 0°$ plane is shown in Fig. 5.21 (note that, in the latter case, the field is

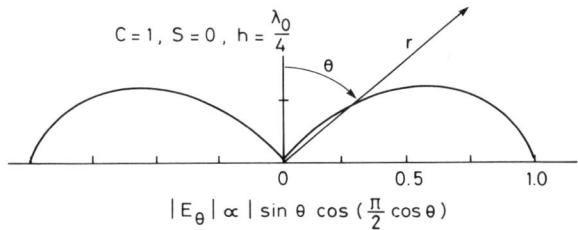

Fig. 5.17 E-plane pattern of vertically oriented dipole

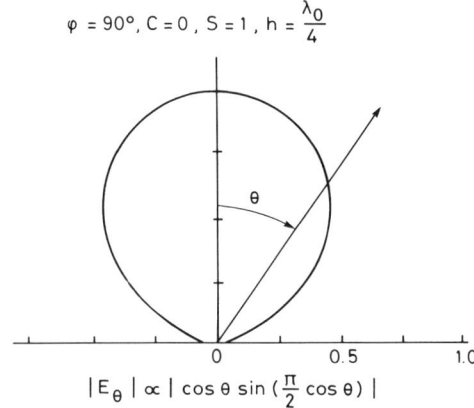

Fig. 5.18 E-plane pattern of horizontally oriented dipole

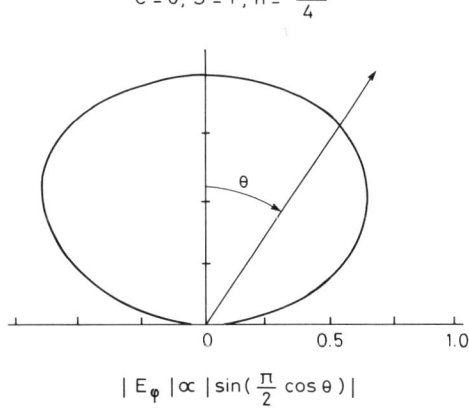

Fig. 5.19 H-plane pattern of horizontally oriented dipole

actually elliptically polarised). Finally, we observe that the H-plane pattern of the dipole, with any tilt angle (for $z = 0$), is merely a circle, which hardly needs to be plotted.

Exercise

Consider a homogeneous half-space ($z > 0$) model of the earth, as indicated in Fig. 5.22. On the vertical z axis pointing downwards we have an electric current source of I amperes that is terminated in electrodes C_1 and C_2. We then measure the received voltage V between two potential elecrodes P_1 and P_2 that are connected by an insulated

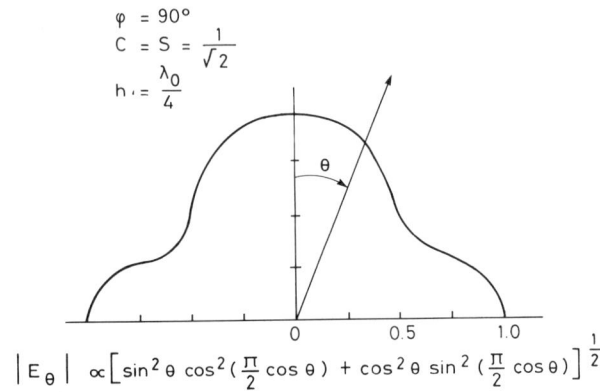

$$|E_\theta| \propto \left[\sin^2\theta \cos^2\left(\frac{\pi}{2}\cos\theta\right) + \cos^2\theta \sin^2\left(\frac{\pi}{2}\cos\theta\right)\right]^{\frac{1}{2}}$$

Fig. 5.20 *E*-plane pattern of tilted dipole

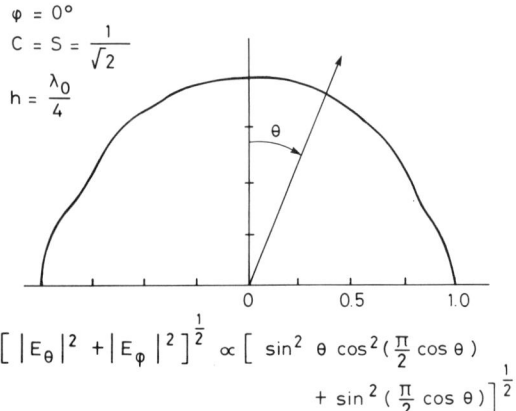

$$\left[|E_\theta|^2 + |E_\varphi|^2\right]^{\frac{1}{2}} \propto \left[\sin^2\theta \cos^2\left(\frac{\pi}{2}\cos\theta\right) + \sin^2\left(\frac{\pi}{2}\cos\theta\right)\right]^{\frac{1}{2}}$$

Fig. 5.21 Total field intensity pattern of tilted dipole in plane $\varphi = 0°$, where field is elliptically polarised

cable along a radial direction from the z axis. We also consider very low frequencies such that all dimensions of the problem are small compared with the free space wavelength. Show that the mutual impedance between the circuits $C_1 C_2$ and $P_1 P_2$ is given by the expression

$$Z(j\omega) = \frac{1}{2\pi\hat{\sigma}(j\omega)}\left(\frac{e^{-\gamma R_{11}}}{R_{11}} + \frac{e^{-\gamma R_{22}}}{R_{22}} - \frac{e^{-\gamma R_{12}}}{R_{12}} - \frac{e^{-\gamma R_{21}}}{R_{21}}\right)$$

where

$$\gamma(j\omega) = [j\mu_0\omega\hat{\sigma}(j\omega)]^{1/2}$$

is the propagation constant. $\hat{\sigma}$ is the complex admittivity of the conducting medium; that is, $\hat{\sigma}(j\omega) = \sigma(\omega) + j\omega\varepsilon(\omega)$, where $\sigma(\omega)$ is the real conductivity and $\varepsilon(\omega)$ is the real permittivity, both being functions of frequency. The distances R_{11}, R_{12}, R_{21} and R_{22} are the distances between C_1 and P_1, C_1 and P_2, C_2 and P_1, and C_2 and P_2, respectively, as indicated in Fig. 5.22.

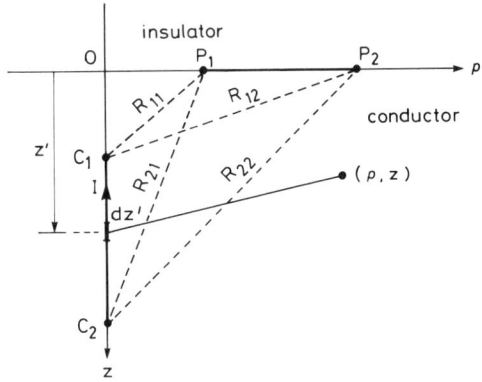

Fig. 5.22 Vertical current-carrying wire between electrodes C_1 and C_2 located in a homogeneous conducting half-space model of the earth, and the potential circuit connected by electrodes P_1 and P_2 on the surface

Solution

We begin by writing down the expression for the potential δA_z produced by the dipole element $I(z')dz'$ at an observer point (ϱ, z) anywhere in the conducting half-space. Thus we have the direct and image contributions:

$$\delta A_z = -\frac{Idz'}{4\pi\hat{\sigma}} \left(\frac{e^{-\gamma r_1}}{r_1} - \frac{e^{-\gamma r_2}}{r_2} \right) \tag{5.163}$$

The negative sign on the second term is needed to ensure that E_z or $J_z = 0$ at $z = 0$. Here the distances r_1 and r_2 are defined by

$$r_1 = [\varrho^2 + (z - z')^2]^{1/2}$$

$$r_2 = [\varrho^2 + (z + z')^2]^{1/2}$$

The horizontal electric field is obtained from

$$\delta E_\varrho = \frac{\partial^2}{\partial\varrho\partial z}(\delta A_z) \tag{5.164}$$

which works out to

$$\delta E_\varrho = \frac{Idz'}{4\pi\hat{\sigma}} \frac{\partial}{\partial\varrho} \left[\frac{1}{r_1^3}(1 + \gamma r_1) e^{-\gamma r_1}(z - z') - \frac{1}{r_2^3}(1 + \gamma r_2) e^{-\gamma r_2}(z + z') \right] \tag{5.165}$$

which, again, holds anywhere in the conductor. Now as we approach the interface,

$$\delta E_\varrho]_{z=0} = -\frac{Idz'}{2\pi\hat{\sigma}} \frac{\partial}{\partial\varrho} \left[\frac{1}{r^3}(1 + \gamma r) e^{-\gamma r} z' \right] \tag{5.166}$$

where

$$r = [\varrho^2 + (z')^2]^{1/2}$$

But an equivalent statement of this result is

$$\delta E_\varrho]_{z=0} = -\frac{I dz'}{2\pi\hat{\sigma}} \frac{\partial^2}{\partial\varrho\partial z'}\left(\frac{e^{-\gamma r}}{r}\right) = \delta E_\varrho(\varrho, 0) \qquad (5.167)$$

which involves a derivative with respect to the z' source coordinate. The voltage developed between P_1 and P_2 is then given by

$$\delta V = -\int_{P_1}^{P_2} \delta E_\varrho(\varrho, 0) \, d\varrho \qquad (5.168)$$

where we neglect any phase shift along the (insulated) cable between P_1 and P_2. Then

$$V = \int_{C_1}^{C_2} \delta V \, dz' = \frac{I}{2\pi\hat{\sigma}} \int_{C_1}^{C_2}\int_{P_1}^{P_2} \frac{\partial^2}{\partial\varrho\partial z'} \frac{e^{-\gamma r}}{r} d\varrho \, dz' \qquad (5.169)$$

which immediately leads to the closed form result

$$V = \frac{I}{2\pi\hat{\sigma}}\left(\frac{e^{-\gamma R_{11}}}{R_{11}} + \frac{e^{-\gamma R_{22}}}{R_{22}} - \frac{e^{-\gamma R_{12}}}{R_{12}} - \frac{e^{-\gamma R_{21}}}{R_{21}}\right) \qquad (5.170)$$

and, of course, the desired impedance is the ratio V/I.

Exercise

To illustrate the quantitative aspects of the problem in the previous exercise, choose the co-ordinates of the electrodes in Fig. 5.22 such that $OC_1 = C_1C_2 = OP_1 = P_1P_2 = a$. Then also let $\hat{\sigma} = \sigma_m \exp(+j\Delta)$, where σ_m and Δ are real. For this case, plot the magnitude of the mutual impedance and the corresponding phase as functions of frequency, using dimensionless parameters.

Solution

For the special geometry indicated, it is evident that $R_{11} = (\sqrt{2})a$, $R_{22} = (2\sqrt{2})a$, and $R_{12} = R_{21} = (\sqrt{5})a$. Then using the basic definition for the propagation constant, we see that

$$\gamma a = (j\mu_0\omega\sigma_m)^{1/2} e^{i\Delta/2} a \qquad (5.171)$$

which can also be written

$$\gamma a = A\left[\cos\left(\frac{\pi}{4} + \frac{\Delta}{2}\right) + j \sin\left(\frac{\pi}{4} + \frac{\Delta}{2}\right)\right] \qquad (5.172)$$

where A is a real dimensionless parameter, proportional to the spacing a, given by

$$A = (\mu_0\omega\sigma_m)^{1/2} a \qquad (5.173)$$

We also find it convenient to normalise the mutual impedance by the zero frequency limit:

$$\lim_{\omega\to 0} Z = R_0 = \frac{1}{2\pi\sigma_m}\left(\frac{1}{R_{11}} + \frac{1}{R_{22}} - \frac{1}{R_{12}} - \frac{1}{R_{21}}\right) \qquad (5.174)$$

which is actually the real transfer or mutual resistance between the two circuits. Thus we plot the ratio $Z(j\omega)/R_0$ as a function of A (as shown in Fig. 5.23) for Δ values ranging from 0 to 0·3. The amplitude and phase scales are identified on the left- and right-hand ordinates, respectively. The shape of these curves can be diagnostic of the electrical properties of the medium, and such schemes are used in geophysical well logging in areas of petroleum resources. (These calculations were performed by Denise Siemers as part of her undergraduate senior honours project.)

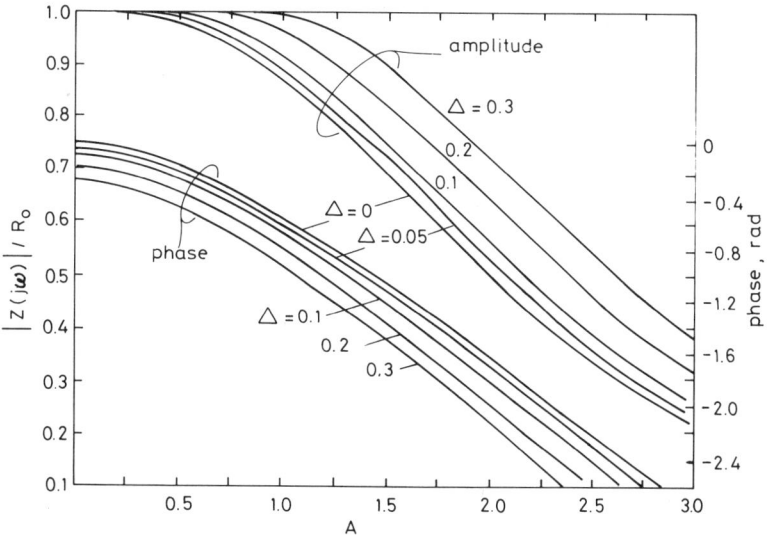

Fig. 5.23 Amplitude and phase of the normalised mutual impedance between the current and potential circuit as a function of dimensionless parameter A, which is proportional to the spacing a between electrodes

Exercise

Consider a thin wire vertical antenna (of height h) located just above a perfectly conducting flat ground plane of infinite extent, as indicated in Fig. 5.24. The antenna is fed at the base. The ambient medium is assumed to be free space. The current on the structure is specified as follows:

$$I(z') = 0 \qquad \text{for } z' > h$$
$$= I_m \sin[k(ph - z')] \qquad \text{for } 0 < z' < h \qquad (5.175)$$

where I_m is a constant, $k = 2\pi/\text{wavelength} = 2\pi/\lambda_0$, and p is a constant.

Determine an expression, valid everywhere, for the vector potential in the free space region above the ground plane.

Solution

We see immediately that the problem is equivalent to locating a centre-driven antenna in unbounded free space where now the total length of the structure is $2h$. Thus we can immediately draw on the form of eqn. 5.61, where we merely identify $L/2$ with h and $-L/2$ with $-h$. Thus, for $z > 0$,

$$A_z = \frac{I_m}{4\pi} \int_{-h}^{h} \frac{e^{-jkR}}{R} \sin[k(ph - |z'|)] \, dz' \qquad (5.176)$$

where
$$R = [\varrho^2 + (z - z')^2]^{1/2}$$
is the desired expression.

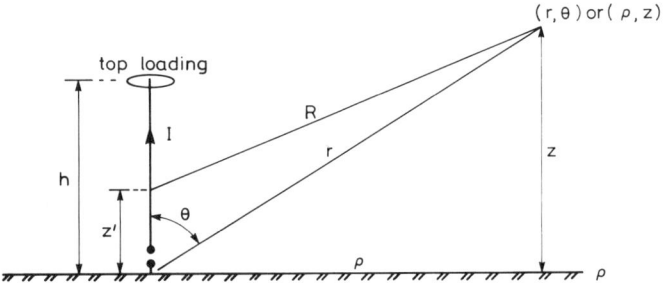

Fig. 5.24 Vertical base-driven antenna of height h over a perfectly conducting ground plane, where we assume a sinusoidally varying current that may be nonzero at the top (e.g. for $p > 1$) such as would occur for a top-loaded structure

Exercise

In the previous exercise, determine useable expressions for the electric and magnetic fields valid in the radiation zone.

Solution

When $kR \gg 1$, we make the usual far field approximation given by
$$R \simeq r - z' \cos \theta \tag{5.177}$$
Then
$$A_z \simeq \left(\frac{I_m e^{-jkr}}{4\pi r}\right) \int_{-h}^{h} \sin [k(ph - |z'|)] e^{jkz'C} dz' \tag{5.178}$$
where $C = \cos \theta$. Furthermore
$$A_z \simeq 2(\cdots) \int_0^h \sin [k(ph - z')] \cos (kCz') dz' \tag{5.179}$$
where the integrand is now real. Then by using the identity
$$2 \sin A \cos B = \sin (A + B) + \sin (A - B) \tag{5.180}$$
we see that
$$A_z \simeq (\cdots) \int_0^h \sin [kph - k(1 - C)z'] dz'$$
$$+ (\cdots) \int_0^h \sin [kph - k(1 + C)z'] dz' \tag{5.181}$$
These are integrals of the type
$$\int_0^h \sin (\alpha - \beta z') dz' = \left.\frac{\cos (\alpha - \beta z')}{\beta}\right|_0^h \tag{5.182}$$
which enables us to write
$$A_z \simeq (\cdots) \left\{\frac{\cos [kph - kh(1 - C)] - \cos (kph)}{k(1 - C)}\right.$$
$$\left. + \frac{\cos [kph - kh(1 + C)] - \cos (kph)}{k(1 + C)}\right\} \tag{5.183}$$

Electromagnetic fields of current distributions

The sole magnetic field component is now obtained from

$$H_\phi = -\frac{\partial A_z}{\partial \varrho} \simeq jk \sin\theta \, A_z \tag{5.184}$$

which allows us to write

$$E_\theta \simeq \eta_0 H_\phi \simeq j\eta_0 \frac{I_m}{2\pi r} e^{-jkr} P(\theta) \tag{5.185}$$

where, by definition, $P(\theta)$ is the radiation pattern. After doing a little algebra, we get the viable form

$$P(\theta) = \{\cos(kh \cos\theta) \cos[k(p-1)h] - \cos(kph)$$
$$- \cos\theta \sin(kh \cos\theta) \sin[k(p-1)h]\}/\sin\theta \tag{5.186}$$

which is an explicit function of the polar angle θ. In the special case $p = 1$, we see that

$$P(\theta) = [\cos(kh \cos\theta) - \cos(kh)]/\sin\theta \tag{5.187}$$

which, not surprisingly, is identical to eqn. 5.63 when h is replaced by $L/2$.

Exercise

Using eqn. 5.186 for $P(\theta)$, show illustrative patterns for antenna height $h = \lambda_0/4$ for the two cases where $p = 1$ and 2. The latter case corresponds to the situation where the current is nonzero at the top of the antenna. The result is shown in Fig. 5.25. As indicated, the patterns are very similar but there is a noticable enhancement of the radiation along the ground plane in the case where $p = 2$ (i.e. the 'high angle radiation' is reduced somewhat).

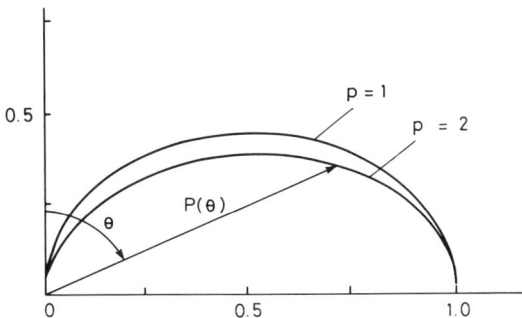

Fig. 5.25 *E*-plane pattern of the vertical base-driven antenna whose physical height is $\lambda_0/4$

5.9 Bibliography

H. BACH and J. E. HANSEN, Uniformly spaced arrays, Chapter 5 in *Antenna Theory*, Part 1 (eds R. E. Collin and F. J. Zucker), McGraw Hill, 1969.

C. A. BALANIS, *Antenna Theory, Analysis and Design*, Harper and Row (especially Chapter 6), 1982.

C. L. DOLPH, A current distribution for broadside arrays which optimises the relationship between beamwidth and side lobe level, *Proc. IRE*, 1946, **34**, pp. 335–48.

M. T. MA, *Theory and Applications of Antenna Arrays*, Wiley, 1974.

A. W. RUDGE, K. MILNE, A. D. OLVER and P. KNIGHT (eds), *The Handbook of Antenna Design*, Peter Peregrinus, 1982.

S. A. SCHELKUNOFF, A mathematical theory of linear arrays, *Bell Syst. Tech. J.* 1943, **22**, pp. 80–107.

S. A. SCHELKUNOFF and H. T. FRIIS, *Antennas, Theory and Practice*, Wiley, 1952.

J. R. WAIT, Pattern of a flush mounted microwave antenna, *J. Res. Natl. Bur. Stand.*, 1957, **59**, pp. 225–9 (considers an end-fire array of slot antennas on a cylindrical surface).

J. R. WAIT and A. M. CONDA, The patterns of a slot array antenna on a finite imperfect ground plane, *L'Onde Electrique* (special supplement), 1959, 376, pp. 21–30.

J. R. WAIT, *Electromagnetic Radiation from Cylindrical Structures*, Pergamon, 1959 (see Chapter 15, dealing with slot arrays on a half-plane).

J. R. WAIT, Characteristics of antennas over lossy earth, Chapter 23 in *Antenna Theory*, Part 2 eds R. E. Collin and F. J. Zucker), McGraw Hill, 1969.

H. P. WILLIAMS, *Antenna Theory and Design*, Pitman, 1966 (see Chapter 6 for an excellent account of directional arrays).

Chapter 6

Guided waves

6.1 Coaxial cable and wire lines

In many telecommunication systems, the desired signal is conveyed by pairs of conductors. Geometrically, the simplest of such transmission lines are coaxial cables. In the ideal coaxial cable, such as that shown in Fig. 6.1, the energy is contained in the concentric region between the centre conductor of radius a and the outer conducting surface of radius b. Using quasi-static arguments, which are valid when $b-a$ is electrically small, we deduce the shunt admittance $\hat{Y}$ and the series impedance $\hat{Z}$, both per unit length of the cable.

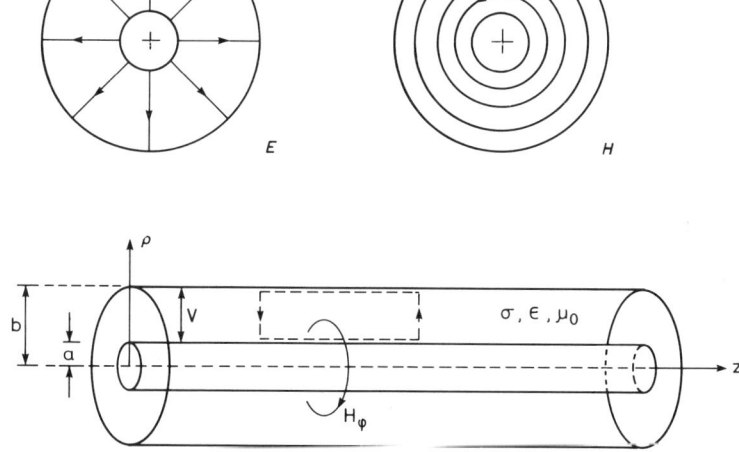

Fig. 6.1 Ideal coaxial cable showing electric field (top left) and magnetic field (top right)

First of all we apply a voltage between the inner and the outer conductor. Then, except near the cable ends, the electric field vector will be radial around the centre conductor. As usual we treat V as a (complex) phasor for an angular frequency ω. Thus the real physical quantity is the real part of $V \exp(j\omega t)$. The corresponding current per unit length between the conductors is denoted by the phasor j_0. Now the radial current density J_ϱ at radius ϱ, with respect to a cylindrical co-ordinate system (ϱ, ϕ, z), is

$$J_\varrho = j_0/2\pi\varrho \tag{6.1}$$

On using Ohm's law, the electric field that has only a radial component is given by

$$E_\varrho = \frac{1}{\sigma + j\varepsilon\omega}\left(\frac{j_0}{2\pi\varrho}\right) \tag{6.2}$$

Then, clearly,

$$V = \int_a^b E_\varrho \, d\varrho = \frac{j_0}{2\pi(\sigma + j\varepsilon\omega)} \ln\frac{b}{a} \tag{6.3}$$

By definition, the admittance $\hat{Y}$ per unit length of the cable is j_0/V. Thus

$$\hat{Y} = 2\pi(\sigma + j\varepsilon\omega)/\ln(b/a) \tag{6.4}$$

or, if we write

$$\hat{Y} = \hat{G} + j\omega\hat{C} \tag{6.5}$$

we see that

$$\hat{G} = 2\pi\sigma/\ln(b/a) \tag{6.6}$$

is the conductance per unit length and

$$\hat{C} = 2\pi\varepsilon/\ln(b/a) \tag{6.7}$$

is the capacitance per unit length, assuming perfect conductors.

Now we consider the longitudinal current I that flows along the inner conductor and returns on the outer conductor. A direct application of Ampère's law tells us that the magnetic field H_ϕ in the region $a < \varrho < b$ is given by

$$H_\phi = I/2\pi\varrho \tag{6.8}$$

Then, if we choose a rectangular circuit of unit length and width $b - a$, as shown in Fig. 6.1, the voltage v induced is given by Faraday's law as follows:

$$v = -j\mu_0\omega \int_a^b H_\phi \, d\varrho = -\frac{j\mu_0\omega}{2\pi} I \ln\frac{b}{a} \tag{6.9}$$

where H is assumed to be essentially constant over the unit length of the assumed rectangle. This voltage v must be overcome by the impressed voltage between the inner and outer conductors. Thus, in the case of perfectly conducting surfaces at $\varrho = a$ and $\varrho = b$, we would have $v = -j\omega\hat{L}I$, where

$$\hat{L} = (\mu_0/2\pi)\ln(b/a) \tag{6.10}$$

is the series inductance per unit length. More generally, we would write

$$v = -(\hat{R} + j\omega\hat{L})I \tag{6.11}$$

where $\hat{R}$, the series resistance of the cable per unit length, would account for the finite conductivity of the walls.

Guided waves

Exercise

Consider a two-wire line that can be idealised as two slender cylindrical conductors, each of a radius a that is small compared with their separation d. Show that the expressions for the series admittance and series impedance per unit length are

$$\hat{Y} \simeq \pi(\sigma + j\varepsilon\omega)/\ln(d/a) \tag{6.12}$$

$$\hat{Z} \simeq (j\mu_0\omega/\pi)\ln(d/a) \tag{6.13}$$

respectively. Assume that the wires are perfectly conducting and that the ambient medium is homogeneous with properties $(\sigma, \varepsilon, \mu)$.

Solution

We regard each wire conductor as a line source, giving rise to radial currents j_0 and $-j_0$, respectively, per unit length. We then evaluate the potential on one conductor due to the current on the other. The inductance calculation is carried out in the same manner as for the coaxial cable.

Exercise

Consider a single wire of radius a located at height h over a perfectly conducting ground plane. Show that the series admittance and the series impedance per unit length of the structure are

$$\hat{Y} = 2\pi(\sigma + j\varepsilon\omega)/\ln(2h/a) \tag{6.14}$$

$$\hat{Z} = (j\mu_0\omega/2\pi)\ln(2h/a) \tag{6.15}$$

Solution

The structure of the fields is identical to that in the preceding exercise. here $2h$ is the distance from the wire to its image at depth h below the ground plane. But note that the conductor voltage, relative to the ground plane, is one-half of the voltage between the two wire conductors in the previous example.

6.2 Simple transmission line theory

The propagation of voltages and currents along coaxial cables and wire conductors can be described adequately by a one-dimensional transmission line theory. This can be developed via circuit concepts in the following manner.

The transmission line structure (e.g. the coaxial cable) is assumed to be of infinite length, and it is characterized by a series admittance $\hat{Y}$ and a series impedance $\hat{Z}$, both per unit length. A small section of this transmission line of length δz is depicted in Fig. 6.2. For conceptual purposes, we show this section as a T network where the shunt element has an admittance $\hat{Y}\delta z$ siemens. For symmetry we divide the series arm into two portions that each have an impedance of $(\hat{Z}/2)\delta z$ ohms. By making δz arbitrarily small, such a network representation becomes increasingly accurate.

Now the voltage drop across the section must be balanced by the current times the incremental impedance. Similarly, the current change must be balanced by the voltage times the incremental admittance. Thus we deduce the basic forms

$$\frac{dV}{dz} = -\hat{Z}I \tag{6.16}$$

$$\frac{dI}{dz} = -\hat{Y}V \tag{6.17}$$

where the common factors δz have conveniently cancelled on both sides of these two differential equations. If we combine eqns. 6.16 and 6.17, and remember that $\hat{Z}$ and $\hat{Y}$ do not depend on z, we obtain the uncoupled second-order equations:

$$\frac{d^2 V}{dz} - \Gamma^2 V = 0 \tag{6.18}$$

$$\frac{d^2 I}{dz} - \Gamma^2 I = 0 \tag{6.19}$$

where

$$\Gamma^2 = \hat{Z}\hat{Y} \tag{6.20}$$

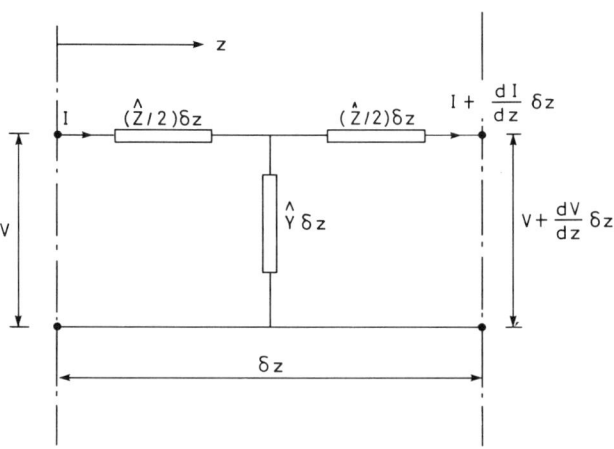

Fig. 6.2 Elemental section δz of simple transmission line with series impedance $\hat{Z}$ and shunt admittance $\hat{Y}$ per unit length

To avoid ambiguity we define

$$\Gamma = (\hat{Z}\hat{Y})^{1/2} \tag{6.21}$$

such that $\text{Re } \Gamma > 0$.

The general solution for the voltage V is

$$V = A e^{-\Gamma z} + B e^{\Gamma z} \tag{6.22}$$

which is verified by direct substitution into eqn. 6.18., assuming A and B are constants. Then, in view of eqn. 6.16, we see that

$$I = -\frac{1}{\hat{Z}}\frac{dV}{dz} = \frac{\Gamma}{\hat{Z}} A e^{-\Gamma z} - \frac{\Gamma}{\hat{Z}} B e^{\Gamma z} \tag{6.23}$$

An equivalent form is

$$I = K^{-1} A e^{-\Gamma z} - K^{-1} B e^{\Gamma z} \tag{6.24}$$

where

$$K = Z/\Gamma = (\hat{Z}/\hat{Y})^{1/2} = \Gamma/\hat{Y} \tag{6.25}$$

is, by definition, the *characteristic impedance* of the transmission line.

Guided waves

Now clearly, if the line were of infinite length (i.e. $0 < z < \infty$) and if there were no sources for $z > 0$, the wave solutions would have the forms

$$V = A e^{-\Gamma z} \tag{6.26}$$

$$I = K^{-1} A e^{-\Gamma z} = K^{-1} V \tag{6.27}$$

which hold for all $z \geqslant 0$. In particular we note that, at $z = 0$, the input impedance of the line is K ohms. In the context of this one-dimensional problem, we could have written $I = MV$, where $M = K^{-1}$ is the characteristic admittance of the line. However, as we have already indicated, there are analogous multidimensional problems where the parameters K and M are not reciprocals of one another.

To illustrate the close analogies between transmission line theory and the plane wave propagation of electromagnetic signals, we consider a simple example. As indicated in Fig. 6.3, we assume that a wave of voltage V and current I is incident on a uniform transmission line of characteristic impedance K_0 and propagation constant Γ_0 (for $z < 0$). Appropriate expressions for the voltage and current functions, for $z < 0$, are

$$V = V_0(e^{-\Gamma_0 z} + R e^{\Gamma_0 z}) \tag{6.28}$$

$$I = K_0^{-1} V_0(e^{-\Gamma_0 z} - R e^{\Gamma_0 z}) \tag{6.29}$$

where V_0 is the specified (complex) value of the incident voltage wave at $z = 0$, and R is a *reflection coefficient* yet to be determined. For $z > 0$, we have

$$V = V_0 T e^{-\Gamma z} \tag{6.30}$$

$$I = V_0 K^{-1} T e^{-\Gamma z} \tag{6.31}$$

where T is a *transmission coefficient*.

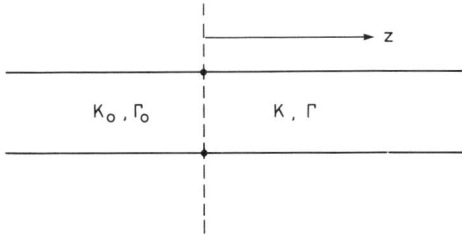

Fig. 6.3 Junction of two ideal transmission lines with characteristic impedance K_0 and K and propagation constants Γ_0 and Γ

Now in such an idealised transmission line it can be assumed that voltage and current are individually continuous at the junction $z = 0$. Then we obtain the algebraic equations

$$1 + R = T \tag{6.32}$$

and

$$K_0^{-1}(1 - R) = K^{-1} T \tag{6.33}$$

These yield the desired forms

$$R = (K - K_0)/(K + K_0) \tag{6.34}$$

$$T = 2K/(K + K_0) \tag{6.35}$$

As can be readily verified, eqns. 6.34 and 6.35 have precisely the same form as eqns. 4.91 and 4.92, respectively. In fact R and T can be written in the forms

$$R = (M_0 - M)/(M_0 + M) \tag{6.36}$$

$$T = 2M_0/(M_0 + M) \tag{6.37}$$

where $M_0 = K_0^{-1}$ and $M = K^{-1}$ are admittances. The analogy is then complete, and demonstrates *that the transmission line model is a convenient way to describe reflection of plane electromagnetic waves at an interface of a homogeneous conducting half-space.*

Exercise

Show that eqns. 6.16 and 6.17 can be decoupled, even in the case where Z and Y are functions of z, to yield

$$\frac{d^2 V}{dz^2} - \frac{1}{\hat{Z}}\frac{d\hat{Z}}{dz}\frac{dV}{dz} - \hat{Y}\hat{Z}V = 0 \tag{6.38}$$

and a similar equation for I.

Solution

Differentiate both sides of eqn. 6.16 with respect to z to give

$$\frac{d^2 V}{dz^2} = -\hat{Z}\frac{dI}{dz} - I\frac{d\hat{Z}}{dz} \tag{6.39}$$

Then substitute eqn. 6.16 for I and eqn. 6.17 for dI/dz into eqn. 6.39, which immediately gives eqn. 6.38. A similar process is carried out to obtain the corresponding form for I. Note that solutions for eqn. 6.38 are not simple exponentials.

Exercise

Consider the same model as shown in Fig. 6.3, but imagine that the line on the right is terminated by an impedance Z_L at $z = l > 0$. Show that the voltage and current for $z < 0$ have the same form as eqns. 6.28 and 6.29, but now R is defined by

$$R = (Z - K_0)/(Z + K_0) \tag{6.40}$$

where

$$Z = K\frac{Z_L + K\tanh \Gamma l}{K + Z_L \tanh \Gamma l} \tag{6.41}$$

is the input impedance at $z = 0$.

Solution

The solution proceeds by assuming that V and I are of the form of eqns. 6.22 and 6.24 in the region $0 \leq z \leq l$. At $z = l$ we then apply $V = Z_L I$, and match V and I individually at $z = 0$ to solve for A, B and R. Also we use the identity $\tanh x/2 = (1 - e^{-x})/(1 + e^{-x})$.

Guided waves

Exercise

In the example of the preceding exercise, show that for lossless transmission lines

$$Z = K\frac{Z_L + jK \tan \beta l}{K + jZ_L \tan \beta l} \qquad (6.42)$$

where $\Gamma = j\beta$ and $\beta = \omega(\hat{L}\hat{C})^{1/2}$.

Solution

Use the identity $\tanh(jy) = j \tan y$.

Exercise

In the above situation, show that $R = 0$ (i.e. perfect matching) is possible if Z_L is real, $\beta l = \pi/2$, and $K^2 = K_0 Z_L$.

Solution

Note that $\tan \pi/2 = \infty$, so that $Z = K^2/Z_L$. This, if equated to K_0, gives no reflection. This is the basis of the so-called quarter-wave transformer to match an impedance K_0 to a load Z_L.

6.3 Rectangular waveguides

In dealing with transmission lines we have ignored the longitudinal electric field in the air space or within the insulation. This is justified when the cross-sectional dimension of the structure are small compared with a wavelength. Such a mode of propagation is called *transverse electromagnetic* (TEM) to signify that both electric and magnetic field vectors are predominantly transverse to the direction of propagation. These guiding structures require two separated conductors to support a TEM mode. Now there is an important class of wave guiding structures, aptly called *waveguides*, that operate on a somewhat different principle. An example that we will consider is the hollow rectangular metal tube, as illustrated in Fig. 6.4.

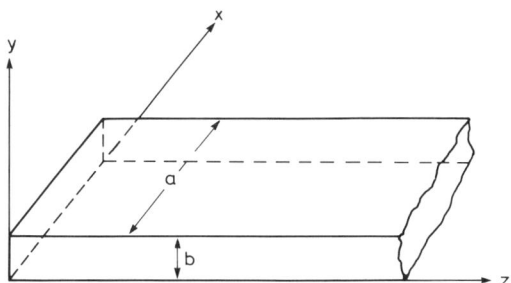

Fig. 6.4 Rectangular waveguide model that extends indefinitely in the positive and negative z directions

To analyse the rectangular structure we obviously choose a rectangular co-ordinate system (x, y, z), where the walls (assumed perfectly conducting) are at $x = 0, a$ and at $y = 0, b$. The region bounded by these walls is a homogeneous medium with properties $(\sigma, \varepsilon, \mu)$.

We now postulate that the propagation of the electromagnetic fields in the structure varies with z in the manner $\exp(-\Gamma z)$, where Γ is an effective propagation constant. We also seek a solution, in the first instance, where $H_z = 0$. Maxwell's equations for these assumptions are

$$(\sigma + j\varepsilon\omega)E_x = \Gamma H_y \qquad (6.43)$$

$$(\sigma + j\varepsilon\omega)E_y = -\Gamma H_x \qquad (6.44)$$

$$(\sigma + j\varepsilon\omega)E_z = \frac{\partial H_y}{\partial x} - \frac{\partial H_x}{\partial y} \qquad (6.45)$$

$$-j\mu\omega H_x = \frac{\partial E_z}{\partial y} + \Gamma E_y \qquad (6.46)$$

$$j\mu\omega H_y = \frac{\partial E_z}{\partial x} + \Gamma E_x \qquad (6.47)$$

It is useful to note that the four transverse components E_x, E_y, H_x and H_y can all be derived from E_z. For example, to show this, we insert H_y in eqn. 6.43 into eqn. 6.47, giving

$$\frac{\gamma^2}{\Gamma} E_x = \frac{\partial E_z}{\partial x} + \Gamma E_x \qquad (6.48)$$

where

$$\gamma^2 = j\mu\omega(\sigma + j\varepsilon\omega)$$

Then from eqn. 6.48 we see that

$$E_x = -\frac{\Gamma}{v^2} \frac{\partial E_z}{\partial x} \qquad (6.49)$$

where

$$v = (\Gamma^2 - \gamma^2)^{1/2} \qquad (6.50)$$

Using eqns. 6.43 and 6.49 we obtain

$$H_y = -\frac{\sigma + j\varepsilon\omega}{v^2} \frac{\partial E_z}{\partial x} \qquad (6.51)$$

Working with eqns. 6.44 and 6.46, we show in a similar fashion that

$$E_y = -\frac{\Gamma}{v^2} \frac{\partial E_z}{\partial y} \qquad (6.52)$$

$$H_x = \frac{\sigma + j\varepsilon\omega}{v^2} \frac{\partial E_z}{\partial y} \qquad (6.53)$$

Furthermore, we can show directly, on inserting H_y given by eqn. 6.51 and H_x given by eqn. 6.53 into eqn. 6.45, that

$$\left(\frac{\partial^2}{\partial x^2} + \frac{\partial^2}{\partial y^2} + v^2\right)E_z = 0 \qquad (6.54)$$

This equation is consistent with the Helmholtz form

$$(\nabla^2 - \gamma^2)E_z = 0 \qquad (6.55)$$

where

$$\nabla^2 = \frac{\partial^2}{\partial x^2} + \frac{\partial^2}{\partial y^2} + \Gamma^2 \qquad (6.56)$$

is the Laplacian operator.

Guided waves

To solve eqn. 6.54 we write

$$E_z = f(x, y) e^{-\Gamma z} \qquad (6.57)$$

where $f(x, y)$ must be selected so that it vanishes at $x = 0, a$ and at $y = 0, b$. A suitable form for the solution is

$$f(x, y) = C \sin\left(\frac{m\pi x}{a}\right) \sin\left(\frac{n\pi y}{b}\right) \qquad (6.58)$$

where m and n are integers and C is a constant. But if eqn. 6.57 is to satisfy eqn. 6.54 we must insist that

$$v^2 = \left(\frac{m\pi}{a}\right)^2 + \left(\frac{n\pi}{b}\right)^2 \qquad (6.59)$$

or, equivalently,

$$\Gamma = \left[\gamma^2 + \left(\frac{m\pi}{a}\right)^2 + \left(\frac{n\pi}{b}\right)^2\right]^{1/2} \qquad (6.60)$$

These discrete values of Γ (denoted $\Gamma_{m,n}$) form a doubly infinite set. To be definite we choose the radical such that Re $\Gamma_{m,n} > 0$.

We are now in the position to write a general solution of eqn. 6.54 with the proviso that no sources exist in the semi-infinite domain $0 < z < \infty$. This step amounts to superimposing all possible solutions of the form eqn. 6.57. Thus the desired general solution, for $z > 0$, is

$$E_z = \sum_{m=1}^{\infty} \sum_{n=1}^{\infty} C_{m,n} \sin\left(\frac{m\pi x}{a}\right) \sin\left(\frac{n\pi y}{b}\right) e^{-\Gamma_{m,n} z} \qquad (6.61)$$

where $C_{m,n}$ are coefficients and are unknown. Explicit expressions for the transverse field components are readily obtained by performing the operations indicated by eqns. 6.49, 6.51, 6.52 and 6.53. These fields satisfy the conditions of the problem, and the configuration is called transverse magnetic (TM) because the magnetic field vector $H = i_x H_x + i_y H_y$ is everywhere transverse or perpendicular to the axis of the waveguide (see Chapter 4). Of course, this is a consequence of assuming $H_z = 0$ at the outset.

Exercise Show that, if we assume $E_z = 0$ at the outset, the transverse field components, for a homogeneous medium, can be obtained from H_z in the manner

$$H_x = -\frac{\Gamma}{v^2} \frac{\partial H_z}{\partial x} \qquad (6.62)$$

$$E_y = \frac{j\mu\omega}{v^2} \frac{\partial H_z}{\partial x} \qquad (6.63)$$

$$H_y = -\frac{\Gamma}{v^2} \frac{\partial H_z}{\partial y} \qquad (6.64)$$

$$E_x = -\frac{j\mu\omega}{v^2} \frac{\partial H_z}{\partial y} \qquad (6.65)$$

Solution

Here we write Maxwell's equations in the same form as eqns. 6.43 to 6.47. By duality, $\sigma + j\varepsilon\omega$ is exchanged with $j\mu\omega$, $\mathbf{E}$ with $\mathbf{H}$, and $\mathbf{H}$ with $-\mathbf{E}$. An analogous process leads to eqns. 6.62 to 6.65, which are the duals of eqns. 6.49, 6.51, 6.52 and 6.53.

Exercise

Adopting the same waveguide model as depicted in Fig. 6.4, show that the general solution for H_z, in the region $0 < x < a$ and $0 < y < b$, for no sources, is

$$H_z = \sum_{p=0}^{\infty} \sum_{q=0}^{\infty} D_{p,q} \cos\left(\frac{p\pi x}{a}\right) \cos\left(\frac{q\pi y}{b}\right) e^{-\Gamma_{p,q} z} \quad (6.66)$$

where

$$\Gamma_{p,q} = \left[\gamma^2 + \left(\frac{p\pi}{a}\right)^2 + \left(\frac{q\pi}{b}\right)^2\right]^{1/2} \quad (6.67)$$

and $D_{p,q}$ are unknown coefficients.

Solution

We find in an analogous fashion that H_z satisfies eqn. 6.54, but now we seek solutions where $E_x = 0$ for $y = 0$ and b and $E_y = 0$ for $x = 0$ and a. Thus eqn. 6.66 is formally correct, where p and q can be any integer including 0. We note here that the transverse fields become identically zero if both p and q are zero. The modes that propagate in the waveguide when $E_z = 0$ are called transverse electric (TE) (see Chapter 4).

We see that the generic equation for the propagation constant for both TM and TE modes is of the same form (i.e. compare eqns. 6.60 and 6.67). To simplify the discussion we will assume that material inside the waveguide is free space; that is, $\sigma = 0$, $\varepsilon = \varepsilon_0$, $\mu = \mu_0$. Then we write

$$\Gamma = \bar{\alpha} + j\bar{\beta} = j\left[k^2 - \left(\frac{m\pi}{a}\right)^2 - \left(\frac{n\pi}{b}\right)^2\right]^{1/2} \quad (6.68)$$

where

$$k = (\varepsilon_0 \mu_0)^{1/2} \omega = 2\pi/\lambda_0 = \omega/c$$

Now clearly the attenuation rate $\bar{\alpha}$ for the modes in the waveguide is zero if

$$\left(\frac{m\pi}{a}\right)^2 + \left(\frac{n\pi}{b}\right)^2 < \left(\frac{\omega}{c}\right)^2 \quad (6.69)$$

An equivalent statement is that the modes propagate freely for angular frequencies such that

$$\omega > \omega_c \quad (6.70)$$

where

$$\omega_c = c\left[\left(\frac{m\pi}{a}\right)^2 + \left(\frac{n\pi}{b}\right)^2\right]^{1/2} \quad (6.71)$$

is by definition the *cutoff frequency*. Now it is clear that the mode with the lowest cutoff frequency is the one labelled TE_{10}. That is, $p = 1$ and

Guided waves

$q = 0$, where we have assumed $b < a$. Thus for the TE_{10} mode, $\omega_c = c\pi/a$. The corresponding cutoff wavelength is $\lambda_c = 2\pi c/\omega_c = 2a$. The TM mode with the lowest cutoff frequency is when $m = n = 1$. For this TM_{11} mode, the cutoff frequency is given by

$$\omega_c = c\left[\left(\frac{\pi}{a}\right)^2 + \left(\frac{\pi}{b}\right)^2\right]^{1/2} \tag{6.72}$$

In general we can see from eqn. 6.68 that, if $\omega < \omega_c$, the phase factor $\bar{\beta} = 0$, but then the attenuation rate is nonzero and given by

$$\bar{\alpha} = \left[\left(\frac{m\pi}{a}\right)^2 + \left(\frac{n\pi}{b}\right)^2 - \frac{\omega^2}{c^2}\right]^{1/2} \text{Np/m} \tag{6.73}$$

Such modes are heavily damped, and the fields associated with them are evanescent in the sense that *no real power is being transmitted*. These cutoff modes, however, play an important role in characterising waveguide junctions and discontinuities in the guiding structure. Such modes will not be discussed further here.

The qualitative behaviour of the electromagnetic field, for the TE_{10} mode, is sketched in Fig. 6.5. End, side and top views are shown, and the directions of both the **E** field lines and the **H** field lines are indicated. The relative closeness of the lines is a measure of the field strength. As expected, the electric field lines terminate at and are perpendicular to the assumed perfectly conducting walls.

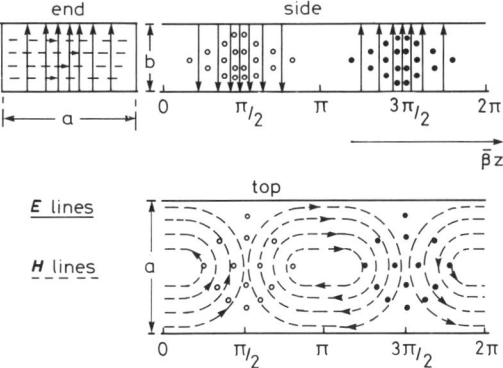

Fig. 6.5 Field lines for TE_{10} mode in rectangular waveguide

The surface current density **j** at the guide walls is a quantity of interest in deducing eddy current losses in the metal. For example,

$$j_z = -H_x \text{ at } y = 0 \quad \text{and} \quad j_z = +H_x \text{ at } y = b$$

$$j_x = +H_z \text{ at } y = 0 \quad \text{and} \quad j_x = -H_z \text{ at } y = b$$

as can be seen from a direct application of Ampère's law. The nature of the surface currents, for the TE_{10} mode, is shown in Fig. 6.6

Exercise

Deduce a useable formula for the attenuation rate $\bar{\alpha}$ for the rectangular waveguide shown in Fig. 6.4 when the guide is filled with a slightly lossy homogeneous nonmagnetic dielectric (i.e. $\mu = \mu_0$).

Solution

Here we can work with eqn. 6.60, which is written in the form

$$\bar{\alpha} + j\bar{\beta} = j\left[\beta^2(1 - j\Delta) - \left(\frac{m\pi}{a}\right)^2 - \left(\frac{n\pi}{b}\right)^2\right]^{1/2} \quad (6.74)$$

where

$$\beta = (\varepsilon\mu_0)^{1/2}\omega = (\varepsilon/\varepsilon_0)^{1/2}k$$

$$\Delta = \sigma/\varepsilon\omega$$

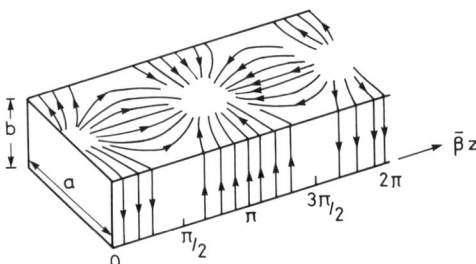

Fig. 6.6 Surface currents on walls of waveguide for TE_{10} mode

By squaring both sides of eqn. 6.74 and equating real and imaginary parts we can solve for $\bar{\alpha}$ and $\bar{\beta}$. The rather complicated algebra can be simplified when the losses are small. For example we may note that

$$\bar{\alpha} + j\bar{\beta} = j\left[(\bar{\beta}_0)^2 - j\beta^2\Delta\right]^{1/2} \quad (6.75)$$

where

$$j\bar{\beta}_0 = j\left[\beta^2 - \left(\frac{m\pi}{a}\right)^2 - \left(\frac{n\pi}{b}\right)^2\right]^{1/2}$$

is seen to be the propagation constant for the loss-free waveguide. Then we find that

$$\bar{\alpha} + j\bar{\beta} = j\bar{\beta}_0\left(1 - j\frac{\beta^2}{\bar{\beta}_0^2}\Delta\right)^{1/2}$$

$$\simeq j\bar{\beta}_0\left(1 - j\tfrac{1}{2}\frac{\beta^2}{\bar{\beta}_0^2}\Delta\right) \quad (6.77)$$

provided $(\beta^2/\bar{\beta}_0^2)\Delta \ll 1$. Finally we obtain the desired result

$$\bar{\alpha} \simeq (1/2)(\beta^2/\bar{\beta}_0)\Delta \quad (6.78)$$

which holds for propagating modes provided they are not near cutoff (i.e. where $\bar{\beta}_0$ becomes small).

Exercise

Consider the rather hypothetical situation where the axial field E_z was specified at the surface $z = z_0$ over the cross-section of the rectangular guide shown in Fig. 6.4. Deduce the field E_z for the region $z_0 < z < \infty$, assuming no other sources exist.

Guided waves

Solution

We require that eqn. 6.61 holds at $z = z_0$. Thus

$$\sum_{m=1}^{\infty}\sum_{n=1}^{\infty} C_{m,n} \sin\left(\frac{m\pi x}{a}\right) \sin\left(\frac{n\pi y}{b}\right) \exp(-\gamma_{m,n}z_0) = E_z(x, y, z_0) \quad (6.79)$$

for $0 < x < a$ and $0 < y < b$, where the right-hand side is known. To solve for $C_{m,n}$ we multiply both sides by $\sin(m'\pi x/a)\sin(m'\pi y/b)$. where m' and n' are integers, and integrate over x from 0 to a and over y from 0 to b. Here we utilise the results

$$\int_0^a \sin\left(\frac{m\pi x}{a}\right)\sin\left(\frac{m'\pi x}{a}\right) dx = \begin{cases} 0 & \text{if } m \neq m' \\ a/2 & \text{if } m = m' \end{cases} \quad (6.80)$$

for $m = 1, 2, 3\ldots$.

and

$$\int_0^b \sin\left(\frac{n\pi y}{b}\right)\sin\left(\frac{n'\pi y}{b}\right) dy = \begin{cases} 0 & \text{if } n \neq n' \\ b/2 & \text{if } n = n' \end{cases} \quad (6.81)$$

for $n = 1, 2, 3\ldots$.

Then we find that the desired expression for the unknown coefficient is

$$C_{m',n'} = \frac{4}{ab} \exp(\Gamma_{m',n'} z_0) \int_0^a \int_0^b E_z(x, y, z_0) \sin\left(\frac{m'\pi x}{a}\right) \sin\left(\frac{n'\pi y}{b}\right) dx\, dy \quad (6.82)$$

Here, of course, m' and n' can be replaced by m and n, respectively, and the resulting $C_{m,n}$ inserted into eqn. 6.61.

Exercise

This exercise is not for the faint hearted. Using the same approach as in the previous exercise, deduce the complete electromagnetic field in the waveguide for $z > z_0$ when E_x and E_y are both specified functions of x and y at $z_0 = 0$.

Solution

Assume that eqns. 6.61 and 6.66 hold everywhere in the region $z_0 < z < \infty$. Then use eqns. 6.49, 6.52, 6.63 and 6.65 to obtain complete modal expansions for E_x and E_y for $z_0 \leqslant z < \infty$. These expressions must reduce to the specified forms $E_x(x, y, z_0)$ and $E_y(x, y, z_0)$ as $z \to z_0$. Using the same method as in the preceding exercise, the appropriate forms for $C_{m,n}$ in eqn. 6.61 and $D_{p,q}$ in eqn. 6.66 are obtained.

6.4 Wave circuit parameters

Although we may view the rectangular waveguide and similar hollow cylindrical structures as boundary value problems, it is useful to examine the circuit equivalences. Such analogies are useful in developing design procedures (see, for example, [1]).*

In the case of TM waves (i.e. $H_z = 0$), in rectangular geometry, we obtained the field equations 6.43 to 6.47 for a homogeneous medium $(\sigma, \varepsilon, \mu)$ where $\Gamma = -\partial/\partial z$. Now from eqns. 6.49 and 6.52, it is clear

---

* In this and the following chapters, references are given by numbers in [ ].

that we can write, for a given mode

$$E_x = -\frac{\partial V}{\partial x} \tag{6.83}$$

$$E_y = -\frac{\partial V}{\partial y} \tag{6.84}$$

where

$$V = (\Gamma/v^2)E_z = -(\partial/\partial z)(E_z/v^2) \tag{6.85}$$

is a scalar potential. Also we can write eqn. 6.47, with the help of eqn. 6.51, in the form

$$\frac{\partial E_x}{\partial z} - \frac{\partial E_z}{\partial x} = \frac{\gamma^2}{v^2}\frac{\partial E_z}{\partial x} \tag{6.86}$$

or, equivalently

$$\frac{\partial V}{\partial z} = -\left(\frac{\gamma^2}{v^2}+1\right)E_z = -\left(j\mu\omega + \frac{v^2}{\sigma + j\varepsilon\omega}\right)\left(\frac{\sigma + j\varepsilon\omega}{v^2}E_z\right) \tag{6.87}$$

where we have used eqn. 6.83. Recall that $v^2 = (m\pi/a)^2 + (n\pi/b)^2$.

Now the equation pair 6.85 and 6.87 are the same as

$$\frac{\partial I}{\partial z} = -(\sigma + j\varepsilon\omega)V \tag{6.88}$$

and

$$\frac{\partial V}{\partial z} = -\left(j\mu\omega + \frac{v^2}{\sigma + j\varepsilon\omega}\right)I \tag{6.89}$$

where

$$I = (\sigma + j\varepsilon\omega)E_z/v^2 \tag{6.90}$$

has the dimensions of amperes. Clearly, eqns. 6.89 and 6.90 have the same form as the generic transmission line equations given by eqns. 6.16 and 6.17. Thus, in the case of a TM waveguide mode, the effective series admittance is given by

$$\hat{Y} = \sigma + j\varepsilon\omega \text{ S/m} \tag{6.91a}$$

while the effective series impedance is given by

$$\hat{Z} = \left(j\mu\omega + \frac{v^2}{\sigma + j\varepsilon\omega}\right) \Omega/\text{m} \tag{6.91b}$$

The equivalent transmission line circuit is identical to that shown in Fig. 6.2 for an elemental section of length δz of the waveguide.

Exercise

Show that in the case of TE wave transmission in the rectangular waveguide, eqns. 6.16 and 6.17 are still applicable with the following identifications:

$$I = -(\Gamma/v^2)H_z \qquad \text{A (amps)} \tag{6.92}$$

$$V = (j\mu\omega/v^2)H_z \qquad \text{V (volts)} \tag{6.93}$$

$$\hat{Z} = j\mu\omega \qquad \Omega/m \qquad (6.94)$$

$$\hat{Y} = (\sigma + j\varepsilon\omega) + \frac{v^2}{j\mu\omega} \quad S/m \qquad (6.95)$$

Solution

In this case we note that, according to eqns. 6.62 and 6.64, $H_x = -\partial I/\partial x$ and $H_y = -\partial I/\partial y$ when I is defined as above. Then we deduce that

$$\frac{\partial I}{\partial z} = -\left(\frac{v^2}{j\mu\omega} + \sigma + j\varepsilon\omega\right)\left(\frac{j\mu\omega}{v^2} H_z\right) \qquad (6.96)$$

which has the required form.

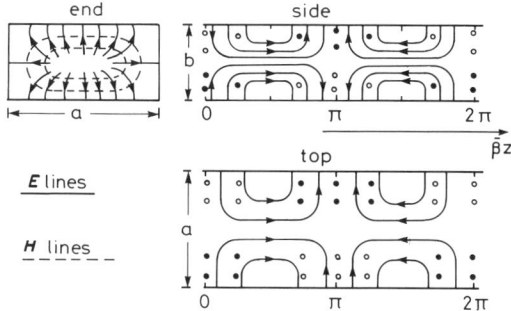

Fig. 6.7 Field lines for TM_{11} mode in a rectangular guide. Dots represent H lines coming out of paper, and small circles represent H lines going into paper

The transmission line circuits are quite general and they hold for all TM and TE modes in the uniform guiding structure. But the appropriate value of v for the mode in question must be employed. In the case of lossless material filling the waveguide (i.e. $\sigma = 0$), the equivalent transmission line circuits can be depicted in terms of equivalent capacitances and inductances. These are shown in Figs. 6.8 and 6.9, respectively, for the TM and TE mode types. The labelling shows impedances per unit length for the series elements and admittance per unit length for shunt elements.

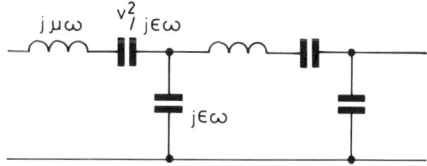

Fig. 6.8 Equivalent transmission line circuit for TM waves in guide

The theory of guided electromagnetic waves in cylindrical tubes with perfectly conducting walls can be developed in a manner that is fully analogous to that for the rectangular waveguide (e.g. see [2], [3].)

In the case of circular cross-sections the wave solutions involve Bessel functions, whereas for the elliptical cross-sections the wave solutions involve Mathieu functions (e.g. see [4]). We will not treat these cases here. Actually there is now a vast literature on waveguide theory and applications (e.g. see [5]–[9]). See also Chapter 7.

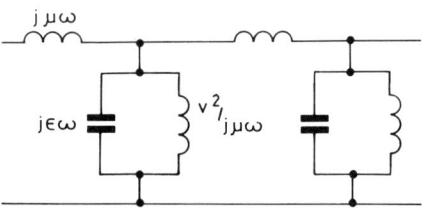

Fig. 6.9 Equivalent transmission line circuit for TE waves in guide

6.5 General slab model

To complete our brief survey of guided electromagnetic wave phenomena we will consider a two-dimensional slab model. The geometry is illustrated in Fig. 6.10 with reference to a co-ordinate system. The central homogeneous slab with properties $(\sigma, \varepsilon, \mu_0)$ is bounded by the surfaces $x = 0$ and $x = d$. The homogeneous region above the slab (i.e. for $x > d$) has properties $(\sigma_1, \varepsilon_1, \mu_0)$, and the homogeneous region below the slab (i.e. for $x < 0$) has properties $(\sigma_2, \varepsilon_2, \mu_0)$. As indicated, the magnetic permeability μ_0 is assumed to be the same everywhere. We now wish to examine the types of waves that can be guided by the slab structure as described. The derivation will be restricted to modes of the TM type, but the extension to the TE mode types will be covered in the exercises. Throughout we will assume that there is no variation of the fields in the y direction. Thus we regard the positive z axis as the direction of propagation.

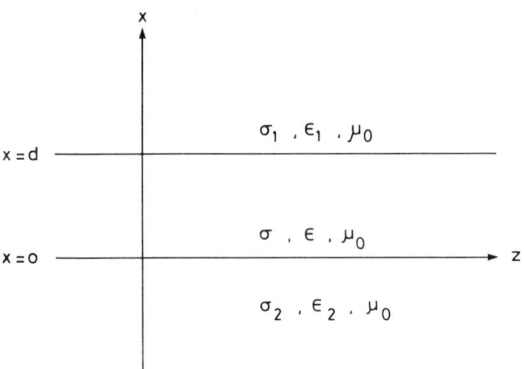

Fig. 6.10 Idealised slab model for a waveguide of width d bounded by arbitrary homogeneous media

For the slab region $(0 < x < d)$ the field equations have the same form as for the rectangular waveguide, but here $\partial/\partial y = 0$. Thus for TM waves that vary as $\exp(-\Gamma z)$ we have

$$(\sigma + j\varepsilon\omega)E_x = \Gamma H_y \tag{6.97}$$

$$(\sigma + j\varepsilon\omega)E_z = dH_y/dx \tag{6.98}$$

Guided waves

$$j\mu\omega H_y = dE_z/dx + \Gamma E_x \tag{6.99}$$

Then, of course

$$\left(\frac{d^2}{dx^2} + v^2\right)H_y = 0 \tag{6.100}$$

where

$$v^2 = -\gamma^2 + \Gamma^2$$

or

$$v = (\Gamma^2 - \gamma^2)^{1/2} \tag{6.101}$$

The same equations apply in the upper region ($x > d$) if we merely add a subscript 1 to the quantities γ and v, but, of course, Γ remains the same. Similarly, for the lower region ($x < 0$) we add a subscript 2 to the quantities γ and v. In all cases Re $\gamma > 0$ and Re $v > 0$.

In the slab region our solution takes the form

$$H_y = (A\,e^{jvx} + B\,e^{-jvx})\,e^{-\Gamma z} \tag{6.102}$$

where A and B are constants. Then

$$E_z = \frac{jv}{\sigma + j\varepsilon\omega}(A\,e^{jvx} - B\,e^{-jvx})\,e^{-\Gamma z} \tag{6.103}$$

In the upper region ($x > d$)

$$H_y = C\,e^{-jv_1 x} \tag{6.104}$$

$$E_z = -\frac{jv_1}{\sigma_1 + j\varepsilon_1\omega}C\,e^{-jv_1 x}\,e^{-\Gamma z} \tag{6.105}$$

where C is a constant.

In the lower region ($x < 0$)

$$H_y = D\,e^{jv_2 x}\,e^{-\Gamma z} \tag{6.106}$$

$$E_z = \frac{jv_2}{\sigma_2 + j\varepsilon_2\omega}D\,e^{jv_2 x}\,e^{-\Gamma z} \tag{6.107}$$

Now, for an acceptable solution, H_y and E_z should both be continuous at the interfaces $x = 0$ and $x = d$. The implementation of this requirement leads to four algebraic equations. For a nontrivial solution the determinant of the coefficients, A, B, C and D should be zero. This condition leads to an equation to solve for the propagation constant.

A simpler and more desirable approach is to note that, according to eqns. 6.104 and 6.105, an impedance may be defined as

$$K_1 = -E_z/H_y]_{x=d} = \frac{jv_1}{\sigma_1 + j\varepsilon_1\omega} \tag{6.108}$$

Similarily, using eqns. 6.106 and 6.107, we define an impedance

$$K_2 = E_z/H_y]_{x=0} = \frac{jv_2}{\sigma_2 + j\varepsilon_2\omega} \tag{6.109}$$

But, according to eqns. 6.102 and 6.103, at $x = d$,

$$K_1 = -K\left(\frac{A\,e^{jvd} - B\,e^{-jvd}}{A\,e^{jvd} + B\,e^{-jvd}}\right) \qquad (6.110)$$

where

$$K = \frac{jv}{\sigma + j\varepsilon\omega} \qquad (6.111)$$

From eqn. 6.110,

$$\frac{A}{B} = \frac{K - K_1}{K + K_1}\,e^{-2jvd} \qquad (6.112)$$

But, also according to eqns. 6.102 and 6.103, at $x = 0$.

$$K_2 = K\frac{A - B}{A + B} \qquad (6.113)$$

in which case

$$\frac{A}{B} = \frac{K + K_2}{K - K_2} \qquad (6.114)$$

Then, on equating eqns. 6.112 and 6.114, we obtain the mode equation

$$\left(\frac{K - K_1}{K + K_1}\right)\left(\frac{K - K_2}{K + K_2}\right)e^{-2jvd} = 1 \qquad (6.115)$$

which, in principle, can be solved for the propagation constant Γ. There can be an infinite number of these solutions, and we denote them by Γ_n where $n = 0, 1, 2, 3, \ldots$. To facilitate the discussion, we replace 1 on the right-hand side of eqn. 6.115 by $\exp(-j\pi 2n)$.

6.6 Limit of perfect wall conductivity

The limiting case when medium (1) and medium (2) are perfect conductors is discussed first. Then, for example, if $\sigma_1 \to \infty$ and $\sigma_2 \to \infty$ we see that $K_1 = 0$ and $K_2 = 0$. The mode equation now reduces to

$$e^{-2jvd} = e^{-j2\pi n} \qquad (6.116)$$

or

$$v = \pi n/d$$

Then

$$\Gamma_n = [\gamma^2 + (\pi n/d)^2]^{1/2} \qquad (6.117)$$

which has the same form as the propagation constant for the rectangular waveguide (see eqn. 6.67) in the limit where $a \to \infty$, $b = d$.

For the zero-order mode ($n = 0$), $\Gamma = \Gamma_0 = \gamma$, in which case the field configuration is the same as that for a plane wave propagating in the positive z direction. For the higher-order modes ($n = 1, 2, 3, \ldots$) we can deduce the attenuation rate $\bar{\alpha}_n$ and the phase factor $\bar{\beta}_n$ from

$$\bar{\alpha}_n + j\bar{\beta}_n = [(\alpha + j\beta)^2 + (\pi n/d)^2]^{1/2} \qquad (6.118)$$

where

$$\alpha + j\beta = \gamma$$

Guided waves

In the lossless case ($\sigma = 0$ and $\alpha = 0$) we see that, for $n = 0$, we have simply

$$\bar{\alpha}_0 = 0$$

$$\bar{\beta}_0 = \beta = (\varepsilon\mu)^{1/2}\omega = \omega/v_0$$

for all frequencies. For $n = 1, 2, 3, \ldots,$

$$\bar{\alpha}_n = 0$$

$$\bar{\beta}_n = [\beta^2 - (\pi n/d)^2]^{1/2} \tag{6.119}$$

provided $\beta > \pi n/d$; and

$$\bar{\alpha}_n = [(\pi n/d)^2 - \beta^2]^{1/2}$$

$$\bar{\beta}_n = 0 \tag{6.120}$$

when $\beta < \pi n/d$. In other words the higher-order modes are cut off if the frequency ω is less than ω_c where $\omega_c = \pi n v_0/d$, $v_0 = (\varepsilon\mu)^{-1/2}$.

6.7 Effect of finite wall conductivity

Here we will examine the transmission of TM waves in the slab model under the assumption that the bounding regions (1) and (2) have *high but not infinite* conductivities. Thus we argue that if $|\sigma_1 + j\varepsilon_1\omega| \gg |\sigma + j\varepsilon\omega|$ and if $|\sigma_2 + j\varepsilon_2\omega| \gg |\sigma + j\varepsilon\omega|$, the following approximations are valid for the important modes:

$$K_1 \simeq \gamma_1/(\sigma_1 + j\varepsilon_1\omega) = \eta_1 \tag{6.121}$$

$$K_2 \simeq \gamma_2/(\sigma_2 + j\varepsilon_2\omega) = \eta_2 \tag{6.122}$$

Then, on noting that

$$K = j\eta v/\gamma \tag{6.123}$$

we have the approximated mode equation

$$\left(\frac{C - \Delta_1}{C + \Delta_1}\right)\left(\frac{C - \Delta_2}{C + \Delta_2}\right) e^{-2\gamma dC} = e^{-j2\pi n} \tag{6.124}$$

where

$$C = jv/\gamma$$

$$\Delta_1 = \frac{\eta_1}{\eta} = \frac{\mu_1\gamma}{\mu\gamma_1}$$

$$\Delta_2 = \frac{\eta_2}{\eta} = \frac{\mu_2\gamma}{\mu\gamma_2}$$

The mode equation in the form given by eqn. 6.124 can be further simplified for the modes of practical importance. For example, if $|C| \gg |\Delta_1|$ and $|C| \gg |\Delta_2|$ we see that

$$(C - \Delta_1)/(C + \Delta_1) \simeq \exp(-2\Delta_1/C) \tag{6.125}$$

Then eqn. 6.124 is reduced to

$$C^{-1}(\Delta_1 + \Delta_2) + \gamma dC = j\pi n$$

This is a quadratic in C, which is solved to give

$$C = C_n = \frac{j\pi n}{2\gamma d} \pm \left[\left(\frac{j\pi n}{2\gamma d}\right)^2 - \frac{\Delta_1 + \Delta_2}{\gamma d}\right]^{1/2} \tag{6.126}$$

The corresponding propagation constant is then

$$\Gamma_n = \gamma S_n \tag{6.127}$$

where

$$S_n = (1 - C_n^2)^{1/2}$$

The + sign in front of the radical is to be chosen because we have stipulated that Re $\Gamma_n > 0$. Also we note that, for $n = 0$,

$$C_0^2 = -(\Delta_1 + \Delta_2)/\gamma d \tag{6.128}$$

and thus

$$\Gamma_0 = \gamma S_0 = \gamma \left(1 + \frac{\Delta_1 + \Delta_2}{\gamma d}\right)^{1/2} \tag{6.129}$$

It is also worth noting that, if $|(\Delta_1 + \Delta_2)\gamma d| \ll 1$, the radical in eqn. 6.126 can be simplified in the manner (for $n \neq 0$):

$$\left(\frac{j\pi n}{2\gamma d}\right)\left[1 - \frac{4(\Delta_1 + \Delta_2)\gamma d}{(j\pi n)^2}\right]^{1/2}$$

$$\simeq \frac{j\pi n}{2\gamma d}\left[1 - \frac{2(\Delta_1 + \Delta_2)\gamma d}{(j\pi n)^2}\right] \simeq \frac{j\pi n}{2\gamma d} - \frac{\Delta_1 + \Delta_2}{j\pi n} \tag{6.130}$$

Then

$$C_n^2 = \left[\frac{j\pi n}{\gamma d} - \frac{(\Delta_1 + \Delta_2)}{j\pi n}\right]^2 \tag{6.131}$$

and finally

$$S_n \simeq \left[1 + \left(\frac{\pi n}{\gamma d}\right)^2 + \frac{2(\Delta_1 + \Delta_2)}{\gamma d}\right]^{1/2} \tag{6.132}$$

which holds for $n = 1, 2, 3, \ldots$. The corresponding propagation constant is obtained from $\Gamma_n = \gamma S_n$. A further simplification, which follows from eqns. 6.129 and 6.132, is

$$S_n \simeq \left[1 + \left(\frac{\pi n}{\gamma d}\right)^2\right]^{1/2} + \frac{\hat{\varepsilon}_n(\Delta_1 + \Delta_2)}{2\gamma d\left[1 + \left(\frac{\pi n}{\gamma d}\right)^2\right]^{1/2}} \tag{6.133}$$

where $\hat{\varepsilon}_0 = 1$ and $\hat{\varepsilon}_n = 2$ for $n = 1, 2, 3, \ldots$. Clearly eqn. 6.133 is restricted by the further condition that the second term on the right-hand side be small compared with the first term.

6.8 Earth–Ionosphere waveguide model

An interesting and important example for which the idealised model can be used is to describe VLF radio wave transmission in the earth–ionosphere waveguide (e.g. see [9]–[11]). Here we neglect earth curvature and represent the ground as a homogeneous half-space (for

Guided waves

$x < 0$) with conductivity σ_g, permittivity ε_g and permeability μ_0. The ionosphere is idealised as a homogeneous electron plasma (for $x > h$) and is described below. The intermediate region (for $0 < x < h$) is assumed to be free space with properties ε_0 and μ_0. The geometry is indicated in Fig. 6.11.

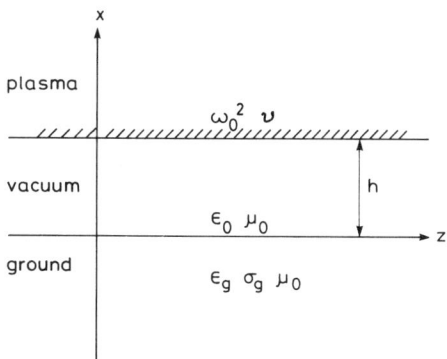

Fig. 6.11 An idealised model of the earth–ionosphere waveguide

The ionosphere can be crudely represented as a lossy dielectric with an effective complex permittivity ε_i given by

$$\varepsilon_i = \varepsilon_0 \left(1 - j\frac{\omega_r}{\omega}\right) \tag{6.134}$$

where $\omega_r = \omega_0^2/v$; ω_0 is the plasma frequency (see Section 4.3) and v is the collision frequency. Typically $\omega_r \simeq 10^5$, but this value can vary by a factor of 10 either way.
Equivalently we may say that the effective real conductivity is $\varepsilon_0 \omega_r$, which is of the order of 10^{-6} S/m.

Our oversimplified model of the earth–ionosphere waveguide shown in Fig. 6.11 is a special case of the general slab model shown in Fig. 6.10. The following substitutions into the general slab model are appropriate: $d \to h$, $\varepsilon \to \varepsilon_0$, $\sigma \to 0$, $\mu \to \mu_0$, $\sigma_1 \to \varepsilon_0 \omega_r$, $\varepsilon_{1,2} \to \varepsilon_0$, $\mu_{1,2} \to \mu_0$, $\varepsilon_2 \to \varepsilon_g$, $\sigma_2 \to \sigma_g$. The general mode equation 6.115 is now expressible in the form

$$R_i(C) R_g(C) \, e^{-j4\pi HC} = e^{-j2\pi n} \tag{6.135}$$

where

$$R_i(C) = \frac{(L-j)C - (C^2 L^2 - jL)^{1/2}}{(L-j)C + (C^2 L^2 - jL)^{1/2}} \tag{6.136}$$

$$R_g(C) = \frac{(KG-j)C - [(K-1)G^2 - jG + C^2 G^2]^{1/2}}{(KG-j)C + [(K-1)G^2 - jG + C^2 G^2]^{1/2}} \tag{6.137}$$

$$K = \frac{\varepsilon_g}{\varepsilon_0}, \quad H = \frac{h}{\lambda_0}, \quad L = \frac{\omega}{\omega_r}, \quad G = \frac{\varepsilon_0 \omega}{\sigma_g}$$

Here λ_0 is the free space wavelength.

When we write the mode equation in the form given by eqn. 6.135 we may interpret eqn. 6.136 as the reflection coefficient of the ionosphere for a TM plane wave incident from below at an angle whose cosine is C. Similarly, eqn. 6.137 is to be interpreted as the reflection coefficient at the ground for a TM plane wave incident from above at an angle whose cosine is C. Furthermore, we may interpret the exponential factor $\exp(-j4\pi HC)$ as the resultant change in amplitude and phase for a plane wave that has been perfectly reflected once at the top and once at the bottom boundary. The imperfect reflection is accounted for by the inclusion of the reflection coefficients $R_i(C)$ and $R_g(C)$ in the mode equation 6.135.

A graphical interpretation of the multiple reflection process is shown in Fig. 6.12. The values of $C = C_n$ that satisfy the mode equation correspond to a resonance of the ray structure, shown in Fig. 6.12, when the net phase shift for an up and back bounce is $2\pi n$ radians. Only in the case of perfect reflection, where $|R_g| = |R_i| = 1$, would the modal values C_n be real, in which case the angles θ_n ($=$ arc cos C_n) would also be real. Now clearly if $|R_g| \neq 1$ and/or $|R_i| \neq 1$, which is the usual case, the modal parameters C_n and S_n are complex. Then if we define $\theta_n =$ arc cos C_n, the angle θ_n must also be complex.

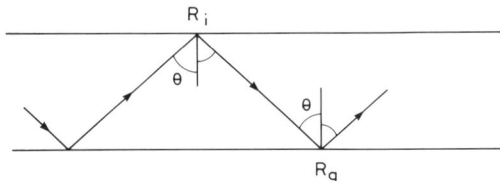

Fig. 6.12 Visualisation of rays that bounce internally within the waveguide. When such a ray is 'resonant', we have a mode

Actually the fields within the waveguide propagate according to $\exp(-\Gamma_n z)$ for transmission in the positive z direction. For the earth–ionosphere waveguide model, this factor can be written in the form

$$\exp(-j2\pi S_n z/\lambda_0) = \exp(-2\pi z u_n/h)\exp(-j2\pi z s_n/\lambda_0) \quad (6.138)$$

where

$$u_n = -(\operatorname{Im} S_n)h/\lambda_0$$

$$s_n = \operatorname{Re} S_n$$

Clearly u_n, as defined, is a measure of the attenuation of the nth mode per unit distance, and s_n is a dimensionless phase factor. In fact the phase velocity of the mode is c/s_n m/s.

The major task is to find solutions of eqn. 6.135 for C_n and thus S_n. Newton's method can be used effectively here when starting values are available. Actually when both G and L are small compared with 1, we may use the approximation given by eqn. 6.133, which in the present context reads

$$\operatorname{Re} S_n \simeq \left[1 - \left(\frac{n\lambda_0}{2h}\right)^2\right]^{1/2} + \frac{\sqrt{G} + \sqrt{L}}{4\pi\sqrt{2}}\varepsilon_n\left[1 - \left(\frac{n\lambda_0}{2h}\right)^2\right]^{-1/2} \quad (6.139)$$

$$\operatorname{Im} S_n \simeq - \frac{\sqrt{G} + \sqrt{L}}{4\pi\sqrt{2}} \hat{\varepsilon}_n \left[1 - \left(\frac{n\lambda_0}{2h}\right)^2\right]^{-1/2} \qquad (6.140)$$

where $\hat{\varepsilon}_0 = 1$ and $\hat{\varepsilon}_n = 2$ for $n = 1, 2, 3, \ldots$.

There is another useful approximate solution that is valid when G is smaller but L is of the order of one (e.g. see [9]). Then, at least for the lower-order modes, say $n = 1$, 2 and 3, we would have

$$\operatorname{Re} S_n \simeq \bar{S}_n + \frac{1}{2\pi\sqrt{2}(h/\lambda_0)\bar{S}_n} \left[\frac{(n-\frac{1}{2})^2}{(2h/\lambda_0)^2}\left(\sqrt{L} - \frac{1}{\sqrt{L}}\right) + \sqrt{G}\right] \qquad (6.141)$$

$$\operatorname{Im} S_n \simeq - \frac{1}{2\pi\sqrt{2}(h/\lambda_0)\bar{S}_n} \left[\frac{(n-\frac{1}{2})^2}{(2h/\lambda_0)^2}\left(\sqrt{L} + \frac{1}{\sqrt{L}}\right) + \sqrt{G}\right] \qquad (6.142)$$

where

$$\bar{S}_n = \left[1 - \frac{(n-\frac{1}{2})^2}{(2h/\lambda_0)^2}\right]^{1/2} \qquad (6.143)$$

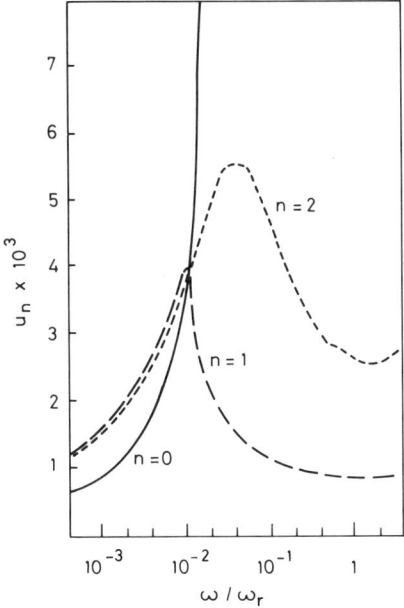

Fig. 6.13 Normalised attentuation rate against frequency parameter for model of Fig. 6.11, where $\omega_r = \omega_0^2/\nu$, $H = h/\lambda_0 = 4$, $G = \varepsilon_0\omega/\sigma_g = 4 \times 10^{-4}$ (for 16 kHz, $\sigma_g = 2.2 \times 10^{-3}$ s/m and $h = 75$ km)

To deal with a concrete case, we plot the normalised attentuation rate u_n in Fig. 6.13 as a function of $L(=\omega/\omega_r)$ for the parameters indicated in the caption. The first three modes only are shown. The behaviour for $\omega/\omega_r \ll 1$ is consistent with the approximate attenuation formula given by eqn. 6.140. Also the broad minima near $L = 1$ for the $n = 1$ and $n = 2$ curves are consistent with eqn. 6.142, which is not applicable to the $n = 0$ mode in the present example.

Finally, to show an example of actual field data, a field strength against range curve is shown in Fig. 6.14 for a transmitter in San Diego. The signal, on an aircraft flying towards Hawaii, was recorded for a frequency of 16·6 kHz (private communication from J. Heritage, US Navy Laboratories, San Diego). The calculated curve was based on summing the propagating modes with an appropriate excitation factor ([9], Chapter 7). For the parameters shown, the fit to the experimental data is quite good. The apparent decrease in the field strength (at the larger ranges) below the calculated curve can be attributed partly to the neglect of earth curvature in this simple model. The mode theory for transmission in the spherical earth–ionosphere waveguide is a great deal more complicated (e.g. see [10] for an updated summary of the general theory). Many other complications arise, such as the influence of the earth's magnetic field, nonuniformity of the ionospheric plasma and mode conversion phenomena at terminations (sunset/sunrise transitions regions).

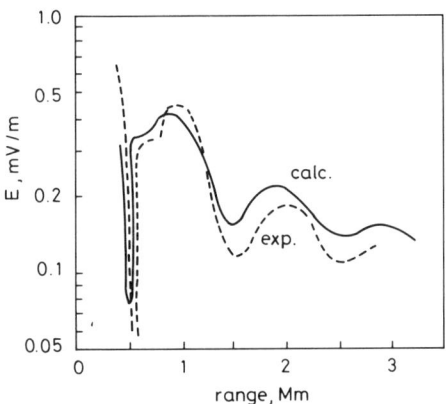

Fig. 6.14 An experimental field strength against distance curve for transmitter at San Diego on daytime flight to Hawaii at 16·6 kHz; parameters for calculated curve are $\omega/\omega_r = 0\cdot52$ or $\omega_r = 2 \times 10^5$, $h = 70$ km and $\sigma_g \geqslant 1$ s/m.

6.9 Surface waves

In this final section on the subject of guided waves, we treat a further special case of the model shown in Fig. 6.11 that exhibits unattenuated wave transmission. Specifically, we consider a perfectly conducting plane that is covered by a lossless dielectric coating. The region above is taken to be free space. This model is shown in Fig. 6.15.

The following specialization to the values in the slab model in Fig. 6.11 are now made: $\sigma = 0$, $\mu = \mu_0$, $\sigma_1 = 0$, $\varepsilon_1 = \varepsilon_0$, $\mu_1 = \mu_0$ and $\sigma_2 \to \infty$. Then the mode equation 6.115 simplifies to

$$R(C)\, e^{-j2NkCd} = e^{-j2\pi n} \tag{6.144}$$

where

$$R(C) = \frac{N^{-2}C - (N^{-2} - S^2)^{1/2}}{N^{-2}C + (N^{-2} - S^2)^{1/2}} \tag{6.145}$$

where $N^2 = \varepsilon/\varepsilon_0$, $S = (1 - C^2)^{1/2}$ and $k = (\varepsilon_0\mu_0)^{1/2}\omega$. Here the modes propagate in the z direction according to $\exp(-\Gamma_n z)$, where $\Gamma_n = jkNS_n$ and where $C_n = (1 - S_n^2)^{1/2}$ is a solution of eqn. 6.144. Now $N^2 > 1$, so we may write

$$R(C) = \frac{C + jN^2(S^2 - N^{-2})^{1/2}}{C - jN^2(S^2 - N^{-2})^{1/2}} \tag{6.146}$$

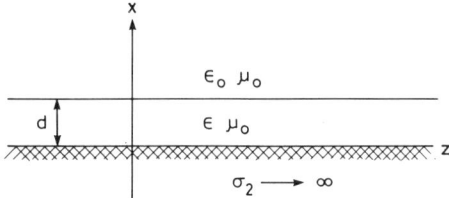

Fig. 6.15 Grounded dielectric slab

For real values of $S^2 > 1/N^2$, $|R(C)| = 1$, and we can define a phase angle $\Phi(C) = \arg R(C)$. Then it is clear that eqn. 6.144 is equivalent to

$$\Phi(C) - 2NkCd = -2\pi n \tag{6.147}$$

There are a finite number of solutions that lead to real S_n (e.g. see [2]) which, in turn, correspond to modes that are unattenuated. In the region above the slab, $x > d$, the fields associated with these modes behave as

$$\exp[-k(S_n^2 N^2 - 1)^{1/2} x] \exp(-jkNS_n z)$$

When $S_n^2 > 1/N^2$ it is clear that the modes are heavily damped in the x direction but unattenuated in the (horizontal) z direction. Thus, as viewed from an observer in the upper region, the propagating wave appears to be bound to the slab. On the other hand, within the slab $(0 < x < d)$ the field structure has the form of a guided wave with multiple bounces between the two boundaries. In this model, the bottom boundary is a perfect electrical conductor with a reflection coefficient equal to $+1$. At the upper boundary the freely guided modes are associated with critical reflection, where the reflection coefficient has a magnitude of 1, but with a variable phase shift.

As an interesting extension we can square both sides of eqn. 6.144 to obtain

$$R^2(C) \exp[-j2NC(2d)k] = \exp[-j2\pi(2n)] \tag{6.148}$$

The modes $2n = 0, 2, 4, \ldots$ that satisfy this equation pertain to the equivalent problem of a dielectric slab of width $2d$ that is bounded top and bottom by free space. The situation is illustrated in Fig. 6.16, where we depict the exponential damping of the field magnitude above and below the slab in the external free space.

It is useful to note that in a grounded dielectric slab of width d or an isolated slab of width $2d$, the phase velocity of the guided mode is $c/S_n N$ which is less than $c = (\mu_0 \varepsilon_0)^{-1/2}$, the velocity of light in vacuum. In this sense the trapped guided modes are often referred to as 'slow waves'.

On the other hand, the phase velocity with reference to the interior of the slab is 'fast' compared with $v = (\mu_0 \varepsilon)^{-1/2} = c/N$ (because $S_n < 1$).

The guiding of electromagnetic waves in a dielectric slab such as in the model shown in Fig. 6.16 has an important counterpart in a cylindrical glass fibre. When the dielectric cylinder is analysed in an analogous fashion, there is a class of unattenuated modes that can be used to convey wide bandwidth signals in miniscule glass threads. In spite of the added complexity of the analysis, the principle of internal critical reflection is the key to the success of this mode of transmission. Actually the theory of fibre optics transmission was developed in the early part of the century, but the implementation came much later. The real breakthrough was the development of glass dielectric materials with very low intrinsic loss. An excellent modern account of the subject is given in [12].

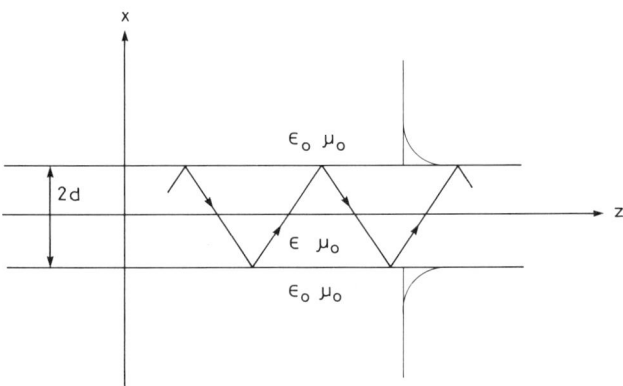

Fig. 6.16 Guided wave mode in dielectric slab, illustrating evanescent field structure in external region

Exercise

Show that a mode equation analogous to eqn. 6.115 can be developed for TE waves in the general slab geometry shown in Fig. 6.10.

Solution

We again assume that the fields vary as $\exp(-\Gamma z)$. The field equations for the region $0 < x < d$ with $\partial/\partial y = 0$ are

$$j\mu\omega H_x = -\Gamma E_y \tag{6.149}$$

$$j\mu\omega H_z = -dE_y/dx \tag{6.150}$$

$$(\sigma + j\varepsilon\omega)E_y = (dH_z/dx) + \Gamma H_x \tag{6.151}$$

E_y also satisfies eqn. 6.101, and solutions for E_y and H_z in $0 < x < d$ are

$$E_y = (A' e^{jvx} + B' e^{-jvx}) e^{-\Gamma z} \tag{6.152}$$

$$H_z = -M(A' e^{jvx} - B' e^{-jvx}) e^{-\Gamma z} \tag{6.153}$$

where

$$M = jv/j\mu\omega = v/\mu\omega$$

… Guided waves …

and $v = (\Gamma^2 - \gamma^2)^{1/2}$ as before. The boundary conditions are now

$$M_1 = H_z/E_y]_{x=d} = \frac{v_1}{\mu_1 \omega} \tag{6.154}$$

$$M_2 = -H_z/E_y]_{x=0} = \frac{v_2}{\mu_2 \omega} \tag{6.155}$$

(i.e. compare with eqns. 6.108 and 6.109). The mode equation is then found to be

$$\left(\frac{M - M_1}{M + M_1}\right)\left(\frac{M - M_2}{M + M_2}\right) e^{-2jvd} = e^{-j2\pi n} \tag{6.156}$$

which is to be solved for $\Gamma = \Gamma_n$.

Exercise

In the limit of perfectly conducting boundaries (i.e. $\sigma_1 = \sigma_2 = \infty$), show that, for the TE waves in the preceding exercise, there is no zero-order mode of propagation.

Solution

In the limit $\sigma_1 \to \infty$ we see that $(M - M_1)/(M + M_1) \to e^{+j\pi}$, and, for $\sigma_2 \to \infty$, $(M - M_2)/(M + M_2) \to e^{+j\pi}$. Thus eqn. 6.156 is equivalent to

$$\exp(-2jvd) = \exp(-j2\pi m) \tag{6.157}$$

where $m = n + 1 = 1, 2, 3, \ldots$.

Exercise

Determine an appropriate closed form TM solution of the mode equation 6.144 for the case when d is very small.

Solution

Here we set $SN = 1 + \delta$ and take δ to be a small parameter. Note that $\delta \to 0$ in the limit $d \to 0$ (i.e. conducting plane with no coating). Now

$$R(C) \simeq \frac{(1 - N^{-2})^{1/2} + jN(2\delta)^{1/2}}{(1 - N^{-2})^{1/2} - jN(2\delta)^{1/2}} \simeq \exp\left[2j \frac{(2\delta)^{1/2} N}{(1 - N^{-2})^{1/2}}\right] \tag{6.158}$$

and

$$e^{-2jNkCd} \simeq \exp[-2jNkd(1 - N^{-2})^{1/2}] \tag{6.159}$$

For $n = 0$, we then find that

$$\delta = \delta_0 \simeq \frac{1}{2}\left(1 - \frac{1}{N^2}\right)^2 (kd)^2$$

and the desired solution for the propagation constant is

$$\Gamma = \Gamma_0 \simeq jk(1 + \delta_0) \tag{6.60}$$

This result can be obtained by other methods (e.g. see Wait, 1970, p. 20). Note that there is no TE mode solution of this type.

6.10 Further exercises

Exercise

Consider the semi-infinite parallel plate waveguide shown in Fig. 6.17. The problem is entirely two dimensional, so that $\partial/\partial y = 0$. The structure consists of two planar surfaces of infinite conductivity at $x = \pm d$ for $0 < z < \infty$. The electric field $E_y(x, z)$ at $z = 0$ is specified in the aperture plane for $-d < x < d$. In other words, $E_y(x, 0) = f(x)$ where $f(x)$ is known. We also restrict attention to even excitation such that $f(x) = f(-x)$. Now do the following:

(a) Deduce $E_y(x, z)$ for $z > 0$ in terms of the known distribution $f(x)$ and, in particular, consider the special form

$$f(x) = \cos \frac{\pi x}{2s} \quad \text{for } |x| < s$$

$$= 0 \quad \text{for } |x| > s$$

(b) Assume $\sigma = 0$, and obtain expressions for the phase velocity of the modes, assuming that at least several of the modes are not near cutoff.
(c) Take $\sigma \neq 0$ but assume $\sigma \ll \varepsilon\omega$, and calculate the attenuation of the propagating modes.
(d) Using Maxwell's equations, deduce the magnetic fields.

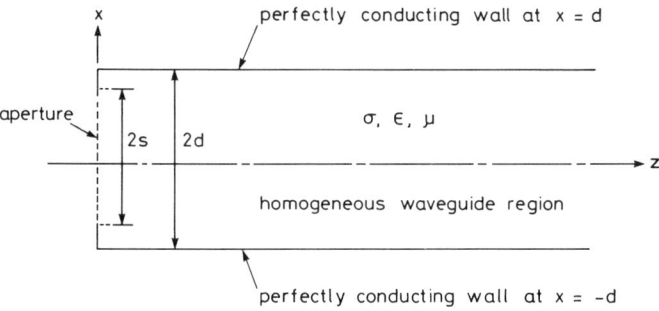

Fig. 6.17 Semi-infinite parallel plate waveguide with aperture excitation

Solution

(a) Bearing in mind that $\mathbf{E}$ has only a y component, Maxwell's equations take the form (for a time factor $\exp(j\omega t)$):

$$j\mu\omega H_x = \frac{\partial E_y}{\partial z} \tag{6.161}$$

$$-j\mu\omega H_z = \frac{\partial E_y}{\partial x} \tag{6.162}$$

$$(\sigma + j\varepsilon\omega)E_y = \frac{\partial H_x}{\partial z} - \frac{\partial H_z}{\partial x} \tag{6.163}$$

Now insert eqns. 6.161 and 6.162 into 6.163 to yield

$$\left(\frac{\partial^2}{\partial x^2} + \frac{\partial^2}{\partial z^2} - \gamma^2\right) E_y = 0 \tag{6.164}$$

where

$$\gamma = [j\mu\omega(\sigma + j\varepsilon\omega)]^{1/2}$$

is the propagation constant.

The desired form of the solution of eqn. 6.164 is

$$E_y = \sum_{m=0}^{\infty} A_m \cos \lambda_m x \, e^{-u_m z} \qquad (6.165)$$

where

$$u_m = (\lambda_m^2 + \gamma^2)^{1/2}$$

and λ_m satisfies $\cos \lambda_m d = 0$. Thus $\lambda_m = (m + \tfrac{1}{2})(\pi/d)$, where m is the order of the mode and $m = 0, 1, 2, \ldots$.

We choose Re $u_m > 0$, so E_y vanishes as $z \to \infty$, as it must. Also we confirm that $E_y = 0$ at $x = d$ and $E_y(-x, z) = E_y(x, z)$.

The source condition stipulates that

$$\sum_{m=0}^{\infty} A_m \cos \lambda_m x = f(x) \qquad (6.166)$$

Now multiply both sides of eqn. 6.166 by $\cos \lambda_n x$ and integrate over x from 0 to d to get

$$\sum_{m=0}^{\infty} A_m \int_0^d \cos \lambda_m x \cos \lambda_n x \, dx = \int_0^d f(x) \cos \lambda_n x \, dx \qquad (6.167)$$

The integral on the left-hand side is equal to zero if $m \neq n$ and equal to $d/2$ if $m = n$. Then all terms on the left-hand side drop out except the one for $m = n$. Thus

$$A_n d/2 = \int_0^d f(x) \cos \lambda_n x \, dx \qquad (6.168)$$

This holds for all n so we can write

$$A_m = \frac{2}{d} \int_0^d f(x) \cos \lambda_m x \, dx \qquad (6.169)$$

which is the desired form of the excitation factor in eqn. 6.165. Using the special cosine distribution in the aperture, we carry out the integrations as follows:

$$A_m = \frac{2E_0}{d} \int_0^s \cos \frac{\pi x}{2s} \cos \lambda_m x \, dx$$

$$= \frac{E_0}{d} \int_0^s \left\{ \cos\left[\left(\frac{\pi}{2s} - \lambda_m\right) x\right] + \cos\left[\left(\frac{\pi}{2s} + \lambda_m\right) x\right] \right\} dx$$

$$= \frac{E_0}{d} \left[\frac{\sin\left(\frac{\pi}{2s} - \lambda_m\right) s}{\frac{\pi}{2s} - \lambda_m} + \frac{\sin\left(\frac{\pi}{2s} + \lambda_m\right) s}{\frac{\pi}{2s} + \lambda_m} \right]$$

$$= E_0(\cos \lambda_m s) \left[\frac{1}{\frac{\pi}{2s} - \lambda_m} + \frac{1}{\frac{\pi}{2s} + \lambda_m} \right]$$

$$= E_0 (\cos \lambda_m s) \frac{\pi/s}{(\pi/2s)^2 - \lambda_m^2} \qquad (6.170)$$

(b) Note that a mode depends on z according to $\exp(-u_m z)$. Now $u_m = \alpha_m + j\beta_m$, where $\alpha_m = \operatorname{Re} u_m$ is the attenuation rate in nepers per metre and $\beta_m = \operatorname{Im} u_m$ is the phase rate in radians per metre. Now $\beta_m = \omega/v_m$, or $v_m = \omega/\beta_m$ is the phase velocity of the mth mode. Then clearly if $\sigma = 0$, $\gamma = jk$, where $k = \sqrt{(\varepsilon\mu)}\omega$ is real, and thus $\alpha_m = \operatorname{Re}(\lambda_m^2 - k^2)^{1/2} = 0$ if $k > \lambda_m$. On the other hand, if $k < \lambda_m$, $\alpha_m > 0$ and the modes are cut off. Also we note that the phase velocities (for the propagating modes) are

$$v_m = \omega/\beta_m = \omega(k^2 - \lambda_m^2)^{-1/2} > \omega/k = 1/\sqrt{(\varepsilon\mu)}$$

for the lossless condition

(c) For $\sigma \neq 0$, $\gamma = jk(1 - j\delta)^{1/2}$, where $\delta = \sigma/\varepsilon\omega$, is exact. Then for small δ we approximate as follows:

$$u_m = j(k^2 - \lambda_m^2 - jk^2\delta)^{1/2} \simeq j(k^2 - \lambda_m^2)^{1/2} + (k^2\delta/2)(k^2 - \lambda_m^2)^{-1/2}$$

The attenuation rate is

$$\alpha_m = \operatorname{Re} u_m \simeq (\mu/\varepsilon)^{1/2}(\sigma/2)[1 - (\lambda_m^2/k^2)]^{-1/2}$$

(d) The magnetic field components are obtained from eqns. 6.161 and 6.162 operating on eqn. 6.165.

Exercise

With reference to Fig. 6.2, derive explicitly the basic coupled transmission line equations given by eqns. 6.16 and 6.17.

Solution

We obtain eqn. 6.16 by noting that the voltage V at the input of the section, of length δz, must be balanced by the sum of the voltages across the two series elements and the voltage at the output of the section. That is

$$V = I\frac{\hat{Z}}{2}\delta z + \left(1 + \frac{dI}{dz}\delta z\right)\frac{\hat{Z}}{2}\delta z + V + \frac{dV}{dz}\delta z \quad (6.171)$$

or

$$0 = I\hat{Z} + \frac{dI}{dz}\frac{\hat{Z}}{2}\delta z + \frac{dV}{dz} \quad (6.172)$$

Then letting $\delta z \to 0$ we obtain eqn. 6.16. To obtain eqn. 6.17 we simply note that the voltage V must be balanced by the sum of the voltages across the first series element and the shunt element. Therefore

$$V = I\frac{\hat{Z}}{2}\delta z - \frac{dI}{dz}\delta z \frac{1}{\hat{Y}\delta z} \quad (6.173)$$

The δz in the latter term nicely cancel; then letting the remaining $\delta z \to 0$, we obtain eqn. 6.17.

Exercise

Show explicitly that the equation for I, corresponding to eqn. 6.38 for V, is

$$\frac{d^2 I}{dz^2} - \frac{1}{\hat{Y}}\frac{d\hat{Y}}{dz}\frac{dI}{dz} - \hat{Y}\hat{Z}I = 0 \quad (6.174)$$

Solution Differentiate both sides of eqn. 6.17 to give

$$\frac{d^2 I}{dz^2} = -\hat{Y}\frac{dV}{dz} - \frac{d\hat{Y}}{dz}V \tag{6.175}$$

Then substitute eqn. 6.16 for dV/dz and eqn. 6.17 for V into eqn. 6.175 to yield eqn. 6.174.

6.11 References

1. E. C. JORDAN and K. BALMAIN, *EM Waves and Radiating Systems*, Prentice Hall, 1968.
2. R. E. COLLIN, *Field Theory of Guided Waves*, McGraw Hill, 1961.
3. J. R. WAIT, *EM Wave Theory*, Harper and Row, 1985.
4. L. LEWIN, *Theory of Waveguides*, Newnes-Butterworth, 1975.
5. S. A. SCHELKUNOFF, *Electromagnetic Fields*, Blaisdell, 1963.
6. M. S. SODHA and A. K. GHATAK, *Inhomogeneous Optical Waveguides*, Plenum, 1977.
7. L. LEWIN, D. C. CHANG and E. F. KUESTER, *EM Waves and Curved Structures*, Peter Peregrinus, 1977.
8. F. SPORLEDER and UNGER, *Waveguide Tapers, Transitions, and Couplers* Peregrinus, 1979.
9. J. R. WAIT, *EM Waves in Stratified media*, Pergamon, 1970.
10. J. R. WAIT, *(Lectures on) Wave Propagation Theory*, Pergamon, 1981.
11. J. GALEJS, *Terrestial Propagation of Long EM Waves*, Pergamon, 1972.
12. G. K. KAO, *Optical Fiber Systems*, McGraw Hill, 1982.

Other reference H. R. L. LAMONT, *Wave Guides*, Methuen, 1950.

Chapter 7

Cylindrical waves and scatter

7.1 General formulation

There is a broad class of problems in electromagnetics where the geometry is cylindrical [1–7]. An example is a long conductor of circular cross-section. Here we are faced immediately with the prospect of solving Maxwell's equations in cylindrical co-ordinates. We shall present a straightforward development which is sufficiently general to encompass many situations of practical interest in telecommunications and radar.

The first step, as always, is to write out Maxwell's equations in a source-free region for a time-harmonic factor exp (jωt). Here we make use of the curl operator in cylindrical co-ordinates (ϱ, ϕ, z) as given in Sec. 1.17. Thus without difficulty we may write

$$-j\mu\omega H_\varrho = \frac{1}{\varrho}\frac{\partial E_z}{\partial \phi} - \frac{\partial E_\phi}{\partial z} \tag{7.1}$$

$$-j\mu\omega H_\phi = \frac{\partial E_\varrho}{\partial z} - \frac{\partial E_z}{\partial \varrho} \tag{7.2}$$

$$-j\mu\omega H_z = \frac{1}{\varrho}\frac{\partial}{\partial \varrho}(\varrho E_\phi) - \frac{1}{\varrho}\frac{\partial E_\varrho}{\partial \phi} \tag{7.3}$$

$$(\sigma + j\varepsilon\omega)E_\varrho = \frac{1}{\varrho}\frac{\partial H_z}{\partial \phi} - \frac{\partial H_\phi}{\partial z} \tag{7.4}$$

$$(\sigma + j\varepsilon\omega)E_\phi = \frac{\partial H_\varrho}{\partial z} - \frac{\partial H_z}{\partial \varrho} \tag{7.5}$$

$$(\sigma + j\varepsilon\omega)E_z = \frac{1}{\varrho}\frac{\partial}{\partial \varrho}(\varrho H_\phi) - \frac{1}{\varrho}\frac{\partial H_\varrho}{\partial \phi} \tag{7.6}$$

We now make a seemingly rash assumption. We specify that the field components all vary with ϕ and z according to exp ($-jm\phi$) exp ($-\Gamma z$), where m is an integer (including zero) and Γ is a generally complex number independent of ϱ, ϕ and z. As we shall see later, suitable superpositions of the basic solutions, for a range of m and ϕ values, will allow us to describe general field configurations. Conceptually it is useful to regard Γ as a propagation constant in the axial direction; m is the order of the Fourier harmonic. Now eqns. 7.1 to 7.6 reduce to

$$-j\mu\omega H_\varrho = -\frac{jm}{\varrho}E_z + \Gamma E_\phi \tag{7.7}$$

Cylindrical waves and scatter

$$-j\mu\omega H_\phi = -\Gamma E_\varrho - \frac{\partial E_z}{\partial \varrho} \tag{7.8}$$

$$-j\mu\omega H_z = \frac{1}{\varrho}\frac{\partial}{\partial \varrho}(\varrho E_\phi) + \frac{jm}{\varrho} E_\varrho \tag{7.9}$$

$$(\sigma + j\varepsilon\omega)E_\varrho = -\frac{jm}{\varrho} H_z + \Gamma H_\phi \tag{7.10}$$

$$(\sigma + j\varepsilon\omega)E_\phi = -\Gamma H_\varrho - \frac{\partial H_z}{\partial \varrho} \tag{7.11}$$

$$(\sigma + j\varepsilon\omega)E_z = \frac{1}{\varrho}\frac{\partial}{\partial \varrho}(\varrho H_\phi) + \frac{jm}{\varrho} H_\varrho \tag{7.12}$$

To proceed further, we now decompose the total field into two parts. In the first case we set $H_z = 0$. Then it is possible to show from eqns. 7.7 to 7.12 that

$$(\nabla^2 - \gamma^2) E_z = 0 \tag{7.13}$$

where

$$\gamma^2 = j\mu\omega(\sigma + j\varepsilon\omega) \tag{7.14}$$

and where

$$\nabla^2 = \frac{1}{\varrho}\frac{\partial}{\partial \varrho}\left(\varrho \frac{\partial}{\partial \varrho}\right) - \frac{1}{\varrho^2} m^2 + \Gamma^2 \tag{7.15}$$

is the Laplacian operator in cylindrical co-ordinates. On the other hand, in the second case we set $E_z = 0$ and show in a similar fashion that

$$(\nabla^2 - \gamma^2) H_z = 0 \tag{7.16}$$

The total field is the superposition of the two partial fields. Thus we deduce that

$$E_\varrho = \frac{1}{u^2}\left(\Gamma \frac{\partial E_z}{\partial \varrho} + \frac{m\omega\mu}{\varrho} H_z\right) \tag{7.17}$$

$$E_\phi = \frac{1}{u^2}\left(\frac{-j\Gamma m}{\varrho} E_z - j\omega\mu \frac{\partial H_z}{\partial \varrho}\right) \tag{7.18}$$

$$H_\varrho = \frac{1}{u^2}\left(\frac{jm\hat{\sigma}}{\varrho} E_z + \Gamma \frac{\partial H_z}{\partial \varrho}\right) \tag{7.19}$$

$$H_\phi = \frac{1}{u^2}\left(\hat{\sigma} \frac{\partial E_z}{\partial \varrho} - \frac{j\Gamma m}{\varrho} H_z\right) \tag{7.20}$$

where $\hat{\sigma} = \sigma + j\varepsilon\omega$ and $u^2 = \gamma^2 - \Gamma^2$.

Exercise

Verify that an equivalent version of eqns. 7.17 to 7.20 is given by the following equations:

$$E_\varrho = \frac{\partial^2 U}{\partial \varrho \partial z} - \frac{j\mu\omega}{\varrho}\frac{\partial V}{\partial \phi} \tag{7.21}$$

$$E_\phi = \frac{\partial^2 U}{\varrho \partial \phi \partial z} + j\mu\omega \frac{\partial V}{\partial \varrho} \qquad (7.22)$$

$$H_\varrho = \frac{\partial^2 V}{\partial \varrho \partial z} + \frac{\sigma + j\varepsilon\omega}{\varrho} \frac{\partial U}{\partial \phi} \qquad (7.23)$$

$$H_\phi = \frac{1}{\varrho} \frac{\partial^2 V}{\partial \phi \partial z} - (\sigma + j\varepsilon\omega) \frac{\partial U}{\partial \varrho} \qquad (7.24)$$

provided U and V are defined such that

$$E_z = \left(-\gamma^2 + \frac{\partial^2}{\partial z^2}\right) U \qquad (7.25)$$

$$H_z = \left(-\gamma^2 + \frac{\partial^2}{\partial z^2}\right) V \qquad (7.26)$$

The scalar functions U and V are sometimes called Debye potentials, and they serve a useful purpose in formulating and solving boundary value problems in cylindrical geometry.

Exercise

Show that the vector form of eqns. 7.21 to 7.26 is given by

$$\boldsymbol{E} = (-\gamma^2 + \mathrm{grad\ div}) \vec{\Pi} - j\mu\omega\ \mathrm{curl}\ \vec{\Pi}^* \qquad (7.27)$$

$$\boldsymbol{H} = (-\gamma^2 + \mathrm{grad\ div}) \vec{\Pi}^* + (\sigma + j\varepsilon\omega)\ \mathrm{curl}\ \vec{\Pi} \qquad (7.28)$$

where

$$\vec{\Pi} = \boldsymbol{i}_z U \quad \text{and} \quad \vec{\Pi}^* = \boldsymbol{i}_z V$$

Here $\vec{\Pi}$ and $\vec{\Pi}^*$ are electric and magnetic Hertz vectors which, in the present instance, are z directed.

7.2 Fields of line sources

A concrete and relatively simple aspect of cylindrical waves is illustrated when we examine a line source. For example, let us deal with an infinitely thin long straight wire carrying a constant current I. Let the surrounding medium be homogeneous with electric and magnetic properties σ, ε and μ. As usual the assumed time factor is $\exp(j\omega t)$. See Fig. 7.1.

We choose cylindrical co-ordinates (ϱ, ϕ, z) with filamental current I coincident with the z axis. An application of Ampère's law tells us that

$$H_\phi = I/2\pi\varrho \qquad (7.29)$$

in the limiting case where $\varrho \to 0$. In fact there will be no other component of the magnetic field for any value of ϕ. Furthermore, there is obvious symmetry about ϕ. Thus $\partial/\partial\phi = 0$. Then, according to eqn. 7.24,

$$H_\phi = -(\sigma + j\varepsilon\omega) \frac{\partial U}{\partial \varrho} \qquad (7.30)$$

where the Debye potential U satisfies

$$\left[\frac{1}{\varrho} \frac{\partial}{\partial \varrho}\left(\varrho \frac{\partial}{\partial \varrho}\right) - \gamma^2\right] U = 0 \qquad (7.31)$$

Cylindrical waves and scatter

More explicitly we would write

$$\frac{d^2 U}{d\varrho^2} + \frac{1}{\varrho}\frac{dU}{d\varrho} - \gamma^2 U = 0 \tag{7.32}$$

(Here we have replaced the partial ∂ by the total d because there is only one variable.) Now eqn. 7.32 can be converted to the modified Bessel's equation of order zero by making the substitution $x = \gamma\varrho$, whence

$$\frac{d^2 U}{dx^2} + \frac{1}{x}\frac{dU}{dx} - U = 0 \tag{7.33}$$

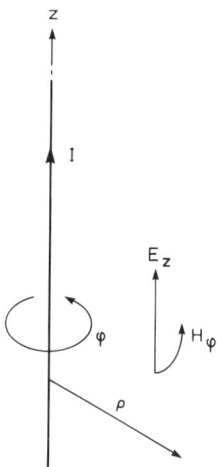

Fig. 7.1 Infinite line source of electric current I, showing the associated electromagnetic field components

Solutions of this equation are the modified *Bessel functions* $I_0(x)$ and $K_0(x)$ of order zero and of argument x (see the Appendix to this chapter). Thus the solutions of eqn. 7.32 are $I_0(\gamma\varrho)$ and $K_0(\gamma\varrho)$. Now in the present instance we are interested in fields that are noninfinite as $\varrho \to \infty$. Only the function $K_0(\gamma\varrho)$ has this property. In fact

$$K_0(\gamma\varrho) \simeq \left(\frac{\pi}{2\gamma\varrho}\right)^{1/2} e^{-\gamma\varrho} \to 0 \tag{7.34}$$

for $|\gamma\varrho| \to \infty$, and

$$I_0(\gamma\varrho) \simeq \left(\frac{1}{2\pi\gamma\varrho}\right)^{1/2} e^{+\gamma\varrho} \to \infty \tag{7.35}$$

in the same asymptotic limit, bearing in mind that Re $\gamma > 0$.

In view of the above discussion we may write

$$U = AK_0(\gamma\varrho) \tag{7.36}$$

for $\varrho > 0$, where A is a constant. Now, in accordance with eqn. 7.30,

$$H_\phi = -A(\sigma + j\varepsilon\omega)\gamma K_0'(\gamma\varrho) \tag{7.37}$$

where the prime indicates differentiation with respect to the argument of the Bessel function. Using a known identity,

$$K_0'(x) = dK_0(x)/dx = -K_1(x) \tag{7.38}$$

where $K_1(x)$ is the modified Bessel function of order one. Thus we are permitted to write

$$H_\phi = A(\sigma + j\varepsilon\omega)\gamma K_1(\gamma\varrho) \tag{7.39}$$

for all ϱ. If we now let $|\gamma\varrho| \to 0$.

$$H_\phi \simeq A(\sigma + j\varepsilon\omega)\frac{1}{\varrho} \tag{7.40}$$

where we have made use of another known result from Bessel function theory, namely $K_1(x) \simeq 1/x$ if $x \to 0$.

The key step is now to equate the right-hand sides of eqns. 7.29 and 7.40 to yield the following expression for the unknown factor:

$$A = \frac{I}{2\pi(\sigma + j\varepsilon\omega)} \tag{7.41}$$

Then eqn. 7.39 is written

$$H_\phi = \frac{I}{2\pi}\gamma K_1(\gamma\varrho) \tag{7.42}$$

The corresponding electric field (which has only a z component) is obtained from eqn. 7.25 to give

$$E_z = -\gamma^2 U = -\frac{j\mu\omega}{2\pi} I K_0(\gamma\varrho) \tag{7.43}$$

An admittance function can now be defined by the ratio

$$\vec{Y} = -\frac{H_\phi}{E_z} = y\frac{K_1(\gamma\varrho)}{K_0(\gamma\varrho)} \tag{7.44}$$

where

$$y = \frac{\gamma}{j\mu\omega} = \left(\frac{\sigma + j\varepsilon\omega}{j\mu\omega}\right)^{1/2} \tag{7.45}$$

is the admittivity of the medium. It is useful to note that this *outward-looking admittance function* has the property that

$$\vec{Y} \simeq y \tag{7.46}$$

in the limit $|\gamma\varrho| \to \infty$.

Exercise

Consider the extension of the preceding analysis for the case where the filamental current has the assumed form $I(z) = I \exp(-\Gamma z)$ for all values of z, where Γ is a constant or fixed parameter. In particular show that the fields can be derived from the electric Debye potential U given by

$$U = \frac{I}{2\pi(\sigma + j\varepsilon\omega)} K_0(u\varrho) e^{-\Gamma z} \tag{7.47}$$

where

$$u = (\gamma^2 - \Gamma^2)^{1/2}$$

is defined such that $\operatorname{Re} u > 0$.

Solution   The method is identical to that used for the uniform line source. We can verify the result by noting that

$$H_\phi = -(\sigma + j\varepsilon\omega)\frac{\partial U}{\partial \varrho}$$

$$= \frac{I}{2\pi} u K_1(u\varrho) e^{-\Gamma z} \qquad (7.48)$$

reduces to

$$H_\phi = \frac{I}{2\pi\varrho} e^{-\Gamma z} \qquad (7.49)$$

if $|\gamma\varrho| \to 0$. The corresponding electric field components are:

$$E_\varrho = \frac{\partial^2 U}{\partial \varrho \partial z} = \frac{I\Gamma u K_1(u\varrho)}{2\pi(\sigma + j\varepsilon\omega)} e^{-\Gamma z} \qquad (7.50)$$

$$E_z = (-\gamma^2 + \Gamma^2)U = -\frac{Iu^2 K_0(u\varrho)}{2\pi(\sigma + j\varepsilon\omega)} e^{-\Gamma z} \qquad (7.51)$$

In the case where $\Gamma \to 0$ we recover eqns. 7.42 and 7.43 from eqns. 7.49 and 7.51, respectively. Note also that $E_\varrho \to 0$, as indicated by eqn. 7.50.

7.3 Internal fields of the cylindrical conductor

As a prelude to the analysis of scattering from a cylinder of arbitrary radius we deal first with the internal fields of a thin cylinder. Specifically we consider a homogeneous region with electric and magnetic properties σ_w, ε_w and μ_w bounded by a cylindrical surface of radius a. This radius a is sufficiently small that only azimuthally symmetric fields need be considered. This 'smallness' assumption is examined more critically below. Some of the notions were first described by Lord Rayleigh [8].

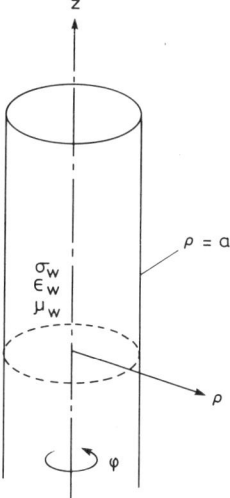

Fig. 7.2 Infinitely long homogeneous circular cylinder

The situation is illustrated in Fig. 7.2, where the cylinder of assumed infinite length is bounded by $\varrho = a$. Again we invoke the symmetry condition $\partial/\partial\phi = 0$ and also we restrict attention to field configurations such that H has only a ϕ component and there are no variations in the z direction (i.e. $\partial/\partial z = 0$). Thus we may derive the fields from an electric Debye potential U_w that satisfies

$$\left[\frac{1}{\varrho}\frac{\partial}{\partial\varrho}\left(\varrho\frac{\partial}{\partial\varrho}\right) - \gamma_w^2\right] U_w = 0 \tag{7.52}$$

being the counterpart of eqn. 7.31 for the external fields. Here

$$\gamma_w = [j\mu_w\omega(\sigma_w + j\varepsilon_w\omega)]^{1/2} \tag{7.53}$$

is the intrinsic propagation constant.

In this problem, the appropriate solution of eqn. 7.52, in the region $0 < \varrho < a$, is

$$U_w = BI_0(\gamma_w\varrho) \tag{7.54}$$

where B is a constant. Here we exclude the solution $K_0(\gamma_w\varrho)$ because it would be singular (i.e. infinite at $\varrho = 0$). The corresponding fields are

$$H_\phi = -(\sigma_w + j\varepsilon_w\omega)\frac{\partial U_w}{\partial\varrho}$$

$$= -B(\sigma_w + j\varepsilon_w\omega)\gamma_w I_1(\gamma_w\varrho) \tag{7.55}$$

$$E_z = -\gamma_w^2 U_w = -B\gamma_w^2 I_0(\gamma_w\varrho) \tag{7.56}$$

(In writing eqn. 7.55 we have employed the Bessel function identity $dI_0(x)/dx = I_1(x)$.)

In analogy to the admittance function given by eqn. 7.44 we have defined here an *internal admittance function* as follows:

$$\overleftarrow{Y} = \frac{H_\phi}{E_z} = y_w \frac{I_1(\gamma_w a)}{I_0(\gamma_w a)} \tag{7.57}$$

where

$$y_w = \left(\frac{\sigma_w + j\varepsilon_w\omega}{j\mu_w\omega}\right)^{1/2} \tag{7.58}$$

is the admittivity. Again we note that, in the limit $|\gamma_w a| \to \infty$,

$$\overleftarrow{Y} \simeq y_w$$

In the present case we also find it convenient to define a series impedance Z_w per unit length as follows:

$$Z_w = \frac{E_z(\varrho = a)}{I_w} \tag{7.59}$$

where $I_w = 2\pi a H_\phi(\varrho = a)$ is clearly the net current carried by the wire. To be explicit,

$$Z_w = \frac{\eta_w}{2\pi a}\frac{I_0(\gamma_w a)}{I_1(\gamma_w a)} \tag{7.60}$$

where

$$\eta_w = \frac{1}{y_w} = \left(\frac{j\mu_w\omega}{\sigma_w + j\varepsilon_w\omega}\right)^{1/2} \tag{7.61}$$

Cylindrical waves and scatter

is the characteristic impedance. In particular we note that, at the low-frequency limit $|\gamma_w a| \to 0$, we see that

$$Z_w = \frac{1}{\pi a^2 (\sigma_w + j\varepsilon_w \omega)} \qquad (7.62)$$

$$\simeq \frac{1}{\pi a^2 \sigma_w}$$

which is the expected DC form for the series resistance per unit length.

Exercise

As in the earlier exercise, now permit the fields inside the cylinder (i.e. $0 < \varrho < a$) to vary with z according to $\exp(-\Gamma z)$. Now show that the series impedance for this case is given by the z-dependent function

$$Z_w(\Gamma) = \frac{E_z(\varrho = a)}{I_w} = \frac{u_w^2 I_0(u_w a)}{2\pi(\sigma_w + j\varepsilon_w \omega) u_w a I_1(u_w a)} \qquad (7.63)$$

where

$$u_w = (\gamma_w^2 - \Gamma^2)^{1/2}$$

and, as before, $I_w = 2\pi a H_\phi(\varrho = a)$.

Solution

The Debye potential now has the form

$$U_w = B I_0(u_w \varrho) e^{-\Gamma z} \qquad (7.64)$$

which clearly is a solution of

$$\left[\frac{1}{\varrho} \frac{\partial}{\partial \varrho} \left(\varrho \frac{\partial}{\partial \varrho} \right) + \frac{\partial^2}{\partial z^2} - \gamma_w^2 \right] U_w = 0 \qquad (7.65)$$

We may check our result by noting that eqn. 7.63 reduces to eqn. 7.60 in the case where $\Gamma \to 0$.

7.4 Scattering from a thin cylinder

Some of the concepts described above can be employed to deal with plane wave scattering from a thin wire that is characterised by a series impedance Z_w.

The surrounding medium is homogeneous with properties σ, ε and μ. The situation is illustrated in Fig. 7.3, where we have located the thin conductor of radius a at the z axis of the cylindrical co-ordinate system (ϱ, ϕ, z). The incident plane wave is chosen with the electric vector $\mathbf{E}^{\text{inc}}$ to have only a z component E_z^{inc}. Also we have assumed that the problem is entirely two dimensional, so that $\partial/\partial z = 0$.

For the assumptions indicated above,

$$E_z^{\text{inc}} = E_0 e^{\gamma \varrho \cos(\phi - \phi_0)} \qquad (7.66)$$

where E_0 is a constant. We now argue that if the wire is sufficiently thin only the axial azimuthally symmetric current on the wire need be considered. Thus the scattered field E_z^{sc} will be azimuthally symmetric and should have the form

$$E_z^{\text{sc}} = -\frac{j\mu\omega I^{\text{ind}}}{2\pi} K_0(\gamma\varrho) \qquad (7.67)$$

in accordance with eqn. 7.43, where I^{ind} is the induced wire current.

Now at the surface of the wire

$$(E_z^{\text{inc}} + E_z^{\text{sc}})_{\varrho=a} = I^{\text{ind}} Z_{\text{w}} \tag{7.68}$$

where Z_{w} is the séries impedance, as given for example by eqn. 7.60. Bearing in mind that $|\gamma a| \ll 1$, we may write eqn. 7.68 in the form

$$E_0 - \frac{j\mu\omega}{2\pi} I^{\text{ind}} K_0(\gamma a) = I^{\text{ind}} Z_{\text{w}} \tag{7.69}$$

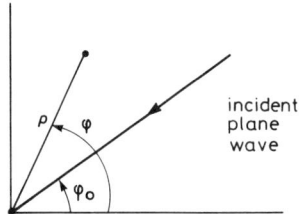

Fig. 7.3 Geometry for studying scattering of a plane wave normally incident on to a thin conductor or 'wire'

Thus

$$I^{\text{ind}} = \frac{E_0}{Z_{\text{w}} + Z_{\text{e}}} \tag{7.70}$$

where

$$Z_{\text{e}} = \frac{j\mu\omega}{2\pi} K_0(\gamma a) \tag{7.71}$$

can be defined as the external series impedance of the wire. Because $|\gamma a| \ll 1$, we can write eqn. 7.71 in equivalent form, in terms of a series resistance and series inductance, according to

$$Z_{\text{e}} \simeq R_{\text{e}} + j\omega L_{\text{e}}$$

$$\simeq \frac{j\mu\omega}{2\pi} [-\ln(\gamma a) - C + \ln 2] \tag{7.72}$$

where $C = 0.5772\ldots$ is Euler's number. If $\gamma = jk$ where k is real (i.e. lossless external medium, $\sigma = 0$), we see that

$$R_{\text{e}} + j\omega L_{\text{e}} = \frac{\mu\omega}{4} + j\omega \frac{\mu}{2\pi} \left(\ln \frac{2}{ka} - C\right) \tag{7.73}$$

Equation 7.70 is an effective Ohm's law. It shows that the applied field E_0, when divided by the total series impedance, yields the induced current. We stress that such a result is only valid if the wire is electrically thin in terms of the effective wavelength in the external medium. This point is discussed in a more quantitative manner below when we deal with a cylinder of nonzero radius.

7.5 Oblique incidence scattering from a thin wire

A nontrivial extension of the preceding scattering problem is when the incoming plane wave is incident at an angle other than perpendicular to the wire (Fig. 7.4). But again we need only consider the z component of the electric field vector denoted E_z^{inc}.

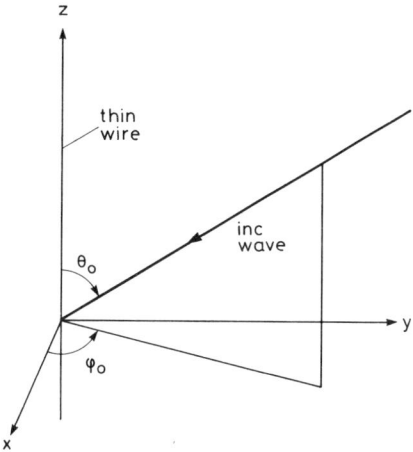

Fig. 7.4 Geometry for scattering of a plane wave at oblique incidence on a thin conductor or 'wire'

Now we write

$$E_z^{inc} = E_{0z}\, e^{(\gamma \sin\theta_0 \cos\phi_0)x}\, e^{(\gamma \sin\theta_0 \sin\phi_0)y}\, e^{(\gamma \cos\theta_0)z} \tag{7.74}$$

or equivalently,

$$E_z^{inc} = E_{0z}\, e^{\gamma\varrho \sin\theta_0 \cos(\phi - \phi_0)}\, e^{(\gamma \cos\theta_0)z} \tag{7.75}$$

where E_{0z} is the strength of the incident electric field (z component) at the origin.

It is useful to note that the incident field can be derived from an electric Debye potential as follows:

$$E_z^{inc} = \left(-\gamma^2 + \frac{\partial^2}{\partial z^2}\right) U^{inc}$$

$$= (-\gamma^2 + \gamma^2 \cos^2\theta_0)\, U^{inc} = -\gamma^2 \sin^2\theta_0\, U^{inc} \tag{7.76}$$

if

$$U^{inc} = \frac{-E_{0z}}{\gamma^2 \sin^2\theta_0} [e^{\gamma\varrho \sin\theta_0 \cos(\phi-\phi_0)}\, e^{\gamma z \cos\theta_0}] \tag{7.77}$$

Again, because of the assumed thinness of the wire, we may adopt only the azimuthally symmetric solution for the Debye potential of the secondary field. Thus we write

$$U^{sec} = D K_0(\gamma\varrho \sin\theta_0)\, e^{\gamma z \cos\theta_0} \tag{7.78}$$

where D is a constant to be determined. Clearly we may confirm that

$$(\nabla^2 - \gamma^2)\, U^{sec} = 0 \tag{7.79}$$

as it must.

The resultant electric fields are now obtained from

$$E_\varrho = \frac{\partial^2}{\partial \varrho \partial z}(U^{\text{inc}} + U^{\text{sec}}) \tag{7.80a}$$

$$E_z = \left(-\gamma^2 + \frac{\partial^2}{\partial z^2}\right)(U^{\text{inc}} + U^{\text{sec}}) \tag{7.80b}$$

while the magnetic field is obtained from

$$H_\phi = -(\sigma + j\varepsilon\omega)\frac{\partial}{\partial \varrho}(U^{\text{inc}} + U^{\text{sec}}) \tag{7.81}$$

The needed boundary condition in the present problem is that

$$E_z]_{\varrho=a} = 2\pi a Z_w H_\phi]_{\varrho=a} \tag{7.82}$$

where Z_w is the specified series impedance, per unit length, of the wire. Equation 7.82 is equivalent to

$$E_{0z} - \gamma^2 \sin^2 \theta_0 \, DK_0(\gamma \sin \theta_0 \, a) = 2\pi a Z_w (\sigma + j\varepsilon\omega)\gamma \sin \theta_0$$
$$\times DK_1(\gamma a \sin \theta_0) \tag{7.83}$$

where we have invoked the condition that $|\gamma a| \ll 1$ and made use of the identity $K_0'(x) = -K_1(x)$. A further simplification is possible here because $K_1(x) \simeq 1/x$ if $x \ll 1$. Thus the right-hand side of eqn. 7.83 can be replaced by $2\pi(\sigma + j\varepsilon\omega)Z_w D$. Then solving for D gives

$$D = E_{0z}/[\gamma^2 \sin^2 \theta_0 K_0(\gamma a \sin \theta_0) + 2\pi Z_w(\sigma + j\varepsilon\omega)] \tag{7.84}$$

The actual induced current on the wire is then given by

$$I^{\text{ind}} = 2\pi a H_\phi]_{\varrho=a} \simeq 2\pi a H_\phi^{\text{sec}}]_{\varrho=a}$$

$$\simeq 2\pi(\sigma + j\varepsilon\omega)\gamma a \sin \theta_0 K_1(\gamma a \sin \theta_0) D \, e^{\gamma z \cos \theta_0}$$

$$\simeq 2\pi(\sigma + j\varepsilon\omega) D \, e^{\gamma z \cos \theta_0} \tag{7.85}$$

Using eqn. 7.84 we obtain

$$I^{\text{ind}} = \frac{E_{0z} \, e^{\gamma z \cos \theta_0}}{Z_w + \dfrac{j\mu\omega}{2\pi}\sin^2 \theta_0 K_0(\gamma a \sin \theta_0)} \tag{7.86}$$

This result, of course, reduces to the form given by eqn. 7.70 in the case of normal incidence (i.e. where $\theta_0 = 90°$).

It is useful to note here that the series impedance can be a function of the angle of incidence θ_0. For example, if the thin cylindrical wire has electrical properties σ_w, ε_w and permeability μ_w we see that Z_w is given explicitly by eqn. 7.63 if we replace Γ by $\gamma \cos \theta_0$.

7.6 Plane wave scattering from a cylinder of any radius

A classical problem in electromagnetic theory first solved by Lord Rayleigh [9] is the scattering of a plane wave from a homogeneous cylinder of infinite length. We shall consider this problem here and later extend it to other conditions such as line source excitation and oblique incidence. But first of all we follow Rayleigh [9] and assume that the electric field vector E^{inc} of the incident wave is parallel to the axis of the cylinder, so that the problem is entirely two dimensional. The situation is shown in Fig. 7.5 where, as indicated, cylindrical co-ordinates are adopted. The homogeneous cylinder is bounded by $\varrho = a$ and it has electrical and magnetic properties σ_c, ε_c and μ_c. The external medium (i.e. for $\varrho > a$) has properties σ, ε and μ.

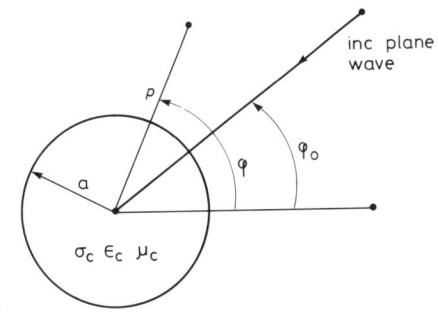

Fig. 7.5 Normally incident plane wave incident on a cylinder with homogeneous properties

The incident wave is given by

$$E_z^{inc} = E_0 \, e^{\gamma \varrho \cos(\phi - \phi_0)} \tag{7.87}$$

which is the same as eqn. 7.66. We now wish to determine the secondary field E_z^{sc} without any restrictions on the magnitude of a.

We note that, for $\varrho > a$,

$$(\nabla^2 - \gamma^2) E_z^{sc} = 0 \tag{7.88}$$

where

$$\nabla^2 = \frac{1}{\varrho} \frac{\partial}{\partial \varrho} \left(\varrho \frac{\partial}{\partial \varrho} \right) + \frac{1}{\varrho^2} \frac{\partial^2}{\partial \phi^2} \tag{7.89}$$

The desired solution of eqn. 7.88 is

$$K_m(\gamma \varrho) \, e^{\pm jm\phi}$$

where K_m is the modified Bessel function of argument $\gamma \varrho$ and order m. This function has the appropriate behaviour as $\varrho \to 0$. For example,

$$K_m(\gamma \varrho) \simeq \left(\frac{\pi}{2\gamma \varrho} \right)^{1/2} e^{-\gamma \varrho} \tag{7.90}$$

for $|\gamma \varrho| \to \infty$. Bearing in mind that the solution for the electric field should be even about $\phi = \phi_0$, we see that the angle dependence should be $\cos m(\phi - \phi_0)$, where m is an integer.

The most general form of the scattered field is

$$E_z^{sc} = \sum_{m=0}^{\infty} A_m K_m(\gamma \varrho) \cos m(\phi - \phi_0) \tag{7.91}$$

where A_m is a coefficient to be determined. The summation need only extend over positive integers because $K_m = K_{-m}$ for any integer m.

To deal with internal field E_z^{int}, for $0 < \varrho < a$, we note that the desired form is

$$E_z^{int} = \sum_{m=0}^{\infty} B_m I_m(\gamma_c \varrho) \cos m(\phi - \phi_0) \qquad (7.92)$$

where B_m is a coefficient and where $I_m(\gamma \varrho)$ is the modified Bessel function of the first type which is finite at $\varrho = 0$. For example

$$I_m(\gamma_c \varrho) \simeq (\gamma_c \varrho)^m \frac{1}{m! 2^m} \qquad (7.93)$$

if $|\gamma_c \varrho| \ll 1$ for all positive integers including zero.

We now recognise that the incident field expression must be cast into a form that involves the angular function $\cos m(\phi - \phi_0)$. The key step here is to employ the addition theorem for modified Bessel functions (see the Appendix to this chapter). In the present context eqn. 7.87 is thus written in the form

$$E_z^{inc} = E_0 \sum_{m=0}^{\infty} \hat{\varepsilon}_m I_m(\gamma \varrho) \cos m(\phi - \phi_0) \qquad (7.94)$$

where $\hat{\varepsilon}_0 = 1$ and $\hat{\varepsilon}_m = 2$ for $m = 1, 2, 3, \ldots$.

Now we find it convenient to change notation slightly and write down the following general forms for the resultant external field (i.e. $\varrho > a$):

$$E_z = E_z^{inc} + E_z^{sc}$$

$$= E_0 \sum_{m=0}^{\infty} \hat{\varepsilon}_m [I_m(\gamma \varrho) + R_m K_m(\gamma \varrho)] \cos m(\phi - \phi_0) \qquad (7.95)$$

where $R_m (= A_m/E_0 \hat{\varepsilon}_m)$ is a dimensionless coefficient. Similarly, for the internal region (i.e. $0 < \varrho < a$), we are led to write

$$E_z = E_z^{int} = E_0 \sum_{m=0}^{\infty} \hat{\varepsilon}_m T_m I_m(\gamma_c \varrho) \cos m(\phi - \phi_0) \qquad (7.96)$$

where $T_m (= B_m/E_0 \hat{\varepsilon}_m)$ is a dimensionless coefficient. As we shall see below, R_m and T_m can be interpreted as generalised reflection and transmission coefficients.

The magnetic field components can be deduced from Maxwell's equations. Thus, for $\varrho > a$, we use

$$H_\varrho = -\frac{1}{j\mu\omega} \frac{\partial E_z}{\varrho \partial \phi} \qquad (7.97)$$

$$H_\phi = \frac{1}{j\mu\omega} \frac{\partial E_z}{\partial \varrho} \qquad (7.98)$$

and, for $\varrho < a$, we use

$$H_\varrho^{int} = -\frac{1}{j\mu_c\omega} \frac{\partial E_z^{int}}{\varrho \partial \phi} \qquad (7.99)$$

$$H_\phi^{int} = \frac{1}{j\mu_c\omega} \frac{\partial E_z^{int}}{\partial \varrho} \qquad (7.100)$$

Cylindrical waves and scatter

The boundary conditions require that the tangential field components be continuous at $\varrho = a$. That is,

$$\lim_{\delta \to 0} \begin{cases} E_z]_{\varrho=a+\delta} = E_z^{\text{int}}]_{\varrho=a-\delta} & (7.101) \\ H_\phi]_{\varrho=a+\delta} = H_\phi^{\text{int}}]_{\varrho=a-\delta} & (7.102) \end{cases}$$

When these conditions are implemented we find that

$$I_m(\gamma a) + R_m K_m(\gamma a) = T_m I_m(\gamma_c a) \qquad (7.103)$$

and

$$y[I'_m(\gamma a) + R_m K'_m(\gamma a)] = T_m y_c I'_m(\gamma_c a) \qquad (7.104)$$

where

$$y = \gamma/j\mu\omega \quad \text{and} \quad y_c = \gamma_c/j\mu_c\omega$$

This algebraic pair may be solved immediately for R_m and T_m. In particular we easily deduce that

$$R_m = -\frac{I_m(\gamma a)}{K_m(\gamma a)} r_m \qquad (7.105)$$

where

$$r_m = \frac{\overset{\leftarrow}{Y}_m^{\text{int}} - \overset{\leftarrow}{Y}_m}{\overset{\leftarrow}{Y}_m^{\text{int}} + \vec{Y}_m} \qquad (7.106)$$

$$\overset{\leftarrow}{Y}_m^{\text{int}} = y_c \frac{I'_m(\gamma_c a)}{I_m(\gamma_c a)} \qquad (7.107)$$

$$\overset{\leftarrow}{Y}_m = y \frac{I'_m(\gamma a)}{I_m(\gamma a)} \qquad (7.108)$$

$$\vec{Y}_m = -y \frac{K'_m(\gamma a)}{K_m(\gamma a)} \qquad (7.109)$$

Clearly in the limits $|\gamma a|$ and $|\gamma_c a| \to \infty$, the admittance functions $\overset{\leftarrow}{Y}$ and $\vec{Y}$ approach y while $\overset{\leftarrow}{Y}^{\text{int}}$ approaches y_c. Thus we may interpret r_m as a reflection coefficient for each harmonic of order m. The half-arrows over the admittance functions signify the direction of the transmission. An attempt to illustrate the situation is shown in Fig. 7.6.

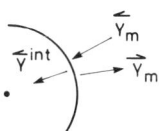

Fig. 7.6 General wave impedances at cylindrical boundary

Some limiting cases are worthy of discussion. For example, if the conductivity σ_c of the cylinder approaches infinity, while the conductivity σ remains finite, we see that $r_m \to 1$, and so

$$R_m = -I_m(\gamma a)/K_m(\gamma a) \qquad (7.110)$$

In this case the resultant field E_z vanishes at $\varrho = a$ as it must.

Another case of interest is when the radius of the cylinder is small in the sense that $|\gamma a| \ll 1$ but no restriction is placed on $\gamma_c a$. Then if we employ the small-argument approximation for the modified Bessel functions it is not difficult to show that

$$\frac{I_m(\gamma a)}{K_m(\gamma a)} \simeq (\gamma a)^{2m} \frac{1}{m!\,(m-1)!\,2^{2m-1}} \tag{7.111}$$

for $m = 1, 2, 3, \ldots$, and

$$\frac{I_0(\gamma a)}{K_0(\gamma a)} \simeq \frac{-1}{\ln(\gamma a/2) + 0.5773} \tag{7.112}$$

Because of the smallness of the right-hand side of eqn. 7.111, we are justified in considering only the $m = 0$ term in characterizing the scattered field. Furthermore, we see that the admittance functions for the $m = 0$ case are

$$\overleftarrow{Y}_0^{\text{int}} = y_c \frac{I_0'(\gamma_c a)}{I_0(\gamma_c a)} = y_c \frac{I_1(\gamma_c a)}{I_0(\gamma_c a)} = (Z_{w,c} 2\pi a)^{-1} \tag{7.113}$$

$$\overleftarrow{Y}_0 = y \frac{I_1(\gamma a)}{I_0(\gamma a)} \simeq y \frac{\gamma a}{2} = (\sigma + j\varepsilon\omega) a \tag{7.114}$$

$$\vec{Y}_0 = -y \frac{K_0'(\gamma a)}{K_0(\gamma a)} = y \frac{K_1(\gamma a)}{K_0(\gamma a)} \simeq \frac{y}{\gamma a} \frac{1}{K_0(\gamma a)} = \frac{1}{j\mu\omega a} \frac{1}{K_0(\gamma a)} \tag{7.115}$$

Now we find that eqn. 7.106 for the $m = 0$ term is well approximated by

$$r_0 \simeq \frac{1/Z_{w,c} 2\pi a}{(1/Z_{w,c} 2\pi a) + [1/j\mu\omega a K_0(\gamma a)]} \tag{7.116}$$

where

$$Z_{w,c} = \frac{\eta_c}{2\pi a} \frac{I_0(\gamma_c a)}{I_1(\gamma_c a)} \tag{7.117}$$

$$\eta_c = 1/y_c = [j\mu_c\omega/(\sigma_c + j\varepsilon_c\omega)]^{1/2}$$

is the characteristic impedance of the cylinder material.

For the case $|\gamma a| \ll 1$, we also readily deduce that eqn. 7.105 for the $m = 0$ term reduces to

$$R_0 \simeq -\frac{1}{K_0(\gamma a)} \quad r_0 \simeq -\frac{1}{\left[K_0(\gamma a) + \dfrac{2\pi}{j\mu\omega} Z_{w,c}\right]} \tag{7.118}$$

Then, without difficulty, we find from eqn. 7.95 that

$$E_z^{\text{sc}} \simeq -E_0 \frac{K_0(\gamma\varrho)}{K_0(\gamma a) + \dfrac{2\pi}{j\mu\omega} Z_{w,c}} \tag{7.119}$$

for the scattered electric field, and

$$I^{\text{ind}} \simeq 2\pi a H_\phi]_{\varrho \to a} \simeq E_0 \frac{1}{\left[Z_{w,c} + \dfrac{j\mu_c\omega}{2\pi} K_0(\gamma a)\right]} \tag{7.120}$$

is the induced current on the cylinder.

Not too surprisingly, eqns. 7.119 and 7.120 are identical in form to 7.67 and 7.70, respectively.

Exercise

Consider the same geometry as in Fig. 7.5 but now assume that the incident plane wave is polarised such that $\boldsymbol{H}^{\text{inc}} = \boldsymbol{i}_z H_z^{\text{inc}}$, where

$$H_z^{\text{inc}} = H_0 \, e^{\gamma \varrho \cos(\phi - \phi_0)} \tag{7.121}$$

Show that the scattered field is given by $\boldsymbol{H}^{\text{sc}} = \boldsymbol{i}_z H_z^{\text{sc}}$, where

$$H_z^{\text{sc}} = H_0 \sum_{m=0}^{\infty} \hat{\varepsilon}_m R_m^* K_m(\gamma \varrho) \cos m(\phi - \phi_0) \tag{7.122}$$

where

$$R_m^* = -\frac{I_m(\gamma a)}{K_m(\gamma a)} r_m^* \tag{7.123}$$

$$r_m^* = \frac{\overleftarrow{Z}_m^{\text{int}} - \overleftarrow{Z}_m}{\overleftarrow{Z}_m^{\text{int}} + \overrightarrow{Z}_m} \tag{7.124}$$

where the impedance functions are

$$\overleftarrow{Z}_m^{\text{int}} = \eta_c \frac{I'_m(\gamma_c a)}{I_m(\gamma_c a)} \tag{7.125}$$

$$\overleftarrow{Z}_m = \eta \frac{I'_m(\gamma a)}{I_m(\gamma a)} \tag{7.126}$$

$$\overrightarrow{Z}_m = -\eta \frac{K'_m(\gamma a)}{K_m(\gamma a)} \tag{7.127}$$

7.7 Scattering from a semi-cylindrical boss on ground plane

There is a class of scattering problems that can be solved exactly by image methods [1]. An excellent example is the semi-cylindrical protuberance on a plane perfectly conducting surface. The situation is illustrated in Fig. 7.7a where the cross-section is shown. The protuberance or boss is infinitely long in the z direction. With reference to a cylindrical co-ordinate system (ϱ, ϕ, z), the semi-cylinder is defined by $0 < \varrho < a$ and $0 < \phi < \pi$ and its electric and magnetic properties are $\sigma_c, \varepsilon_c, \mu_c$. The external medium is defined by $a < \varrho < \infty$ and $0 < \phi < \pi$. The whole ground plane (i.e. $0 < \varrho < \infty$ for $\phi = 0$ and π) is taken to be perfectly conducting.

In our first example we choose the incident wave (defined for $0 < \phi < \pi$) to be given by $\boldsymbol{E}^{\text{inc}} = \boldsymbol{i}_z E_z^{\text{inc}}$, where

$$E_z^{\text{inc}} = E_0 \, e^{\gamma \varrho \cos(\phi - \phi_0)} \tag{7.128}$$

This has precisely the same form as eqn. 7.87, but here $0 < \phi_0 < \pi$. We now wish to obtain the expression for the scattered field.

The key point is that the resultant electric field E_z must vanish everywhere on the ground plane. The next point to note is that, in the absence of the boss, we would have to add a field E_z^{image} to eqn. 7.87 to cancel it for the case where $\phi = 0$ and π. Clearly this image field must be given by

$$E_z^{image} = -E_0 e^{\gamma\varrho \cos(\phi+\phi_0)} \tag{7.129}$$

Now as Lord Rayleigh [1] observed many years ago, the ground plane can be removed and the other half of the cylinder filled in to yield the picture shown in Fig. 7.7b. The desired solution is thus immediate because we need only superimpose the two solutions (i.e. the solutions for two plane waves incident at angles $+\phi_0$ and $-\phi_0$ on an isolated full cylinder).

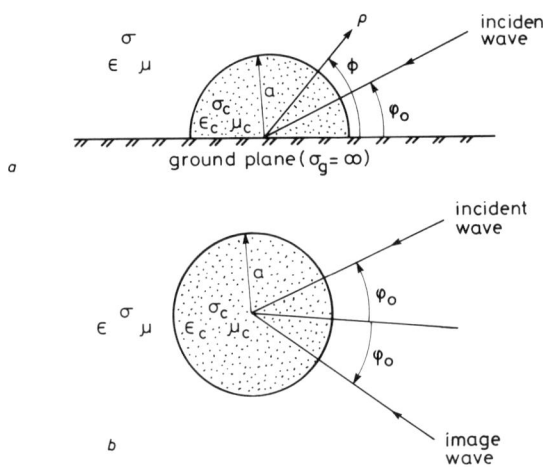

Fig. 7.7 *a* Plane wave scattering from a semi-cylindrical boss *b* The equivalent image problem

The final solution is thus given as follows, for $\varrho > a$ and $0 < \phi < \pi$:

$$E_z = E_0 [e^{\gamma\varrho\cos(\phi-\phi_0)} - e^{\gamma\varrho\cos(\phi+\phi_0)}]$$

$$+ E_0 \sum_{m=0}^{\infty} \hat{\varepsilon}_m R_m K_m(\gamma\varrho) [\cos m(\phi-\phi_0) - \cos m(\phi+\phi_0)] \tag{7.130}$$

while, for $\varrho < a$ and $0 < \phi < \pi$, we have

$$E_z = E_0 \sum_{m=0}^{\infty} \hat{\varepsilon}_m T_m I_m(\gamma_c\varrho) [\cos m(\phi-\phi_0) - \cos m(\phi+\phi_0)] \tag{7.131}$$

The coefficients R_m and T_m are precisely the same as defined by eqns. 7.105 and 7.103 respectively. The corresponding magnetic fields are obtained from eqns. 7.97 to 7.100.

The validity of the image solution can be checked by seeing that the boundary conditions are satisfied.

Cylindrical waves and scatter

Exercise

With reference to the geometry shown in Fig. 7.7a, consider the case where the incident plane wave is polarised such that $\boldsymbol{H}^{\text{inc}} = H_z \boldsymbol{i}_z$, where

$$H_z = H_0 \, e^{\gamma \varrho \cos(\phi - \phi_0)} \quad (7.132)$$

for $0 < \phi < \pi$. Show that the resultant magnetic field for $\varrho > a$ and $0 < \phi < \pi$ is given by

$$H_z = H_0 [e^{+\gamma \varrho \cos(\phi - \phi_0)} + e^{+\gamma \varrho \cos(\phi + \phi_0)}]$$

$$+ H_0 \sum_{m=0}^{\infty} \hat{\varepsilon}_m R_m^* K_m(\gamma \varrho) [\cos m(\phi - \phi_0) + \cos m(\phi + \phi_0)] \quad (7.133)$$

where R_m^* is given by eqn. 7.123. Also show that the field inside the cylindrical boss, for $\varrho < a$ and $0 < \phi < \pi$, is given by

$$H_z = H_0 \sum_{m=0}^{\infty} \hat{\varepsilon}_m T_m^* I_m(\gamma_c \varrho) [\cos m(\phi - \phi_0) + \cos m(\phi + \phi_0)] \quad (7.134)$$

where T_m^* is given by a form analogous to eqn. 7.103. Finally, confirm that the boundary conditions are satisfied: that is, H_z and E_ϕ are continuous at $\varrho = a$ and E_ϱ is zero for $\phi = 0$ and $\phi = \pi$ for all $\varrho < \infty$.

7.8 Application to scattering of a radio ground wave by a ridge

The semi-cylindrical boss can be employed as a model to study the influence of a ridge for a radio ground wave [10, 11]. We will examine explicitly a very idealistic case. With reference to Fig. 7.8, we assume that the semi-cylinder representing the ridge is perfectly conducting. The ground plane is also perfectly conducting, which is a prerequisite for the application of image theory in the present context. The objective is now to deduce the resultant magnetic field both on the ridge and on the ground surface for an incident plane wave polarised with H parallel to the axis of the ridge. Furthermore, the incoming wave is at grazing incidence on the ground plane.

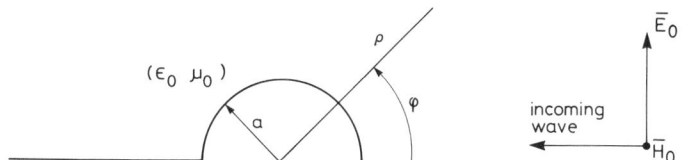

Fig. 7.8 Geometry for incoming (ground) wave and semi-cylindrical model of ridge

The problem, as posed, is really a special case of the configuration shown in Fig. 7.7a when $\sigma_c \to \infty$ and $\phi_0 \to 0$ for the case where H^{inc} is z directed. Thus eqn. 7.134 is applicable with the appropriate substitutions. However, we will examine the problem afresh for pedagogical reasons. Also, we deal with a free space environment, so $\gamma = jk$ where $k = (\varepsilon_0 \mu_0)^{1/2} \omega$ in terms of the electrical properties ε_0 and μ_0. The incoming or primary wave is clearly given by

$$H_z^{\text{pr}} = H_0 \, e^{jk\varrho \cos \phi} \quad (7.135a)$$

which is the total field in the absence of the semi-cylindrical ridge. Here H_0 is a constant reference field. The resultant field is now to be determined and, in particular, we wish to examine the quantitative aspects for a ridge of wavelength dimensions.

Using the addition theorem for the (garden variety) Bessel functions, we now re-express eqn. 7.135 in the form

$$H_z^{\text{pr}} = \bar{H}_0 \sum_{m=0}^{\infty} \hat{\varepsilon}_m e^{jm\pi/2} J_m(k\varrho) \cos m\phi \qquad (7.135b)$$

where $\hat{\varepsilon}_0 = 1$, $\hat{\varepsilon}_m = 2$ for $m = 1, 2, 3, \ldots$. Here $J_m(k\varrho)$ is related to $I_m(jk\varrho)$ differing only by a constant.

The total scattering or secondary magnetic field is now found to be

$$H_z^{\text{sec}} = H_0 \sum_{m=0}^{\infty} \hat{\varepsilon}_m P_m H_m^{(2)}(k\varrho) \cos m\phi \qquad (7.136)$$

where $H_m^{(2)}(k\varrho)$ is the Hankel function of the second type, which is related to $K_m(jk\varrho)$, differing only by a constant. The coefficient P_m is determined by imposing the boundary condition that $E_\phi^{\text{pr}} + E_\phi^{\text{sec}}$ vanishes at $\varrho = a$ for $0 < \phi < \pi$. An equivalent statement is

$$\frac{\partial}{\partial \varrho}(H_z^{\text{pr}} + H_z^{\text{sec}})|_{\varrho=a} = 0 \qquad (7.137)$$

for $0 < \phi < \pi$. Invoking this condition, we deduce immediately that

$$P_m = -[J_m'(ka)/H_m^{(2)\prime}(ka)] e^{jm\pi/2} \qquad (7.138)$$

We also readily confirm that $E_\varrho^{\text{pr}} + E_\varrho^{\text{sec}}$ vanishes at $\varrho > a$ for $\phi = 0$ and π. For example, note that

$$\frac{\partial}{\partial \phi}(H_z^{\text{pr}} + H_z^{\text{sec}})|_{\phi=0,\pi} = 0 \qquad (7.139)$$

for $\varrho > a$. Why? Because $\sin m\pi = 0$ for all integers m.

A useful and meaningful parameter for this problem is the *relative amplitude*, defined by the ratio

$$|H_z^{\text{pr}} + H_z^{\text{sec}}|/\bar{H}_0 \qquad (7.140)$$

where $\bar{H}_0$ is assumed to be real. A related quantity is the *phase lag*, defined by

$$k\varrho \cos \phi - \arg \frac{H_z^{\text{pr}} + H_z^{\text{sec}}}{\bar{H}_0} \qquad (7.141)$$

In words, the relative amplitude is the magnitude of the total magnetic field for an incoming wave of unit amplitude. The phase lag, as defined above, is the argument of the incoming wave field minus the argument of the total field. The amplitude and phase so described are plotted in Figs. 7.9a and 7.9b respectively for points on the semi-cylinder (i.e. $\varrho = a$) for $0 < \phi < 180°$. Various values of ka (= radius $\times 2\pi/$ wavelength) are indicated. As can be seen, the amplitude of the magnetic field is approaching 2·0 on the illuminated side of the cylinder. On the other hand, the amplitude is decreased below 1·0 on the shadow side. Also we see that the phase lag is negative (i.e. relative phase lead) on the illuminated side. But the phase lag rises sharply on the shadow

Cylindrical waves and scatter

side. Actually, geometrical optics would predict an amplitude of 2·0 and zero phase lag for $0 < \phi < 90°$ on the surface of the semi-cylinder, whereas it would predict zero or vanishing field in the shadow region. The amplitude curves in Fig. 7.9a are beginning to show this behaviour for $ka = 3$. Actually, as ka becomes very large (e.g. $ka > 10$) the geometrical optical description is appropriate but there are still problems in the transition region (i.e. at $\phi = 90°$). Further discussion of this topic is deferred to a later chapter.

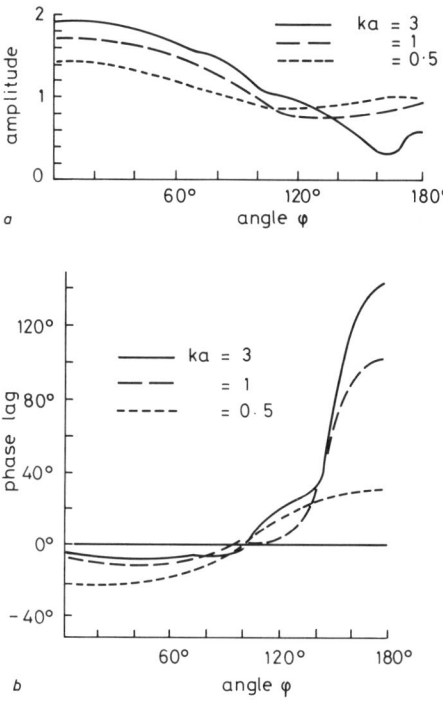

Fig. 7.9 *a* Amplitude and *b* phase of magnetic field on boss

The amplitude (normalised as indicated) on the ground plane in front of the boss ($\varrho > a$, $\phi = 0°$) is shown in Fig. 7.10a for $ka = 0.5$, 1 and 3. The abscissa is the distance measured from the centre of the boss and expressed in radians. Here it is evident that the secondary or scattered wave is interfering with the incoming wave to form a standing wave pattern. As would be expected, the period of the oscillation is approximately π radians or an actual linear distance of one-half of a wavelength. The phase lag also shows an oscillatory pattern, but we do not show the results here.

The amplitude on the ground plane to the rear of the boss ($\varrho > a$, $\phi = 180°$) is shown in Fig. 7.10b. As indicated, the field very close to the boss is reduced significantly, but as we move away the field recovers to its original state.

Exercise

Instead of the perfectly conducting assumption for the boss, assume that $E_\phi = -Z_s H_z$ for $\varrho = a$ and $0 < \phi < \pi$, where Z_s is a fixed surface impedance. The ground plane remains perfectly conducting, and the other conditions of the problem are the same. Using this so-called Leontovich boundary condition, show that the expression for the secondary field is now given by

$$H_z^{sec} = -\bar{H}_0 \sum_{m=0}^{\infty} \hat{\varepsilon}_m e^{jm\pi/2} \frac{J'_m(x) + GJ_m(x)}{H_m^{(2)'}(x) + GH_m^{(2)}(x)} H_m^{(2)}(k\varrho) \cos m\phi \tag{7.142}$$

where

$$G = -jZ/120\pi \tag{7.143}$$

Note that $Z/120\pi \simeq (1 + j)(\varepsilon_0 \omega/2\sigma_c)^{1/2}$ if $\sigma_c \gg \varepsilon_c \omega$ in terms of the conductivity σ_c and permittivity ε_c of the boss material. Thus $G \simeq (1 - j)(\varepsilon_0 \omega/2\sigma_c)^{1/2}$.

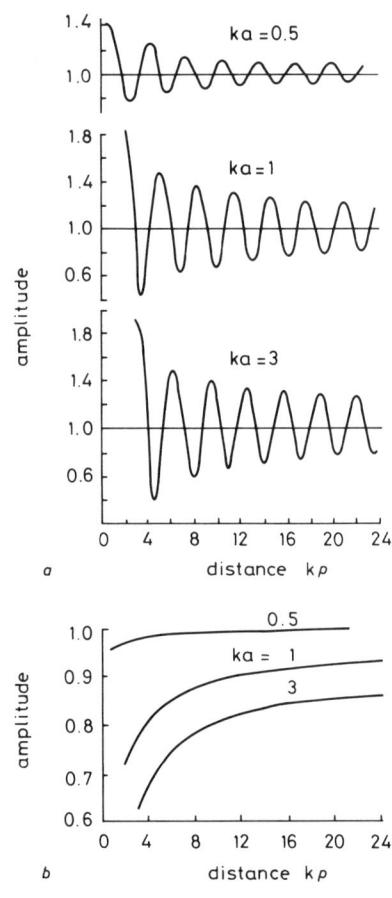

Fig. 7.10 Amplitude of magnetic field a in front and b at rear of boss

7.9 Scattering by two parallel wires

There are many important situations where the scattering takes place in a multiplicative fashion [5, 6, 12]. A good example is when the electromagnetic wave impinges on two parallel thin cylinders. To deal

Cylindrical waves and scatter

with such problems, the mathematics becomes quite difficult. Our purpose here is to employ a relatively simple model that permits an understandable solution. The interested reader is referred to a very general analysis [13] if the present approach is not to his liking.

The first case we deal with is illustrated in Fig. 7.11. A cylindrical conductor or wire of radius a_1 is coaxial with the z axis of the (ϱ, ϕ, z) system. A cylindrical conductor or wire of radius a_2 is centred at $\varrho = d$ and $\phi = 0$. The radii a_1 and a_2 are assumed to be small compared with the separation d.

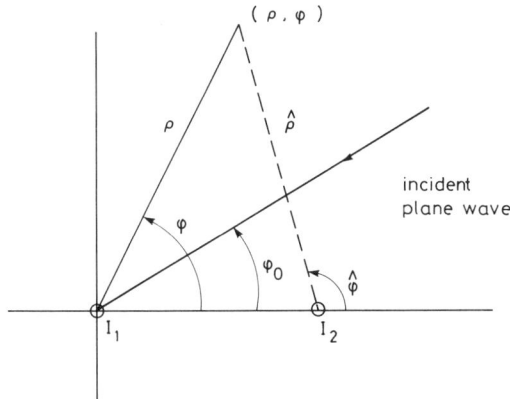

Fig. 7.11 Geometry for scattering of a normally incident plane wave by two parallel wires

We now specify that a plane wave with the electric vector E^{inc}, parallel to the two wires, is incident at an angle ϕ_0 with respect to the $\phi = 0$ plane. Thus the problem is entirely two dimensional (i.e. $\partial/\partial z = 0$). Again assuming that the surrounding medium is homogeneous, with properties σ, ε and μ, we can write

$$E_z^{inc} = E_0 \, e^{\gamma \varrho \cos(\phi - \phi_0)} \tag{7.144}$$

where

$$\gamma = [j\mu\omega(\sigma + j\varepsilon\omega)]^{1/2}$$

This incident field will now induce axial currents I_1 and I_2 on wire 1 and wire 2, respectively.

It is evident that the secondary or scattered field is merely the superposition of the individual contributions from the wires. Thus we see that

$$E_z^{sc} = -\frac{j\mu\omega}{2\pi} [I_1 K_0(\gamma\varrho) + I_2 K_0(\gamma\hat{\varrho})] \tag{7.145}$$

where

$$\hat{\varrho} = (\varrho^2 + d^2 - 2\varrho d \cos \phi)^{1/2}$$

and where we have made use of eqn. 7.43.

Now at this stage we do not know I_1 and I_2. To illustrate the procedure to determine these induced currents, we assume that the wires are perfectly conducting. The requisite boundary conditions are that E_z^{inc} and E_z^{sc} should vanish at $\varrho = a_1$ and $\varrho = (d^2 + a_2^2 + 2da_2 \cos \hat{\phi})^{1/2}$, where $0 < \hat{\phi} < 2\pi$. The latter condition can be replaced by $\varrho = d$ because we have already assumed $d \gg a_2$ and certainly $|\gamma a_2| \ll 1$. Furthermore, we note that when $\varrho = a_1$ we must have $\hat{\varrho} \simeq d$ because $d \gg a_1$ and $|\gamma a_1| \ll 1$.

On applying the boundary conditions as simplified above, we obtain the following two algebraic equations:

$$E_0 - \frac{j\mu\omega}{2\pi}[I_1 K_0(\gamma a_1) + I_2 K_0(\gamma d)] = 0 \quad (7.146)$$

$$E_0 e^{-\gamma d \cos \phi_0} - \frac{j\mu\omega}{2\pi}[I_1 K_0(\gamma d) + I_2 K_0(\gamma a_2)] = 0 \quad (7.147)$$

Solving for I_1 and I_2 gives

$$I_1 = E_0 \frac{2\pi}{j\mu\omega}\left[\frac{K_0(\gamma a_2) - e^{-\gamma d \cos \phi_0} K_0(\gamma d)}{K_0(\gamma a_1)K_0(\gamma a_2) - K_0^2(\gamma d)}\right] \quad (7.148)$$

$$I_2 = E_0 \frac{2\pi}{j\mu\omega}\left[\frac{K_0(\gamma d) - e^{-\gamma d \cos \phi_0} K_0(\gamma a_1)}{K_0^2(\gamma d) - K_0(\gamma a_1)K_0(\gamma a_2)}\right] \quad (7.149)$$

To interpret the above results it is useful to rewrite the formulas in the following form:

$$I_1 \simeq I_1^{(0)}\left[1 - \frac{K_0(\gamma d)}{K_0(\gamma a_2)} e^{-\gamma d \cos \phi_0}\right](1 - \chi)^{-1} \quad (7.150)$$

$$I_2 \simeq I_2^{(0)}\left[1 - \frac{K_0(\gamma d)}{K_0(\gamma a_1)} e^{-\gamma d \cos \phi_0}\right](1 - \chi)^{-1} \quad (7.151)$$

where

$$I_1^{(0)} = E_0 \frac{2\pi}{j\mu\omega} \frac{1}{K_0(\gamma a_1)} \quad (7.152)$$

$$I_2^{(0)} = E_0 \frac{2\pi}{j\mu\omega} \frac{1}{K_0(\gamma a_2)} e^{-\gamma d \cos \phi_0} \quad (7.153)$$

$$\chi = \frac{K_0^2(\gamma d)}{K_0(\gamma a_1)K_0(\gamma a_2)} \quad (7.154)$$

It is clear that $I_1^{(0)}$ and $I_2^{(0)}$ are the expressions for the induced currents that one would derive if all interactions between the wires were neglected. Then, if for the moment we ignored the $(1 - \chi)^{-1}$ factor on the right-hand sides of eqns. 7.150 and 7.151, we can interpret the square bracket factors as first-order interactions between the wire currents. Then we may assert that the factor $(1 - \chi)^{-1}$ represents the totality of the higher-order interactions. In fact, we may expand the factor as a geometrical series as follows:

$$(1 - \chi)^{-1} = 1 + \chi + \chi^2 + \chi^3 + \ldots \quad (7.155)$$

where, for example, the χ^p term is the contribution of the $2p$th-order interaction.

Employing the expressions for I_1 and I_2 given by eqns. 7.150 and 7.151, respectively, in 7.145 we can compute the resultant scattered field at any point $P(\varrho, \phi)$. All interactions are considered implicitly in such a scheme. However, we are free to expand the secondary field into zero-, first-, second- and higher-order scatter contributions. Schematically the sequence is illustrated in Fig. 7.12, where zero-, first- and second-order contributions are shown with reference to wire 1.

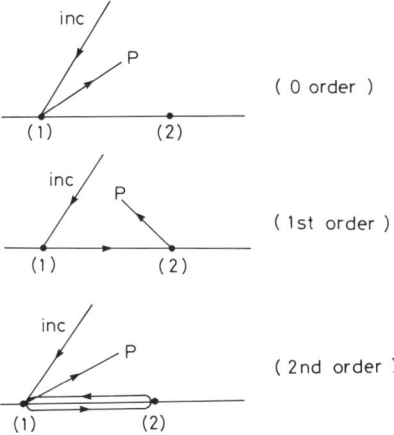

Fig. 7.12 Ray interpretation of plane wave scattering from parallel wires showing typical zero order and first- and second-order multiple scatter contributions

From a physical standpoint, it is not difficult to see that the higher-order interaction or scattering is decreased when the interwire separation d is increased. Also, from a qualitative standpoint it is not difficult to see that the higher-order interaction or scattering is decreased when the interwire separation d is increased. Also from a quantitative standpoint, we note that the higher-order or multiple scatter is negligible if χ, as defined by eqn. 7.154, has a small magnitude (i.e. $|\chi| \ll 1$). To exhibit this condition we assume that $|\gamma d| \gg 1$ and note that $|\gamma a_1|$ and $|\gamma a_2| \gg 1$, as already specified. Then the condition for the justified neglect of multiple scatter is

$$\left| \frac{\pi}{2\gamma d} e^{-2\gamma d} \frac{1}{\left(\ln \frac{\gamma a_1}{2} - C\right)\left(\ln \frac{\gamma a_2}{2} - C\right)} \right| \ll 1 \qquad (7.156)$$

In the case of a lossless medium $\gamma = jk = 2\pi/\lambda_0$, the inequality can be effectively replaced by $kd \gg 1$. In lossy media the condition is less stringent.

7.10 Scattering from a wire located over a plane conductor

A closely related problem that yields to the above approach is scattering from a thin conductor or wire of radius a located over a perfectly conducting plane. The situation is illustrated in Fig. 7.13, where the ground plane is located at $y = 0$ and the overhead wire (of assumed infinite length) is located at $y = h$. With reference to the Cartesian co-ordinate system (x, y, z), the incident wave is

$$E_z^{inc} = E_0 \, e^{\gamma x \cos \phi_0} \, e^{\gamma y \sin \phi_0} \tag{7.157}$$

where ϕ_0 is the angle subtended by the wave normal and the ground plane.

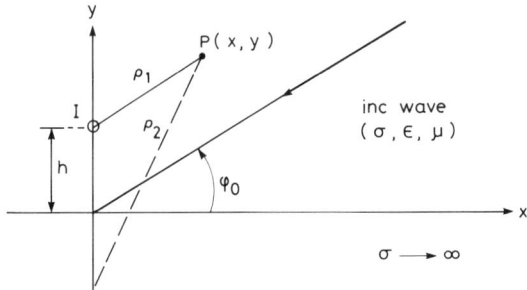

Fig. 7.13 Scattering of a plane wave from an overhead wire parallel to a perfectly conducting ground plane

The filamental current on the wire is designated I, which as yet is not known. But now we can see that field E_z^{sc} scattered from the wire must have the form

$$E_z^{sc} = -\frac{j\mu\omega}{2\pi} I[K_0(\gamma \varrho_1) - K_0(\gamma \varrho_2)] \tag{7.158}$$

where

$$\varrho_1 = [x^2 + (y - h)^2]^{1/2}$$

$$\varrho_2 = [x^2 + (y + h)^2]^{1/2}$$

Clearly we may verify that

$$\left(\frac{\partial^2}{\partial x^2} + \frac{\partial^2}{\partial y^2} - \gamma^2 \right) E_z^{sc} = 0 \tag{7.159}$$

and also note that $E_z^{sc} = 0$ at $y = 0$. Thus we may interpret E_z^{sc} as the field of a line source at $y = h$ and an image line source, of opposite sign, at $y = -h$.

To complete the picture, we need to add a specular reflected field E_z^{refl}, for $y > 0$, given by

$$E_z^{refl} = -E_0 \, e^{\gamma x \cos \phi_0} \, e^{-\gamma y \sin \phi_0} \tag{7.160}$$

which, of course, is present even if the thin wire were not.

Now we need to find I. This task is accomplished by requiring that the total field E_z vanishes at the surface of the wire (which we assume is perfectly conducting). An explicit statement is

$$(E_z^{inc} + E_z^{refl} + E_z^{sc}) = 0 \tag{7.161}$$

for $\varrho_1 = a$. Because we can assume that $h \gg a$, it is permissible to make the following substitutions in the boundary conditions:

$$(E_z^{inc} + E_z^{refl})_{\substack{x=0 \\ y=h}} + (E_z^{sc})_{\substack{\varrho_1 = a \\ \varrho_2 = 2h}} = 0 \tag{7.162}$$

Then, on using eqns. 7.157, 7.160 and 7.158 together with 7.162, it follows that

$$E_0(e^{\gamma h \sin \phi_0} - e^{-\gamma h \sin \phi_0}) - \frac{j\mu\omega}{2\pi} I[K_0(\gamma a) - K_0(2\gamma h)] = 0 \quad (7.163)$$

Then, of course,

$$I = \frac{2\pi}{j\mu\omega} E_0 \left[\frac{e^{\gamma h \sin \phi_0} - e^{-\gamma h \sin \phi_0}}{K_0(\gamma a) - K_0(2\gamma h)} \right] \quad (7.164)$$

is the desired expression for the induced current.

In the case of a lossless medium $\gamma = jk$, where k is real, we may write

$$I = \frac{4\pi}{\mu\omega} E_0 \left[\frac{\sin(kh \sin \phi_0)}{K_0(jka) - K_0(2jkh)} \right] \quad (7.165)$$

Not surprisingly this result shows that the induced current is zero if $kh \sin \phi_0 = m\pi$ (for $m = 0, 1, 2, \ldots$), in which case the incident and reflected field are cancelling and forming a null at the wire location.

There is a nice simplification of eqn. 7.165 when $2kh \ll 1$, corresponding to a small electrical height of the elevated wire. Then, using the small-argument approximation for both K_0 functions, we find that

$$I \simeq \frac{4\pi}{\mu\omega} E_0 \frac{\sin(kh \sin \phi_0)}{\ln(2h/a)}$$

$$\simeq \frac{4\pi}{\mu\omega} E_0 \, kh \, [\ln(2h/a)]^{-1} \sin \phi_0 \quad (7.166)$$

while, for free space,

$$I \simeq (E_0 h/30) [\ln(2h/a)]^{-1} \sin \phi_0 \quad (7.167)$$

There is a very simple extension of the foregoing analysis to the case where the overhead wire has a finite conductivity or is otherwise modified so that the tangential electric field is nonzero. We can deal with this problem by using a surface impedance boundary condition such as stated by eqn. 7.82. What this amounts to is that the right-hand side of eqn. 7.162 is replaced by IZ_w, where Z_w is the series impedance of the wire per unit length. Then it is a simple matter to show that eqn. 7.164 is modified to

$$I = E_0 \frac{e^{\gamma h \sin \phi_0} - e^{-\gamma h \sin \phi_0}}{Z_w + \frac{j\mu\omega}{2\pi} [K_0(\gamma a) - K_0(2\gamma h)]} \quad (7.168)$$

In the case of a free space environment and for small wire heights (i.e. $kh \ll 1$), we see that

$$I \simeq E_0 \frac{2jkh \sin \phi_0}{Z_w + \frac{j\mu\omega}{2\pi} \ln \frac{2h}{a}} \quad (7.169)$$

The numerators of the expressions above can be identified as the effective voltage driving the current I, and the denominators are equivalent impedances.

7.11 Line source excitation of a homogeneous cylinder

An intermediate solution, which we need for future use, is for a uniform line source in the vicinity of a cylinder of any radius. The situation is illustrated in Fig. 7.14, where we have located a homogeneous cylinder of radius a to be coaxial with the cylindrical co-ordinate system (ϱ, ϕ, z). A uniform line source of current I is located at (ϱ_1, ϕ_1), where $\varrho_1 > a$. The external region $\varrho > a$ is homogeneous with properties σ, ε and μ, and the internal region is also homogeneous with properties σ_c, ε_c and μ_c. As indicated, the problem is entirely two dimensional and thus $\partial/\partial z = 0$ throughout.

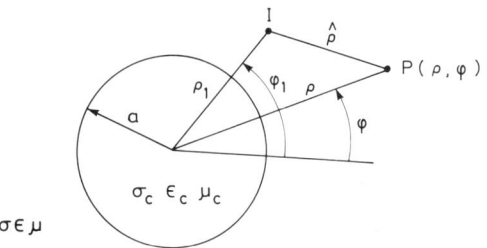

Fig. 7.14 Line source excitation of homogeneous cylinder

Now at the observer point $P(\varrho, \phi)$ we may write the primary electric field that has only a z component as follows:

$$E_z^{\text{prim}} = -\frac{j\mu\omega}{2\pi} I K_0(\gamma\hat{\varrho}) \qquad (7.170)$$

where

$$\hat{\varrho} = [\varrho^2 + \varrho_1^2 - 2\varrho\varrho_1 \cos(\phi - \phi_1)]^{1/2}$$

Clearly eqn. 7.170 has the same form as 7.43. But now we wish to re-express eqn. 7.170 with respect to the origin at $\varrho = 0$, as indicated in Fig. 7.14. This task is accomplished by making use of an addition theorem for modified Bessel functions, so that we may write

$$E_z^{\text{prim}} = -\frac{j\mu\omega}{2\pi} I \sum_{m=0}^{\infty} \hat{\varepsilon}_m K_m(\gamma\varrho_1) I_m(\gamma\varrho) \cos m(\phi - \phi_1) \quad (7.171)$$

which is applicable for $a < \varrho < \varrho_1$.

The problem is now seen to be closely analogous to scattering of a plane wave from the cylinder. Then, taking a lead from the form given by eqns. 7.91 and 7.92, we surmise that the secondary field E_z^{sc}, for $\varrho > a$, should have the form

$$E_z^{\text{sc}} = -\frac{j\mu\omega}{2\pi} I \sum_{m=0}^{\infty} \hat{\varepsilon}_m K_m(\gamma\varrho_1) R_m K_m(\gamma\varrho) \cos m(\phi - \phi_1) \qquad (7.172)$$

and the internal field E_z for $\varrho < a$, would have the form

$$E_z = -\frac{j\mu\omega I}{2\pi} \sum_{m=0}^{\infty} \hat{\varepsilon}_m K_m(\gamma\varrho_1) T_m I_m(\gamma_c\varrho) \cos m(\phi - \phi_1) \qquad (7.173)$$

The definitions of the coefficients R_m and T_m are given precisely by eqns. 7.105 and 7.103, respectively.

7.12 Application to subsurface radio

There is an interesting application of the present formulation to subsurface radio communication and source location [14, 15]. To this end we are interested in the fields inside the cylinder as given by eqn. 7.173. The coefficient T_m is given explicitly by

$$T_m = -\{\gamma a[I_m(\gamma_c a)K'_m(\gamma a) - (\gamma_c/\gamma)I'_m(\gamma_c a)K_m(\gamma a)]\}^{-1} \quad (7.174)$$

where we have assumed that $\mu_c = \mu = \mu_0$; thus

$$\gamma_c = [j\mu_0\omega(\sigma_c + j\varepsilon_c\omega)]^{1/2}$$

$$\gamma = [j\mu_0\omega(\sigma + j\varepsilon\omega)]^{1/2}$$

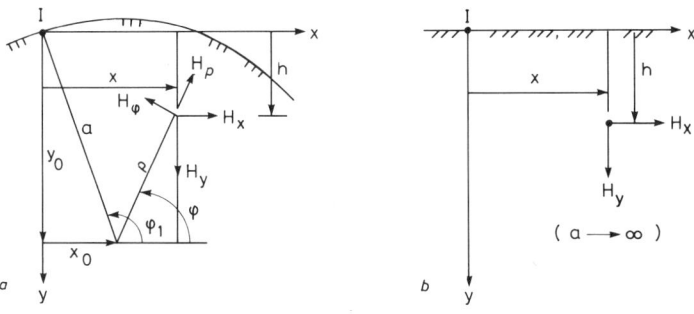

Fig. 7.15 *a* Line source located on a cylindrical topographic feature *b* Idealised half-space model showing location of subsurface observer

Here we have also made use of the Wronskian relationship

$$K_m(Z)I'_m(Z) - K'_m(Z)I_m(Z) = 1/Z \quad (7.175)$$

We now apply our exact formulation to the special case where $\varrho_1 = a$ so that the line source is on the cylindrical surface. The situation is illustrated in Fig. 7.15a, where we also show a rectangular co-ordinate system centred at the line source and positioned such that (x_0, y_0) is the $\varrho = 0$ axis in the cylindrical system. Now we are principally interested in the magnetic field at the subsurface point (x, h) with reference to the rectangular co-ordinate system. As indicated in Fig. 7.15b, the half-space or 'flat earth' model is achieved in the limit $a \to \infty$. The practical question is: how do the field components H_x and H_y, for a fixed x and h, depend on the radius of curvature? To simplify the calculations we make the following approximations, which are certainly justified at audio frequencies and distances of the order of 100 m. Specifically $|\gamma a| \ll 1$, which in effect is saying that a is much less than a wavelength in the outer region (i.e. the air). Now we use the small-argument limiting forms to show that

$$K'_m(Z)/K_m(Z) \simeq -m/Z \quad (7.176)$$

and thus eqn. 7.173 is simplified to

$$E_z = -\frac{j\mu_0\omega I}{2\pi}\sum_{m=0}^{\infty}\hat{\varepsilon}_m\frac{I_m(\gamma_c\varrho)\cos m(\phi - \phi_1)}{mI_m(\gamma_c a) + \gamma_c aI'_m(\gamma_c a)} \quad (7.177)$$

The corresponding expressions for the magnetic fields are

$$H_\varrho = -\frac{1}{j\mu_0\omega\varrho}\frac{\partial E_z}{\partial \phi}$$

$$= \frac{I}{2\pi\varrho}\sum_{m=0}^{\infty}\hat{\varepsilon}_m \frac{mI_m(\gamma_c\varrho)\sin m(\phi-\phi_1)}{mI_m(\gamma_c a)+\gamma_c aI'_m(\gamma_c a)} \quad (7.178)$$

$$H_\phi = \frac{1}{j\mu_0\omega}\frac{\partial E_z}{\partial \varrho}$$

$$= -\frac{I}{2\pi\varrho}\sum_{m=0}^{\infty}\hat{\varepsilon}_m \frac{(\gamma_c\varrho)I'_m(\gamma_c\varrho)\cos m(\phi-\phi_1)}{mI_m(\gamma_c a)+(\gamma_c a)I'_m(\gamma_c a)} \quad (7.179)$$

Now, in the context of the geometry shown in Fig. 7.15a we are interested in the rectangular components H_x and H_y. Clearly

$$H_x = H_\varrho \cos\phi - H_\phi \sin\phi \quad (7.180)$$

$$H_y = -H_\varrho \sin\phi - H_\phi \cos\phi \quad (7.181)$$

To simplify the presentation of numerical data, we define dimensionless parameters

$$A = -(2\pi h/I)H_x \quad (7.182)$$

$$B = (2\pi/I)H_y \quad (7.183)$$

When we apply this model to the earth where the cylinder is a typical topographic feature, we can neglect displacement currents (i.e. $\varepsilon_c\omega \ll \sigma_c$), in which case $\gamma_c \simeq |\gamma_c|e^{j\pi/4}$ where $|\gamma_c| \simeq (\sigma_c\mu_0\omega)^{1/2}$. Then the Bessel function I_m can be represented conveniently in terms of ber and bei functions (see the Appendix to this chapter) in the manner

$$I_m(Z\, e^{j\pi/4}) = e^{-jm\pi/2}(\text{ber}_m Z + j\,\text{bei}_m Z) \quad (7.184)$$

where Z is real.

We show some typical numerical calculations of the response functions $|A|$ and $|B|$ in Figs. 7.16a, b and c, plotted as a function of x/h. The following dimensionless parameters are employed: $D = (\sigma_c\mu_0\omega)^{1/2}h$, $R = |\gamma_c a| = (\sigma_c\mu_0\omega)^{1/2}a$. We choose $R = 5$, 10 and ∞. For the curves in Fig. 7.16a we have $x_0 = 0$ and $D = 0\cdot2$; in Fig. 7.6b we have $x_0 = 0\cdot5h$ and $D = 0\cdot$; and finally in Fig. 7.16c we have $x_0 = 0$ and $D = 1\cdot0$. The qualitative behaviour of the curves is similar in all three cases. The effect of finite curvature of the air/ground interface is to increase the level of the normalised horizontal field (i.e. $|A|$), whereas the maximum response of the vertical field (i.e. $|B|$) is lowered somewhat. Actually, the curves designated $R = \infty$ are based on the computed response for a conducting half-space model as described elsewhere [15]. As we see, if the radius of curvature is sufficiently large it appears that the results for the cylinder approach those for the half-space. This is a reassuring check on the numerical data since the method of calculation is vastly different in the two cases.

To deal with a specific example, we choose the ground conductivity

Cylindrical waves and scatter 181

σ_c to be 1 mS/m and the operating frequency to be 100 Hz. Then the parameter $D = 0{\cdot}20$ if the observer depth $h = 225$ m, and the parameter $D = 11{\cdot}0$ if $h = 1{\cdot}12$ km. Also we note that, for this example, if $R = 10$ we have $a = 11.2$ km. These specific examples would correspond to moderately rough terrain features.

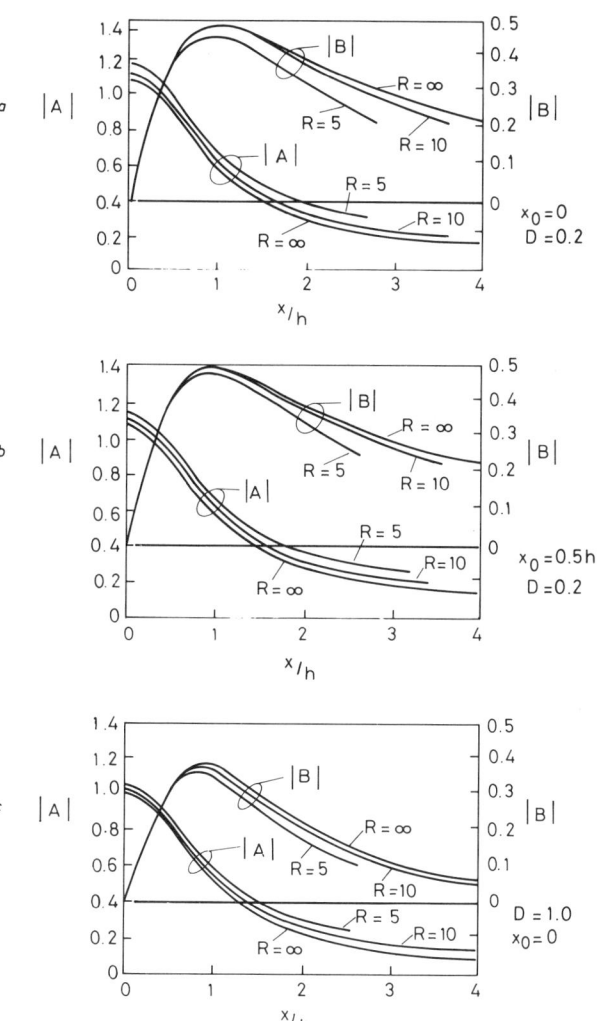

Fig. 7.16 Examples of how the subsurface magnetic field components are modified by the radius of curvature of the air/ground interface $R = \infty$ corresponds to a half-space model

7.13 Scattering of a plane wave from a cylinder and parallel wire combination

In our earlier development for scattering from two thin cylinders, we assumed that radii were both sufficiently small that only axially symmetric currents on each 'wire' needed to be considered. Here we outline a solution for the case where one of the cylinders can have any radius. Again we restrict attention to a purely two-dimensional situation.

The model we adopt is shown in Fig. 7.17. A cylinder with properties

σ_c, ε_c and μ_c, of radius a, is coaxial with the co-ordinate system (ϱ, ϕ, z). A thin cylinder or 'wire' is located at $\varrho = d$ and $\phi = 0$, and it is characterised by a series impedance Z_w per unit length. The incident plane wave is polarised with the electric vector parallel to the axes of the cylinders. Thus we write

$$E_z^{inc} = E_0 \, e^{\gamma \varrho \cos(\phi - \phi_0)} \tag{7.185}$$

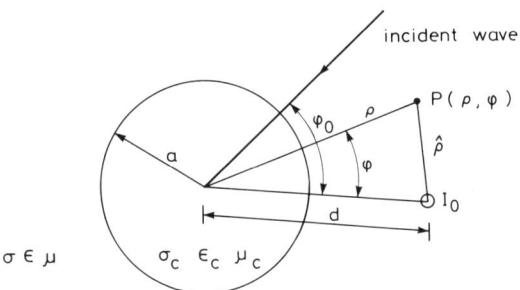

Fig. 7.17 Geometry for scattering of a plane wave from a cylindrical parallel wire combination

Now in the absence of the thin wire conductor, we would have a scattered field E_{z0}^{sc} given, in accordance with eqn. 7.95, by

$$E_{z0}^{sc} = E_0 \sum_{m=0}^{\infty} \hat{\varepsilon}_m R_m K_m(\gamma \varrho) \cos m(\phi - \phi_0) \tag{7.186}$$

where R_m is given by eqn. 7.105. Then we need to recognise that a current I_0, yet to be determined, is induced in the wire at $\varrho = d, \phi = 0$. The resultant field E_z^w produced by this induced current, in the presence of the big cylinder, is given by the general result of the preceding section. Thus, using eqn. 7.172, we write

$$E_z^w = -\frac{j\mu\omega I_0}{2\pi}\left[K_0(\gamma\hat{\varrho}) + \sum_{m=0}^{\infty} \hat{\varepsilon}_m K_m(\gamma d) R_m K_m(\gamma \varrho) \cos m\phi\right] \tag{7.187}$$

where

$$\hat{\varrho} = (\varrho^2 + d^2 - 2\varrho d \cos\phi)^{1/2}$$

Now we invoke the boundary condition on the thin wire which reads

$$(E_z^{inc} + E_{z0}^{sc} + E_z^w)|_{\hat{\varrho}=a_0} = I_0 Z_w \tag{7.188}$$

Invoking the condition that $a_0 \ll d$, we deduce that eqn. 7.188 is given by

$$E_0\left[e^{\gamma d \cos\phi_0} + \sum_{m=0}^{\infty} \hat{\varepsilon}_m R_m K_m(\gamma d) \cos m\phi_0\right]$$

$$- \frac{j\mu\omega I_0}{2\pi}\left[K_0(\gamma a_0) + \sum_{m=0}^{\infty} \hat{\varepsilon}_m R_m [K_m(\gamma d)]^2\right] = I_0 Z_w \tag{7.189}$$

Solving this equation for I_0:

$$I_0 = \frac{E_0\left[e^{\gamma d \cos\phi_0} + \sum_{m=0}^{\infty} \hat{\varepsilon}_m R_m K_m(\gamma d) \cos m\phi_0\right]}{Z_w + \dfrac{j\mu\omega}{2\pi}\left[K_0(\gamma a_0) + \sum_{m=0}^{\infty} \hat{\varepsilon}_m R_m K_m^2(\gamma d)\right]} \quad (7.190)$$

The numerator here is the total applied electric field at the wire and the denominator is the total circuit impedance.

7.14 Scattering by two parallel wires for oblique incidence

As we indicated earlier, oblique incidence of a plane wave on a thin conducting wire could be handled as a straightforward extension of the normal incidence case. The geometry we treated earlier is shown in Fig. 7.4. But now we wish to generalise this situation to allow for the presence of two parallel thin wires of radii a_1 and a_2. The surrounding medium is assumed to be free space. The configuration is shown in Fig. 7.8 with reference to a rectangular co-ordinate system. The wires are located at $x = 0$ and $x = d$ and contained in the $y = 0$ plane. Because of the assumed thinness of the wires, only the z component E_z^{inc} of the incident plane wave interacts with the wires. It is specified to be

$$E_z^{\text{inc}} = E_{0z}\, e^{jkx\cos\phi_0 \sin\theta_0}\, e^{jky\sin\phi_0 \sin\theta_0}\, e^{jkz\cos\theta_0} \quad (7.191)$$

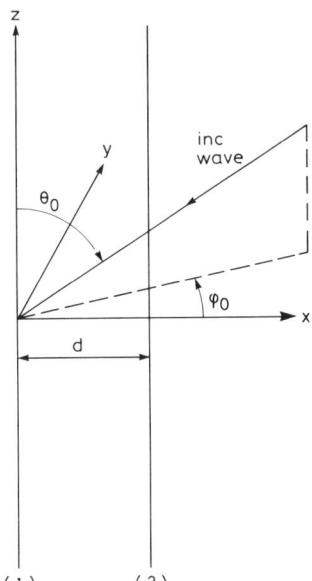

Fig. 7.18 Scattering of a plane wave at oblique incidence from two parallel wires

where θ_0 and ϕ_0 are the polar and azimuthal angles respectively, as indicated in Fig. 7.18, and where E_{0z} is the value of the incident field referred to the origin of the (x, y, z) co-ordinate system. Now eqn. 7.191 can be derived from an electric Hertz vector that has only a z component U^{inc}, which is given by

$$U^{\text{inc}} = \frac{E_{0z}}{k^2 \sin^2\theta_0}\, e^{jkx\cos\phi_0 \sin\theta_0}\, e^{jky\sin\phi_0 \sin\theta_0}\, e^{jkz\cos\theta_0} \quad (7.192)$$

The form of eqn. 7.192 is readily verified by noting that

$$E_z^{inc} = \left(k^2 + \frac{\partial^2}{\partial z^2}\right) U^{inc} \tag{7.193}$$

A key point in the development of the solution is to note that $\exp(jkz \cos \theta_0)$ must be the universal z dependence of the total field. Furthermore, the secondary fields can be derived from a Hertz vector with only a z component U^{sec}, bearing in mind that the induced currents on the wires are essentially axial. Thus we write

$$U^{sec} = f(\varrho, \phi) e^{jkz \cos \theta_0}$$

where $f(\varrho, \phi)$ is some undetermined function of ϱ and ϕ. Then taking the lead from the form of eqn. 7.78, we deduce that

$$U^{sec} = [A_1 K_0(u\varrho) + A_2 K_0(u\hat{\varrho})] e^{jkz \cos \theta_0} \tag{7.194}$$

where

$$\hat{\varrho} = (\varrho^2 + d^2 - 2\varrho d \cos \phi)^{1/2}$$

$$u = jk \sin \theta_0$$

and where A_1 and A_2 are coefficients yet to be determined. The relevant geometry is indicated in Fig. 7.19, where a plan view is shown. In fact, we confirm that U^{sec} as given by eqn. 7.194 satisfies

$$(\nabla^2 + k^2) U^{sec} = 0 \tag{7.195}$$

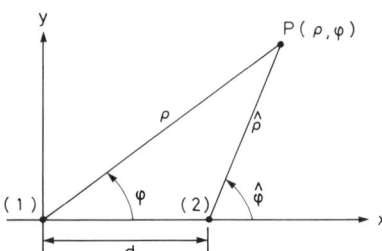

Fig. 7.19 Plan view of the two parallel wires showing local co-ordinate systems

We now assert that the wires have series impedances Z_1 and Z_2 per unit length, so that the boundary conditions are

$$E_z]_{\varrho=a_1} = Z_1 \, 2\pi a_1 H_\phi]_{\varrho=a_1} \tag{7.196}$$

$$E_z]_{\hat{\varrho}=a_2} = Z_2 \, 2\pi a_2 H_{\hat{\phi}}]_{\hat{\varrho}=a_2} \tag{7.197}$$

It is clear that the left-hand sides of eqns. 7.196 and 7.197 are

$$[E_z^{inc} + E_z^{sc}]_{\varrho=a_1} = \{E_{0z} + k^2 \sin^2 \theta_0 [A_1 K_0(jka_1 \sin \theta_0) + A_2 K_0(jkd \sin \theta_0)]\} e^{jkz \cos \theta_0} \tag{7.198}$$

$$[E_z^{inc} + E_z^{sc}]_{\hat{\varrho}=a_2} = \{E_{0z} e^{jkd \cos \phi_0 \sin \theta_0} + k^2 \sin^2 \theta_0 [A_1 K_0(jkd \sin \theta_0) + A_2 K_0(jka_2 \sin \theta_0)]\} e^{jkz \cos \theta_0} \tag{7.199}$$

respectively. On the other hand, the induced wire currents are

$$I_1 = [2\pi a_1 H_\phi]_{\varrho=a_1} = -j\varepsilon_0 \omega 2\pi a_1 \left.\frac{\partial U^{sc}}{\partial \varrho}\right]_{\varrho=a_1} \simeq j\varepsilon_0 \omega 2\pi A_1 \, e^{jkz\cos\theta_0}$$

(7.200)

$$I_2 = [2\pi a_2 H_{\hat{\phi}}]_{\hat{\varrho}=a_2} = -j\varepsilon_0 \omega 2\pi a_2 \left.\frac{\partial U^{sc}}{\partial \hat{\varrho}}\right]_{\hat{\varrho}=a_2} \simeq j\varepsilon_0 \omega 2\pi A_2 \, e^{jkz\cos\theta_0}$$

(7.201)

where we have used the valid approximation $ZK_1(Z) \simeq 1$ when $|Z| \ll 1$. Also we should note that $2\pi a_1 H_\phi^{inc}$ and $2\pi a_2 H_{\hat{\phi}}^{inc}$ are negligible when evaluated at $\varrho = a_1$ and $\hat{\varrho} = a_2$, respectively.

With the indicated substitutions, eqns. 7.196 and 7.197 lead to two algebraic equations that may be solved for A_1 and A_2 explicitly as follows

$$A_1 = -\frac{E_{0z}}{k^2 \sin^2 \theta_0} \frac{K_0(\alpha_2) - K_0(\delta) \, e^{\delta \cos \phi_0}}{K_0(\alpha_1)K_0(\alpha_2) - K_0^2(\delta)}$$

(7.202)

$$A_2 = -\frac{E_{0z}}{k^2 \sin^2 \theta_0} \frac{K_0(\alpha_1) \, e^{\delta \cos \phi_0} - K_0(\delta)}{K_0(\alpha_1)K_0(\alpha_2) - K_0^2(\delta)}$$

(7.203)

where $\alpha_1 = ka_1 \sin \theta_0$, $\alpha_2 = jka_2 \sin \theta_0$ and $\delta = jkd \sin \theta_0$

Using eqns. 7.202 and 7.203 for the coefficients A_1 and A_2, we see that 7.200 and 7.201 are the desired expressions for the induced currents. Furthermore, in the limit of normal incidence (i.e. $\theta_0 \to 90°$), the results for I_1 and I_2 are fully consistent with eqns. 7.148 and 7.149.

Exercise

Derive explicit expressions for the induced currents on the wires in the case where Z_w is nonzero.

Exercise

Show that in the far field (i.e. $k\varrho \sin \theta_0 \gg 1$, and $\varrho \gg d$) we have the simplified form:

$$U^{sc} \simeq \left(\frac{\pi}{2jk\varrho \sin \theta_0}\right)^{1/2} e^{-jk\varrho \sin \theta_0} e^{jkz\cos\theta_0} (A_1 + A_2 \, e^{jkd\sin\theta_0 \cos\phi_0})$$

(7.204)

and then show that

$$H_\phi^{sc} = -j\varepsilon_0 \omega \frac{\partial U^{sc}}{\partial \varrho} \simeq -\varepsilon_0 \omega k \sin \theta_0 \, U^{sc}$$

(7.205)

which can also be expressed as

$$H_\phi^{sc} \simeq \frac{1}{2}\left(\frac{jk}{2\pi r}\right)^{1/2} e^{-jkr} (\hat{I}_1 + \hat{I}_2 \, e^{jkd\sin\theta_0 \cos\phi_0})$$

where $r = (\varrho^2 + z^2)^{1/2}$ in the direction $\theta = \pi - \theta_0$, and where $\hat{I}_1$ and $\hat{I}_2$ are the values of I_1 and I_2 at $z = 0$.

7.15 Scattering at oblique incidence from cylindrical structures

Some interesting complexities arise when we deal fully with an obliquely incident plane wave for a cylinder of arbitrary radius. We shall outline the method of solution, which is described more fully in the literature [16].

The model is shown in Fig. 7.20, where cylindrical co-ordinates

(ϱ, ϕ,z) are chosen to be coaxial with the cylindrical target of radius a with homogeneous properties σ_c, ε_c and μ_c; the surrounding medium has properties σ, ε and μ. There are clearly two cases to consider. In the first, the electric vector of the incident wave has a z component E_{0z}^{inc} but no z component of magnetic field. In the second, the magnetic vector of the incident wave has a z component H_{0z}^{inc} but no component of electric field. Clearly, the general case of arbitrary incidence can always be handled by a linear superposition of these two cases. Of course, in the thin wire limit we would not expect H_{0z}^{inc} to interact with the target in any significant fashion and, in fact, the general solution bears out this conjecture.

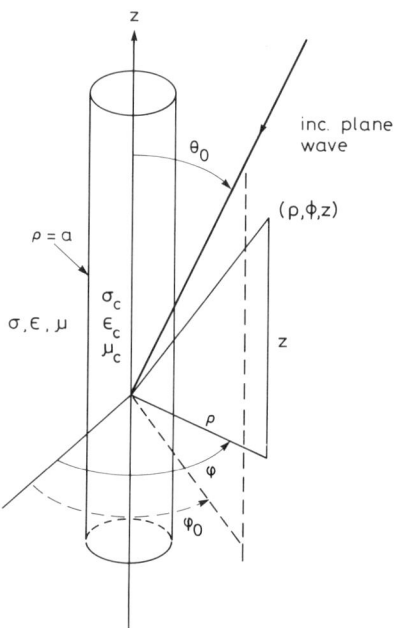

Fig. 7.20 Scattering of a plane wave at oblique incidence from a homogeneous cylinder of any radius but again of infinite length

We shall proceed by treating case 1 explicitly and deducing case 2 by duality. Thus, let us specify that

$$E_{0z}^{inc} = E_0 \, e^{\gamma \varrho \sin \theta_0 \cos(\phi - \phi_0)} \, e^{\gamma z \cos \theta_0} \tag{7.206}$$

where the angles θ_0 and ϕ_0 indicate the direction of the incident wave as noted in Fig. 7.20 and where E_0 is the value of E_{0z}^{inc} referred to the origin. Now we derive eqn. 7.206 from a z-directed Hertz vector or Debye potential U^{inc} given by

$$U^{inc} = -(E_0/u^2) \, e^{u\varrho \cos(\phi - \phi_0)} \, e^{\Gamma z} \tag{7.207}$$

where $u = \gamma \sin \theta_0$ and $\Gamma = \gamma \cos \theta_0$. Then clearly $u^2 = \gamma^2 - \Gamma^2$, where $\gamma = [j\mu\omega(\sigma + j\varepsilon\omega)]^{1/2}$.

Then, on using an addition theorem (e.g. compare with eqn. 7.94), we are able to write eqn. 7.207 in the equivalent form

$$U^{\text{inc}} = -\frac{E_0}{u^2} \sum_{m=0}^{\infty} \hat{\varepsilon}_m I_m(u\varrho) \cos m(\phi - \phi_0) e^{\Gamma z} \qquad (7.208)$$

This result suggests immediately that the secondary electric Debye potential, for $\varrho > a$, should have the form

$$U^{\text{sc}} = -\frac{E_0}{u^2} \sum_{m=0}^{\infty} \hat{\varepsilon}_m K_m(u\varrho) R_m \cos m(\phi - \phi_0) e^{\Gamma z} \qquad (7.209)$$

where R_m is a coefficient to be determined. Here we can check that

$$(\nabla^2 - \gamma^2) U^{\text{sc}} = 0 \qquad (7.210)$$

and also note that $K_m(u\varrho)$ is an outgoing wave function as required.
For the internal region, $\varrho < a$, we are led to write

$$U = -\frac{E_0}{u^2} \sum_{m=0}^{\infty} \hat{\varepsilon}_m I_m(u_c \varrho) T_m \cos m(\phi - \phi_0) e^{\Gamma z} \qquad (7.211)$$

where $u_c = (\gamma_c^2 - \Gamma^2)^{1/2}$ and $\gamma_c = [j\mu_c \omega (\sigma_c + j\varepsilon_c \omega)]^{1/2}$. Again we note that

$$(\nabla^2 - \gamma_c^2) U = 0 \qquad (7.212)$$

and that the wave function $I_m(u_c \varrho)$ is bounded at $\varrho = 0$.

Up to this point the solution is completely analogous to the case treated earlier for normal incidence (i.e. $\Gamma = 0$) where $\partial/\partial z = 0$. Thus again we require that E_z and H_ϕ are continuous at the boundary, $\varrho = a$. But now we have an additional complication because E_ϕ is nonzero. Thus it can be confirmed that we need to allow for the coupling to the magnetic Hertz vector or magnetic Debye potential V. This fact leads us to write, for $\varrho > a$, the following:

$$V^{\text{sc}} = -\frac{E_0}{u^2} \sum_{m=1}^{\infty} K_m(u\varrho) R_m^* \sin m(\phi - \phi_0) e^{\Gamma z} \qquad (7.213)$$

where R_m^* is a coefficient. Similarly, for the internal region, $\varrho < a$, we have

$$V = -\frac{E_0}{u^2} \sum_{m=1}^{\infty} I_m(u_c \varrho) T_m^* \sin m(\phi - \phi_0) e^{\Gamma z} \qquad (7.214)$$

where T_m^* is yet another coefficient. The adoption of the $\sin m(\phi - \phi_0)$ function is necessary if eqns. 7.21 to 7.24 are to be applied consistently to satisfy the boundary conditions.

The set of four boundary conditions are now listed as follows:

$$\begin{bmatrix} E_z^{\text{inc}} + E_z^{\text{sc}} \\ H_z^{\text{sc}} \\ E_\phi^{\text{inc}} + E_\phi^{\text{sc}} \\ H_\phi^{\text{inc}} + H_\phi^{\text{sc}} \end{bmatrix}_{\varrho = a+0} = \begin{bmatrix} E_z \\ H_z \\ E_\phi \\ H_\phi \end{bmatrix}_{\varrho = a-0} \qquad (7.215)$$

Using eqns. 7.21 to 7.24 we now readily transform these boundary conditions into four linear algebraic equations for the unknown coefficients R_m, T_m, R_m^* and T_m^*. The reader is invited to carry out the process to yield the somewhat cumbersome formulas (e.g. see [16]).

Case 2, for the other polarisation such that

$$H_{0z}^{inc} = H_0 \, e^{\gamma\varrho \sin\theta_0 \cos(\phi-\phi_0)} \, e^{\gamma z \cos\theta_0} \qquad (7.216)$$

is carried out in an analogous fashion. Formally we may appeal to duality and make the following exchanges to yield the results directly from case 1 to case 2: $E_0 \to H_0$, $\mathbf{E} \to \mathbf{H}$, $\mathbf{H} \to -\mathbf{E}$, $\sigma + j\varepsilon\omega \to j\mu\omega$ and $j\mu\omega \to \sigma + j\varepsilon\omega$.

The significant find here is that, for both case 1 and for case 2, the electric and magnetic Debye potentials are coupled. This fact has important consequences in determining the polarisation of the scattered field. The coupling vanishes at normal or perpendicular incidence (i.e. $\theta_0 = 90°$), and also the coupling is nonexistent if the conductivity σ_c or the permeability μ_c are effectively infinite.

7.16 Scattering from conducting cylinder of arbitrary cross-section

In many situations one is interested in cylindrical targets of noncircular cross-section. The most direct and useful approach to such problems is to deal with the integral equation formulation by numerical means. Here we follow Richmond [17] who dealt with purely dielectric cylinders. Without difficulty, we may generalise his solutions to lossy structures of infinite length. However, we will assume that the magnetic permeability of the whole space is $\mu_0 = 4\pi \times 10^{-7}$ H/m.

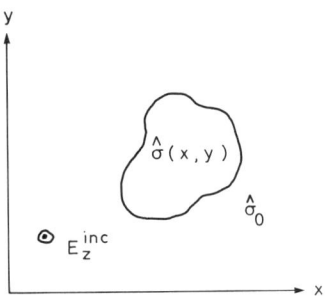

Fig. 7.21 Inhomogeneous cylinder of arbitrary cross-section

The situation is illustrated in Fig. 7.21. The cylinder of arbitrary cross-sectional shape has a conductivity $\sigma(x, y)$ and a permittivity $\varepsilon(x, y)$ that are functions of x and y within the interior. The region exterior to the cylinder is homogeneous with conductivity σ_c and permeability ε_b. To simplify our discussion we define complex conductivities in the manner

$$\hat{\sigma}(x, y) = \sigma(x, y) + j\omega\varepsilon(x, y) \qquad (7.217)$$

and

$$\hat{\sigma}_b = \sigma_b + j\omega\varepsilon_b \qquad (7.218)$$

Furthermore we will keep the problem two dimensional. Thus we choose the incident field to have only a z component $E_z^{inc}(x, y)$ which does not depend on z. Also the resultant electric field in the presence of the cylindrical target will also have only a z component that we designate $E_z(x, y)$. Now we may write

$$E_z(x, y) = E_z^{inc}(x, y) + E_z^{sc}(x, y) \qquad (7.219)$$

where $E_z^{sc}(x, y)$, by definition, is the scattered or secondary field. The objective, of course, is to obtain E_z^{sc} everywhere in terms of E_z^{inc}.

The basic concept underlying the solution is to observe that the scattered field arises from the polarisation currents J_z within the target. In fact, from Ohm's law,

$$J_z(x, y) = (\hat{\sigma}(x, y) - \hat{\sigma}_b)E_z(x, y) \tag{7.220}$$

Now we consider an element of cross-sectional area dA within the target as carrying a current dI that contributes to the scattered field according to eqn. 7.43. Thus

$$dE_z^{sc} = -\frac{j\mu\omega}{2\pi} dI\, K_0(\gamma_b \varrho) \tag{7.221}$$

is the corresponding increment to the total response, where $\gamma_b = (j\mu\omega\hat{\sigma}_b)^{1/2}$. It then follows that

$$dI = (\hat{\sigma} - \hat{\sigma}_b)E_z\, dA \tag{7.222}$$

So eqn. 7.221 becomes

$$dE_z^{sc} = -\frac{j\mu\omega}{2\pi} (\hat{\sigma} - \hat{\sigma}_b) E_z\, dA\, K_0(\gamma_b \varrho) \tag{7.223}$$

The next step is to superimpose the fields of all the elemental line sources and to replace the elemental area dA by $dx'\,dy'$, to yield

$$E_z^{sc} = -\frac{\gamma_b^2}{2\pi} \iint_A C(x', y') E_z(x', y') K_0(\gamma_b \hat{\varrho})\, dx'\, dy' \tag{7.224}$$

where

$$\hat{\varrho} = [(x - x')^2 + (y - y')^2]^{1/2}$$

$$C = \frac{\hat{\sigma}}{\hat{\sigma}_b} - 1$$

The integration in eqn. 7.224 is over the area of the target. The contrast function $C(x', y')$ differs from zero in the case of any material target and, in general, it could be a function of x' and y'. We also note that $E_z(x', y')$, where it occurs in the integrand of eqn. 7.224, is the unknown resultant field. Only in the case where $C(x', y')$ is sufficiently small can we replace E_z by E_z^{inc}. Such a step is analogous to the first-order Born approximation used in quantum mechanics [18]. Before we describe the numerical solution of eqn. 7.224, it is worth while to mention the Born approximation and the higher-order iterations.

In accordance with the preceding comments we designate the 'first approximation' to the resultant scattered field as follows:

$$E_{z,1} = E_z^{inc} + E_{z,1}^{sc} \tag{7.225}$$

where

$$E_{z,1}^{sc} = -\frac{\gamma_b^2}{2\pi} \iint_A CE_z^{inc} K_0\, dx'\, dy' \tag{7.226}$$

with an obvious contraction of notation. Then clearly the second-order approximation is given by

$$E_{z,2} = E_z^{inc} + E_{z,2}^{sc} \tag{7.227}$$

where

$$E_{z,2}^{sc} = -\frac{\gamma_b^2}{2\pi} \iint C(E_z^{inc} + E_{z,1}^{sc}) K_0 \, dx' \, dy' \tag{7.228}$$

The iterative process can be continued any number of times, and we may write, for example, the nth-order approximation for the scattered field as

$$E_{z,n}^{sc} = -\frac{\gamma_b^2}{2\pi} \iint C(E_z^{inc} + E_{z,n-1}^{sc}) K_0 \, dx' \, dy' \tag{7.229}$$

While the iterative procedure is conceptually simple, it often has severe convergence problems. Only in the case where the contrast is small and for bounded targets does the process converge. Normally the condition

$$\left| \gamma_b^2 \iint C \, dx' \, dy' \right| \ll 1$$

will assure reasonable convergence.

As mentioned above, a more generally useful procedure is to solve the integral equation 7.224 by numerical means [17, 19–21]. To begin with we combine eqns. 7.219 and 7.224 so that we may write

$$E_z(x, y) + \frac{1}{2\pi} \gamma_b^2 \iint_A C(x', y') E_z(x', y') K_0(\gamma_b \varrho) \, dx' \, dy' = E^{inc}(x, y) \tag{7.230}$$

To deal with this equation we now divide the region A into P cells that are sufficiently small to regard the complex conductivity and electric field to be constant within an individual cell (see Fig. 7.22). Then we say that eqn. 7.230 is to be enforced at the centre of cell m, so that

$$E_z^{(m)} + \frac{1}{2\pi} \gamma_b^2 \sum_{p=1}^{P} C^{(p)} E_z^{(p)} \iint K_0(\gamma_b \varrho) \, dx' \, dy' = E^{inc}(x_m, y_m) \tag{7.231}$$

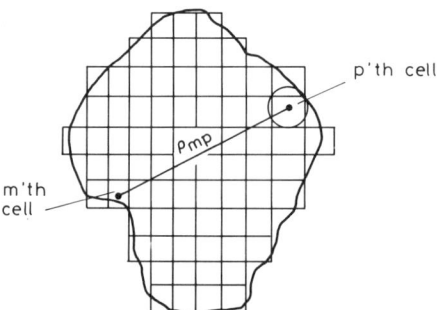

Fig. 7.22 Discretising the inhomogeneous cylinder by rectangular (or square) cells

where $C^{(p)}$ is the conductivity contrast and $E_z^{(p)}$ is the electric field at the centre of cell p. Here we note that

$$\varrho = [(x' - x_m)^2 + (y' - y_m)^2]^{1/2} \tag{7.232}$$

By taking $m = 1, 2, 3, \ldots, P$ we see that eqn. 7.231 yields a system of P linear equations. The task then is to solve this system in order to determine the fields $E_z^{(1)}, E_z^{(2)}, E_z^{(3)}, \ldots, E_z^{(P)}$ at the centre of each cell.

The integral in eqn. 7.231 is not expressible in closed form. But it may be evaluated numerically if care is exercised in dealing with the singularity as $\varrho \to 0$. On the other hand, an analytical approximation is usually adequate. Following Richmond [17] and current practice, the rectangular or square cells are replaced by circular cells of the same area. Then the integral in eqn. 7.231 can be expressed as follows:

$$\frac{1}{2\pi}\gamma_b^2 \int_0^{2\pi}\int_0^{a_p} K_0(\gamma_b \varrho) \varrho' \, d\varrho' \, d\phi'$$

$$= [1 - \gamma_b a_p K_1(\gamma_b a_p)] \quad \text{if } m = p \quad (7.233)$$

$$\simeq \gamma_b a_p I_1(\gamma_b a_p) K_0(\gamma_b \varrho_{mp}) \quad \text{if } m \neq p \quad (7.234)$$

where

$$\varrho_{mp} = [(x_m - x_p)^2 + (y_m - y_p)^2]^{1/2} \quad (7.235)$$

The system of linear equations represented by eqn. 7.231 is now given succinctly by

$$\sum_{p=1}^{P} D_{mp} E_z^{(p)} = E_m^{\text{inc}} \quad (m = 1, 2, \ldots, P) \quad (7.236)$$

where $E_z^{(p)} = E_z(x_p, y_p)$ is the resultant (unknown) electric field at the centre of the pth cell, and the coefficients are given by

$$D_{mp} = 1 + C^{(p)}[1 - \gamma_b a_p K_1(\gamma_b a_p)] \quad \text{if } p = m \quad (7.237)$$

$$= \gamma_b a_p I_1(\gamma_b a_p) K_0(\gamma_b \varrho_{mp}) \quad \text{if } p \neq m \quad (7.238)$$

Once we have solved the linear system of equations given by eqn. 7.236 for $E_z^{(p)}$, we may use eqn. 7.224 to give the scattered field. The form given by eqn. 7.235 is then exploited so that

$$E_z^{\text{sc}} = -\gamma_b \sum_{p=1}^{P} C^{(p)} E_z^{(p)} a_p I_1(\gamma_b a_p) K_0(\gamma_b \varrho_p) \quad (7.239)$$

where $C^{(p)}$ is taken as the average complex conductivity contrast over a square cell of area πa_p^2. Now ϱ_p is the distance from the centre of the square pth cell and the observation point (x, y), i.e.

$$\varrho_p = [(x - x_p)^2 + (y - y_p)^2]^{1/2}$$

When the observation point is sufficiently removed from the target, such that $|\gamma_b \varrho_p| \gg 1$, we may use the asymptotic approximation

$$K_0(\gamma_b \varrho_p) \simeq \left(\frac{\pi}{2\gamma_b \varrho_p}\right)^{1/2} e^{-\gamma_b \varrho_p} \quad (7.240)$$

Some further simplifications are possible if the background medium is a perfect dielectric. Then $\hat{\sigma}_b = \sigma_b + j\varepsilon_b \omega = j\varepsilon_0 \omega$ and $\gamma_b = jk$, where $k = (\varepsilon_0 \mu_0)^{1/2} \omega$ is the real wave number. Then, in terms of cylindrical co-ordinates (ϱ_0, ϕ),

$$\varrho_p \simeq \varrho_0 - x_p \cos \phi - y_p \sin \phi$$

so that

$$K_0(\gamma_b \varrho_p) \simeq e^{-jk\varrho_0} \left(\frac{\pi}{2\gamma_0 \varrho_0}\right)^{1/2} e^{jk(x_p \cos \phi + y_p \sin \phi)} \tag{7.241}$$

Then, following Harrington [22], we may define an 'echo width' parameter $W(\phi)$ according to

$$W(\phi) = \lim_{\varrho_0 \to \infty} (2\pi \varrho_0) \left| \frac{E_z^{sc}(\varrho_0, \phi)}{E_0^{inc}} \right|^2 \tag{7.242}$$

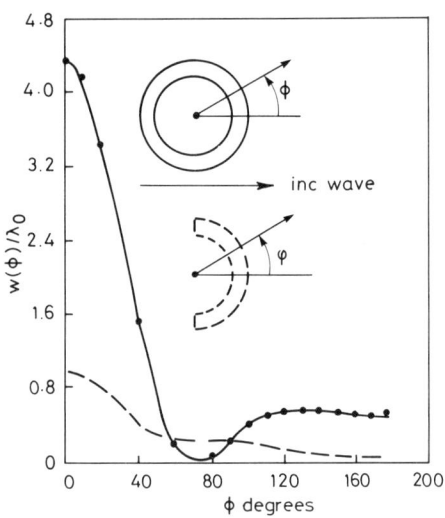

Fig. 7.23 Echo width (normalised by the free space wavelength) for a circular and semi-circular dielectric shell. The parameters are $\varepsilon/\varepsilon_0 = 4$, $\sigma = 0$, inner radius = $0.25 \lambda_0$, outer radius = $0.30 \lambda_0$. The solid curve is based on the classical solution (i.e. harmonic series) for the circular shell. The data points are from the corresponding integral equation solution, and the broken curve is based on the integral equation solution for the semi-circular shell (data from [17]). Plane wave incidence is assumed in each case

where E_0^{inc} can be taken as the amplitude of the incident field at the centre of the cylindrical scatter. Using such a formulation, Richmond [17] has provided a number of interesting examples of scattering from dielectric cylinders of various cross-sections. An example is shown in Fig. 7.23, where $W(p)/\lambda_0$ is plotted as a function of ϕ for a plane wave that is incident on the structure. Here $\phi = 0°$ is the direction of forward scatter. The two structures considered are the circular dielectric shell and the semi-circular dielectric shell of the same radius and thickness. The parameters are indicated in the caption to Fig. 7.23. As a check on the calculation procedure, the results for the circular shell were also obtained using the exact harmonic series form in terms of Bessel functions. As indicated, in Fig. 7.23, the agreement is very close.

Richmond [23] has also extended his formulation to the case where the incident wave is polarised with the magnetic field parallel to the axis of the cylindrical structure. Corresponding investigations for three-dimensional structures have been carried out by Cauterman, Nguyen and Degauque [24].

In a number of situations the analytical and numerical aspects of the problem can be combined. Such a situation would arise, for example, if the target is partly or mostly described by a regular body and the inhomogeneity is localised to some extent. An example is discussed briefly here.

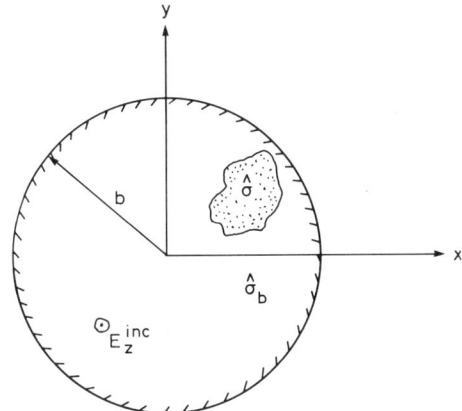

Fig. 7.24 Localised inhomogeneity within a homogeneous cylinder

The new situation is illustrated in Fig. 7.24, which can be regarded as the same model as shown in Fig. 7.21 except that the homogeneous background region of conductivity $\hat{\sigma}_b$ is now bounded by the cylindrical surface $\varrho = b$. For example, this surface could be a perfectly conducting sheath or the external region might be free space. Again we restrict attention to a purely two-dimensional geometry.

The incident electric field, which has only a z component E_z^{inc}, is defined as that field which would exist in the homogeneous cylindrical region in the absence of inhomogeneity whose properties are characterised by a complex conductivity $\hat{\sigma}_b$.

With reference to Fig. 7.25, we are thus interested in the field dE_z at (x, y) produced by the polarisation current dI in the element at (x_m, y_m). This is really the same problem as that found earlier for the line source excitation of a homogeneous cylinder. However, in the present case the line source is inside the cylinder. Without difficulty, we deduce that

$$dE_z = -\frac{j\mu_0\omega}{2\pi} dI\, G(x_m, y_m; x, y) \qquad (7.243)$$

where the Green's function is given by

$$G(x_m, y_m; x, y) = K_0(\gamma_b\hat{\varrho}_m) - \sum_{n=0}^{\infty} \hat{\varepsilon}_n I_n(\gamma_b\varrho_m) r_n \frac{K_n(\gamma_b b)}{I_n(\gamma_b b)} I_n(\gamma_b\varrho)\cos n(\phi - \phi_m)$$

$$(7.244)$$

where

$$\hat{\varrho}_m = [\varrho^2 + \varrho_m^2 - 2\varrho\varrho_m \cos(\phi - \phi_m)]^{1/2}$$

$\hat{\varepsilon}_0 = 1$, $\hat{\varepsilon}_n = 2$ ($n \neq 0$) and r_n is a reflection coefficient at $\varrho = b$. For example, if the surface at $\varrho = b$ is a perfect conductor we may confirm that dE_z vanishes there by noting that

$$K_0(\gamma_b \hat{\varrho}_m) = \sum_{n=0}^{\infty} \hat{\varepsilon}_n I_n(\gamma_b \varrho_m) K_n(\gamma_b \varrho) \cos n(\phi - \phi_m) \quad (7.245)$$

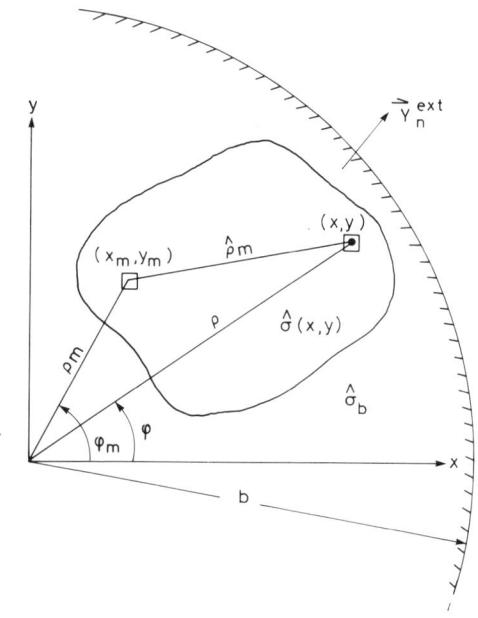

Fig. 7.25 Expanded view of local inhomogeneity and relevant geometry

which is the form of the addition theorem for modified Bessel functions valid for $\varrho > \varrho_m$. On the other hand, if the external region is characterised by an admittance $\vec{Y}_n^{\text{ext}}$ at $\varrho = b$, we would use the form

$$r_n = \frac{\vec{Y}_n^{\text{ext}} - \vec{Y}_n}{\vec{Y}_n^{\text{ext}} + \overleftarrow{Y}_n} \quad (7.246)$$

where

$$\vec{Y}_n = -y_b K_n'(\gamma_b b)/K_n(\gamma_b b) \quad (7.247)$$

$$\overleftarrow{Y}_n = y_b I_n'(\gamma_b b)/I_n(\gamma_b b) \quad (7.248)$$

$$y_b = \gamma_b/j\mu_0\omega = (\hat{\sigma}_b/j\mu_0\omega)^{1/2}$$

Equation 7.246 is completely analogous to 7.106 and the derivation is almost identical. Thus in place of eqn. 7.231 we find that

$$E_z^{(m)} + \frac{1}{2\pi} \gamma_b^2 \sum_{p=1}^{P} C^{(p)} E_z^{(p)} \iint_{\text{cell } p} G(x_m, y_m; x', y') \, dx' \, dy' = E^{\text{inc}}(x_m, y_m)$$
$$(7.249)$$

This system of linear equations has the same form as of eqn. 7.236. Furthermore, it is found that D_{mp}, for $p = m$, is given by eqn. 7.237 because the singularity of the Green's function G is the same as the Bessel function $K_0(\gamma_b \hat{\varrho}_m)$ for $\hat{\varrho}_m \to 0$. But now

$$D_{mp} = \gamma_b a_p I_1(\gamma_b a_p) G(x_m, y_m; x_p, y_p)$$

for $m \neq p$.

The numerical implementation of eqn. 7.249 would be tedious but well worthwhile if one is dealing with bounded systems where the inhomogeneity is localised. A good example is the hyperthermia problem [25] where we wish to focus the electromagnetic energy into a tumor that is contained within a cylindrically shaped torso. The actual local power deposition would be $\sigma(x, y)|E_z^2|$ W/m² at the point (x, y).

7.17 Further exercises

Exercise

We consider a plane wave incident obliquely on an infinitely long thin wire of finite conductivity as indicated in Fig. 26. The wire is coincident with the z axis of a rectangular (x, y, z) or cylindrical (ϱ, ϕ, z) system. The incoming wave is defined by

$$\mathbf{H}_0^{inc} = H_0 \, e^{jk\varrho \sin\theta_0 \cos\phi} \, e^{jkz \cos\theta_0} \, \mathbf{i}_y$$

Deduce the receiving antenna patterns for electric and magnetic probes for a point near the wire (at $\varrho = d$). In particular, deal with the pattern functions defined as follows:

$$P_v(\theta_0) = \left| \frac{E_z(x = d, \phi = 0°, z = 0; \theta_0)}{E_z(x = d, \phi = 0°, z = 0; 90°)} \right| \quad (7.250)$$

$$P_h(\theta_0) = \left| \frac{E_x(x = d, \phi = 0°, z = 0; \theta_0)}{E_z(x = d, \phi = 0°, z = 0; 90°)} \right| \quad (7.251)$$

$$Q(\theta_0) = \left| \frac{H_y(x = d, \phi = 0°, z = 0°; \theta_0)}{H_y(x = d, \phi = 0°, z = 0°; 90°)} \right| \quad (7.252)$$

$$S(\phi) = \left| \frac{E_z(\varrho = d, \theta_0 = 90°, z = 0; \phi)}{E_z(\varrho = d, \theta_0 = 90°, z = 0; 0°)} \right| \quad (7.253)$$

$$T(\phi) = \left| \frac{H_\phi(\varrho = d, \theta_0 = 90°, z = 0; \phi)}{H_\phi(\varrho = d, \theta_0 = 90°, z = 0; 0°)} \right| \quad (7.254)$$

Using the results given by eqns. 7.74 to 7.86, and noting that $\gamma = jk$, show that:

$$P_v(\theta_0) = \left| \frac{\sin\theta_0 \left[e^{jkd \sin\theta_0} - \dfrac{K_0(jkd \sin\theta_0)}{K_0(jka \sin\theta_0) + W(\theta_0)} \right]}{e^{jkd} - \dfrac{K_0(jkd)}{K_0(jka) + W(90°)}} \right| \quad (7.255)$$

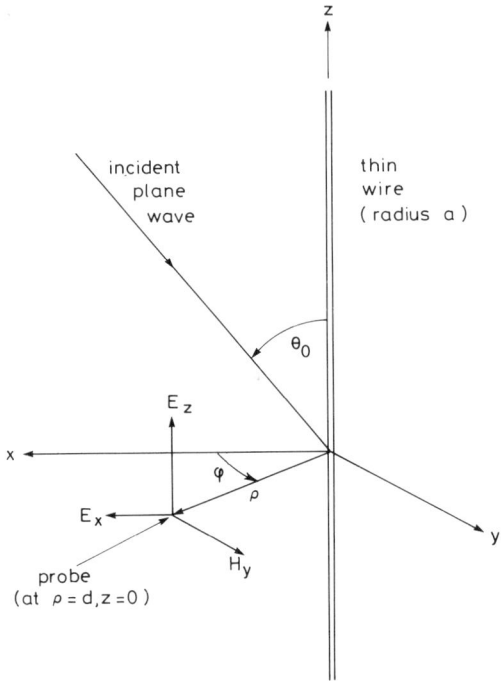

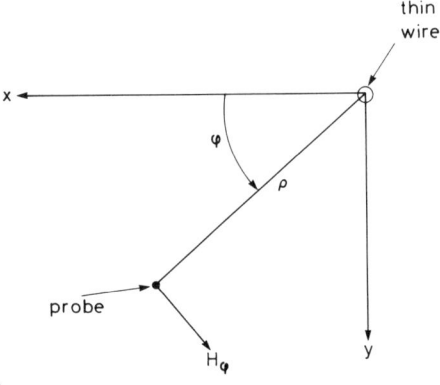

Fig. 7.26 *a* Geometry for scattering of a plane wave at oblique incidence on a thin imperfectly conducting wire of radius *a*. The field is to be probed at (ρ, φ) in the $z = 0$ plane. *b* The plan view of the geometry. The surrounding medium is free space

$$P_h(\theta_0) = \left| \frac{\cos\theta_0 \left[e^{jkd\sin\theta_0} + \dfrac{K_1(jkd\sin\theta_0)}{K_0(jka\sin\theta_0) + W(\theta_0)} \right]}{e^{jkd} - \dfrac{K_0(jkd)}{K_0(jka) + W(90°)}} \right| \qquad (7.256)$$

$$Q(\theta_0) = \left| \frac{e^{jkd\sin\theta_0} + \dfrac{K_1(jkd\sin\theta_0)}{K_0(jka\sin\theta_0) + W(\theta_0)}}{e^{jkd} + \dfrac{K_1(jkd)}{K_0(jka) + W(90°)}} \right| \qquad (7.257)$$

$$S(\phi) = \left| \frac{e^{jkd\cos\phi} - \dfrac{K_0(jkd)}{K_0(jka) + W(90°)}}{e^{jkd} - \dfrac{K_0(jkd)}{K_0(jka) + W(90°)}} \right| \qquad (7.258)$$

$$T(\phi) = \left| \frac{(\cos\phi)\,e^{jkd\cos\phi} + \dfrac{K_1(jkd)}{K_0(jka) + W(90°)}}{e^{jkd} + \dfrac{K_1(jkd)}{K_0(jka) + W(90°)}} \right| \qquad (7.259)$$

where

$$W(\theta) = \frac{2\pi Z_w(jk\cos\theta_0)}{j\mu_0\omega \sin^2\theta_0} \qquad (7.260)$$

and

$$Z_w(jk\cos\theta_0) \simeq \frac{(j\mu_w\omega)^{1/2}}{2\pi a(\sigma_w + j\varepsilon_w\omega)^{1/2}} \frac{I_0(\gamma_w a)}{I_1(\gamma_w a)} \qquad (7.261)$$

where σ_w, ε_w and μ_w are the electrical properties of the wire material. Note that in eqn. 7.261, $I_0/I_1 \simeq 1$ because $|\gamma_w a| \gg 1$.

Solution and numerical results

The above expressions for the pattern functions are obtained by direct substitution of the indicated field components using eqns. 7.80a, 7.70b and 7.81 for the probe location. We also note that

$$K_0(j\beta) = -\frac{\pi}{2}[Y_0(\beta) + jJ_0(\beta)] \qquad (7.262)$$

$$K_1(j\beta) = -\frac{\pi}{2}[J_1(\beta) - jY_1(\beta)] \qquad (7.263)$$

reduce the modified Bessel functions of imaginary argument to the common Bessel functions of real argument. Thus for numerical work we may employ tabulated values of the J and Y functions of order zero and one. To illustrate the pattern functions we choose a metal wire of conductivity $\sigma_w = 10^7$ S/m, magnetic permeability $\mu_w = \mu_0$, and an operating frequency of 100 MHz (i.e. $\omega/2\pi = 3 \times 10^8$). An example of such calculations is shown in Fig. 7.27, where $P_v(\theta_0)$ is plotted as a function of the angle θ_0 for the principal E-plane (i.e. where $\phi = 0°$); $d/\lambda_0 = 0\cdot1$, $0\cdot3$, $0\cdot5$ and $1\cdot0$, and the wire radius $a = 2$ cm. Within graphical accuracy in this figure, the corresponding curves for $\sigma_w = 10^7$ S/m and $\sigma_2 = \infty$ could not be distinguished. As indicated, in the example shown in Fig. 7.28 the difference between the two cases is only discernible at very small angles (i.e. where the incident waves are arriving at highly oblique incidence). This same comment is applicable to the curves shown for the pattern functions plotted in Figs. 7.29 to 7.32.

The influence of wire diameter plays some role, as can be seen in the pattern plots in Fig. 7.33 where both $a = 2$ mm and $a = 2$ cm cases are compared for the same value of d/λ_0. Actually the thinner wire (i.e. $a = 2$ mm) exhibits a slightly stronger dependence on the finite wire conductivity, as shown in the example in Fig. 7.34. But again here the effect is almost undetectable except for small angles.

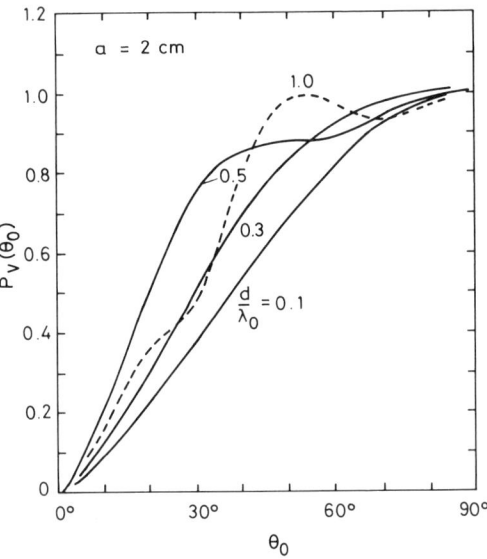

Fig. 7.27 Pattern of an electric field probe oriented in the vertical direction for the E-plane (i.e. $\varphi = 0°$) for an obliquely incident plane wave on the thin wire

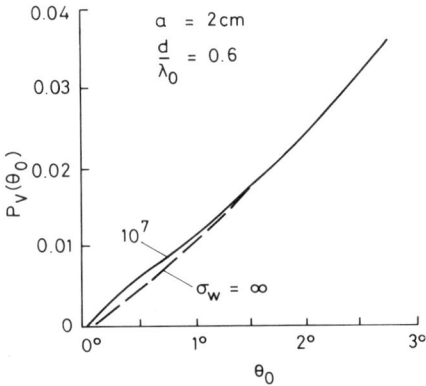

Fig. 7.28 Same conditions as for Fig. 7.27, but two different wire conductivities are shown

The patterns shown in Figs. 7.27, 7.29, 7.30, 7.31 and 7.32 can be scaled to other frequencies (at least by a factor of 10 up or down) for most practical purposes. They provide a good illustration of the field behaviour near a slender metallic structure where the principal scattering mechanism is the induced longitudinal current. By reciprocity, the patterns, of course, also apply directly to the case where an electric or magnetic dipole source is placed at $\varrho = d$ and $z = 0$, in which case the observer is at infinity

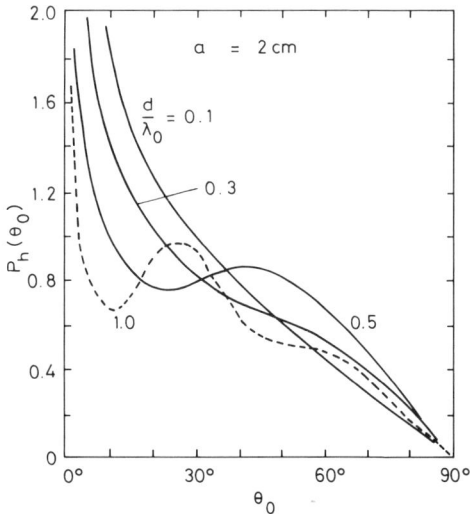

Fig. 7.29 Pattern of an electric field probe oriented in the horizontal or radial direction for the E-plane (i.e. $\varphi = 0°$) for an obliquely incident plane wave on the thin wire

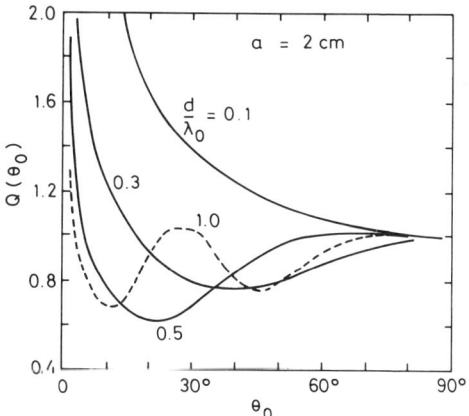

Fig. 7.30 Pattern of a magnetic field probe or small loop with axis oriented for maximum pick up in E-plane (i.e. loop senses H_φ in $\varphi = 0°$ plane)

7.18 Appendix: cylindrical Bessel functions – an outline

This outline quotes liberally from references [26–32] using modern notation [28].

In many electromagnetic problems we are concerned with boundaries that are cylindrical surfaces of constant curvature. Two specific examples are hollow circular waveguides with metal walls, and dielectric optical fibre waveguides. The generic equation that we deal with is the Helmholtz equation, written as follows:

$$(\nabla^2 - \gamma^2)\psi = 0 \tag{7.264}$$

where

$$\nabla^2 = \frac{1}{\varrho}\frac{\partial}{\partial \varrho}\left(\varrho \frac{\partial}{\partial \varrho}\right) + \frac{1}{\varrho^2}\frac{\partial^2}{\partial \phi^2} + \frac{\partial^2}{\partial z^2} \tag{7.265}$$

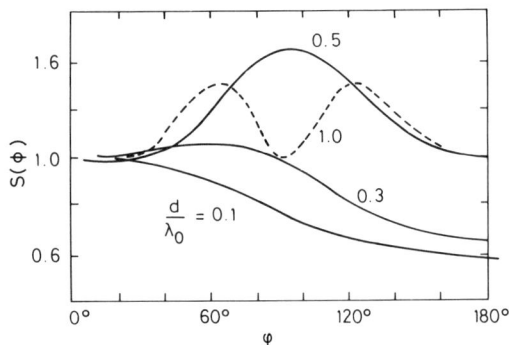

Fig. 7.31 Pattern of an electric field probe oriented in the vertical direction for the H-plane (i.e. $z = 0°$) for a normally incident wave (i.e. $\Theta_0 = 90°$)

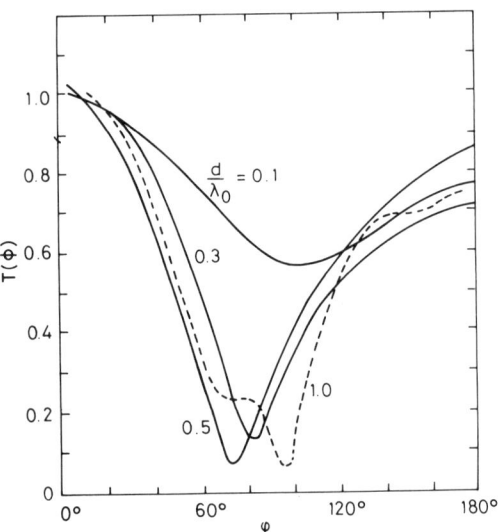

Fig. 7.32 Pattern of a magnetic field probe, oriented in the azimuthal direction about the wire, for the H-plane for a normally incident wave

is the Laplacian operator in cylindrical co-ordinates (ϱ, ϕ, z). The scalar function ψ represents the physical quantity of interest, such as the axial component of the field or some associated quantity. Also we should note that γ is a fixed parameter and defined such that Re $\gamma > 0$ for a common time factor $\exp(j\omega t)$. In the electromagnetic problem, $\gamma = [j\mu\omega(\sigma + j\omega\varepsilon)]^{1/2}$, where σ, ε, μ are the specified properties of the homogeneous region under consideration.

Using a separation of variables approach, we are led to write solutions of eqn. 7.264 in the form of a product of a function of ϱ, a function of ϕ and a function of z. That is,

$$\psi = Z(\varrho)\, e^{-jm\phi}\, e^{-\Gamma z} \tag{7.266}$$

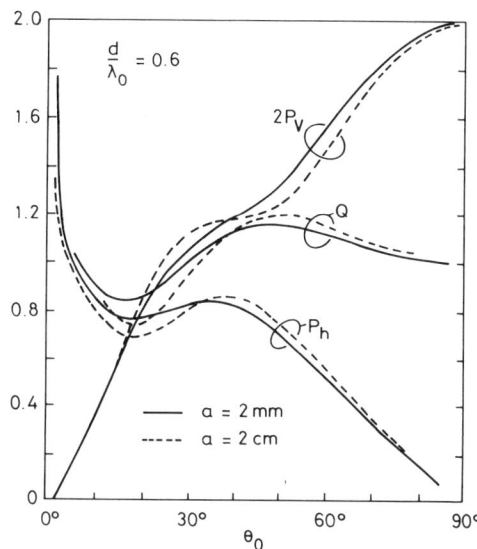

Fig. 7.33 Effect of changing wire diameter on th E-plane patterns

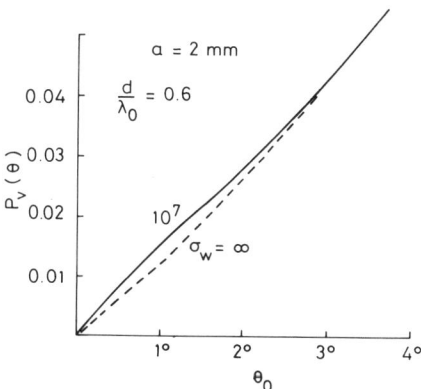

Fig. 7.34 Effect of finite wire conductivity for the thinner wire

where m are Γ are parameters that do not depend on the coordinates and where $Z(\varrho)$ is some function of ϱ yet to be determined. (We follow standard convention here and regard m as an integer, but a more general form of eqn. is obtained if we merely replace m by ν). When eqn. 7.266 is inserted into eqn. 7.264 we obtain the following differential equation:

$$\frac{1}{\varrho}\frac{d}{d\varrho}\left(\varrho\,\frac{dZ}{d\varrho}\right) - \frac{m^2}{\varrho^2}Z + \Gamma^2 Z - \gamma^2 Z = 0 \tag{7.267}$$

for the radial function $Z(\varrho)$. Of course, eqn. 7.267 is equivalent to

$$\varrho^2 \frac{d^2Z}{d\varrho^2} + \varrho \frac{dZ}{d\varrho} - (m^2 + u^2\varrho^2)Z = 0 \tag{7.268}$$

where $u = (\gamma^2 - \Gamma^2)^{1/2}$ is defined such that Re $u > 0$. The latter equation is a form of Bessel's equation that has been studied in the literature for more than a century. Two independent solutions are the functions $I_m(u\varrho)$ and $K_m(u\varrho)$, where m designates the order and $u\varrho$ is the argument. Following G. N. Watson [26], they are defined as follows. The first solution is

$$I_m(u\varrho) = \sum_{n=0,1,2,\ldots}^{\infty} \frac{(u\varrho)^{m+2n}}{n!\,(m+n)!\,2^{m+2n}} \tag{7.269}$$

for $m = 0, 1, 2, 3, \ldots$. Also we have the identity $I_m(u\varrho) = I_{-m}(u\varrho)$ which takes care of the negative integer values of the order. On the other hand, if the order of the Bessel function is a noninteger, we would have

$$I_\nu(u\varrho) = \sum_{n=0,1,2,\ldots}^{\infty} \frac{(u\varrho)^{\nu+2n}}{n!\,(\nu+n)!\,2^{\nu+2n}} \tag{7.270}$$

$$I_{-\nu}(u\varrho) = \sum_{n=0,1,2,\ldots}^{\infty} \frac{(u\varrho)^{-\nu+2n}}{n!\,(-\nu+n)!\,2^{-\nu+2n}} \tag{7.271}$$

which actually are independent solutions of eqn. 7.268 if ν is a noninteger. One may confirm that eqns. 7.270 and 7.271 are solutions of 7.268 by direct substitutions and equating equal powers of $u\varrho$.

To arrive at the definition of the second solution $K_m(u\varrho)$ for integer order m, we proceed as follows. First of all we note that, for noninteger ν, the linear combination of eqns. 7.27 and 7.271 given by

$$K_\nu(u\varrho) = \frac{\pi}{2 \sin \nu\pi} [I_{-\nu}(u\varrho) - I_\nu(u\varrho)] \tag{7.272}$$

is also a solution of eqn. 7.268. Now for integer m we define

$$K_m(u\varrho) = \lim_{\nu \to m} K_\nu(u\varrho)$$

$$= \frac{1}{2 \cos m\pi} \left(\frac{\partial I_{-\nu}}{\partial \nu} - \frac{\partial I_\nu}{\partial \nu} \right) \bigg|_{\nu \to m} \tag{7.273}$$

We will not give the explicit form for $K_m(u\varrho)$ because it is rather complicated and it is not needed here. Actually I_m and K_m are usually called modified Bessel functions of integer order m. The more common Bessel functions [28] $J_m(v\varrho)$ and $Y_m(v\varrho)$, of order m, are independent solutions of the equation

$$\varrho^2 \frac{d^2Z}{d\varrho^2} + \varrho \frac{dZ}{d\varrho} - (m^2 - v^2\varrho^2)Z = 0 \tag{7.274}$$

which is mathematically equivalent to eqn. 7.267 if we set $u = jv$ or $v = -ju = (\Gamma^2 - \gamma^2)^{1/2}$ and define v such that Im $v < 0$. Usually J_m and Y_m are called Bessel functions, of order m, of the first and second type, respectively. Two other closely related functions are defined by

$$H_m^{(1)}(v\varrho) = J_m(v\varrho) + jY_m(v\varrho) \tag{7.275}$$

$$H_m^{(2)}(v\varrho) = J_m(v\varrho) - jY_m(v\varrho) \tag{7.276}$$

Usually $H_m^{(1)}$ and $H_m^{(2)}$ are called Hankel functions, of order m, of the first and second kind, respectively.

The above functions, all of the Bessel type, are closely related. Two key connecting relationships (for $u = jv$) are

$$I_\nu(u\varrho) = \exp(jv\pi/2)\, J_\nu(v\varrho) \tag{7.277}$$

$$K_\nu(u\varrho) = (\pi/2)\exp[-j(\nu+1)\pi/2]\, H_\nu^{(2)}(v\varrho) \tag{7.278}$$

which are also valid when $\nu = m$ is an integer. In particular, we note that

$$I_0(u\varrho) = J_0(v\varrho), \qquad I_1(u\varrho) = jJ_1(v\varrho) \tag{7.279}$$

$$K_0(u\varrho) = \frac{\pi}{2j}[J_0(v\varrho) - jY_0(v\varrho)] = \frac{\pi}{2j} H_0^{(2)}(v\varrho) \tag{7.280}$$

$$K_1(u\varrho) = -\frac{\pi}{2}[J_1(v\varrho) - jY_1(v\varrho)] = -\frac{\pi}{2} H_1^{(2)}(v\varrho) \tag{7.281}$$

The limiting forms of the Bessel functions for small values of the argument follow from the defining series expansions. Thus, for example, we have for $|u\varrho| \ll 1$

$$I_\nu(u\varrho) \simeq \frac{(u\varrho)^\nu}{\nu!\, 2^\nu} \quad \text{and} \quad I_0(u\varrho) \simeq 1 \tag{7.282}$$

where again ν can also be an integer. Furthermore, if $|u\varrho| \ll 1$, we have

$$K_\nu(u\varrho) \simeq \frac{(\nu-1)!\, 2^{\nu-1}}{(u\varrho)^\nu} \quad \text{provided Re } \nu > 0 \tag{7.283}$$

$$K_0(u\varrho) \simeq -[\ln(u\varrho) + C - \ln 2] \tag{7.284}$$

where $C = 0.5772\ldots$ is Euler's constant.

When the arguments of the Bessel functions are sufficiently large, we may employ the asymptotic approximations. For example, when $|u\varrho| \gg 1$ we have

$$K_\nu(u\varrho) \simeq \left(\frac{\pi}{2u\varrho}\right)^{1/2} e^{-u\varrho}\left[1 + \frac{4\nu^2 - 1}{1!\, 8u\varrho} + \frac{(4\nu^2 - 1)(4\nu^2 - 3)}{2!\, (8u\varrho)^2} + \cdots\right] \tag{7.285}$$

On the other hand,

$$I_\nu(u\varrho) \simeq \left(\frac{1}{2\pi u\varrho}\right)^{1/2} e^{u\varrho} \tag{7.286}$$

is exponentially large as $\varrho \to \infty$.

The small-argument approximations for J_ν and Y_ν are given as follows (for $|v\varrho| \ll 1$):

$$J_\nu(v\varrho) \simeq \frac{(v\varrho)^\nu}{2^\nu \nu!} \quad \text{and} \quad J_0(v\varrho) \simeq 1 \tag{7.287}$$

where ν can also be an integer. Also (for $|v\varrho| \ll 1$) we have

$$Y_\nu(v\varrho) \simeq -\frac{2^\nu(\nu-1)!}{\pi(v\varrho)^\nu} \quad \text{provided} \quad \text{Re } \nu > 0 \tag{7.288}$$

$$Y_0(v\varrho) \simeq (2/\pi)[\ln(v\varrho) + C - \ln 2] \tag{7.289}$$

For large values of the argument $v\varrho$, the leading terms of the asymptotic series are useful:

$$J_\nu(v\varrho) \simeq \left(\frac{2}{\pi v\varrho}\right)^{1/2} \cos\left(v\varrho - \frac{\nu\pi}{2} - \frac{\pi}{4}\right) \tag{7.290}$$

$$Y_\nu(v\varrho) \simeq \left(\frac{2}{\pi v\varrho}\right)^{1/2} \sin\left(v\varrho - \frac{\nu\pi}{2} - \frac{\pi}{4}\right) \tag{7.291}$$

$$H_\nu^{(1)}(v\varrho) \simeq \left(\frac{2}{\pi v\varrho}\right)^{1/2} \exp\left[+j\left(v\varrho - \frac{\nu\pi}{2} - \frac{\pi}{4}\right)\right] \tag{7.292}$$

$$H_\nu^{(2)}(v\varrho) \simeq \left(\frac{2}{\pi v\varrho}\right)^{1/2} \exp\left[-j\left(v\varrho - \frac{\nu\pi}{2} - \frac{\pi}{4}\right)\right] \tag{7.293}$$

These forms are the phase of $v\varrho$ is in the range $-\pi$ to $+\pi$, but they are most appropriate and useful when $v\varrho$ is both large and real.

Recurrence relations, for any value of the argument z and order ν, are

$$\frac{d}{dz}[z^\nu I_\nu(z)] = z^\nu I_{\nu-1}(z) \tag{7.294}$$

$$\frac{d}{dz}[z^{-\nu} I_\nu(z)] = z^{-\nu} I_{\nu+1}(z) \tag{7.295}$$

$$\frac{d}{dz}[z^\nu K_\nu(z)] = -z^\nu K_{\nu-1}(z) \tag{7.296}$$

$$\frac{d}{dz}[z^{-\nu} K_\nu(z)] = -z^{-\nu} K_{\nu+1}(z) \tag{7.297}$$

$$I_{\nu-1}(z) = +I_{\nu+1}(z) = 2I_\nu'(z) \tag{7.298}$$

$$I_{\nu-1}(z) - I_{\nu+1}(z) = \frac{2\nu}{z} I_\nu(z) \tag{7.299}$$

$$K_{\nu-1}(z) + K_{\nu+1}(z) = -2K_\nu'(z) \tag{7.300}$$

$$K_{\nu-1}(z) - K_{\nu+1}(z) = -\frac{2\nu}{z} K_\nu(z) \tag{7.301}$$

These, of course, also hold if $v = m$ is an integer.

The following so-called Wronskian relation is also to be noted

$$I_v(z)K_v'(z) - I_v'(z)K_v(z) = -1/z \qquad (7.302)$$

where the prime indicates differentiation with respect to the argument z.

The Bessel functions can often be expressed in terms of integrals. Two examples are [29]

$$I_v(z) = \frac{(z/2)^v}{\sqrt{\pi}\,(v - \tfrac{1}{2})!} \int_{-1}^{+1} (1 - t^2)^{v-\tfrac{1}{2}} \cosh zt\, dt \qquad (7.303)$$

$$|\arg z| < \pi, \qquad \text{Re } v > -\tfrac{1}{2}$$

$$K_v(z) = \int_0^\infty e^{-z\cosh u} \cosh vu\, du \qquad (7.304)$$

$$\text{Re } z > 0, \qquad v \text{ arbitrary}$$

Corresponding results for the 'oscillatory' Bessel functions are obtained from the above by merely replacing z by jy and then using eqns. 7.277 and 7.278. Such results, of course, are listed in many places explicitly [26, 27, 30].

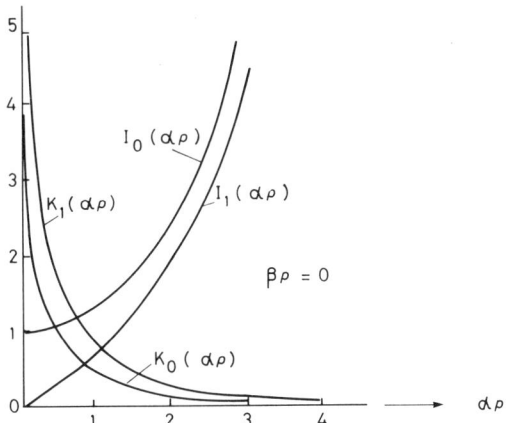

Fig. 7.35 Functions $I_m(\alpha\rho)$ and $K_m(\alpha\rho)$

When the argument of the Bessel functions is $\pi/4$ or $45°$, the representation in terms of the Kelvin functions are often used [27]. These are designated by the symbols ber, bei, ker and kei, with a subscript to denote the order. The essential connecting relationships are

$$\text{ber}_v x + j\,\text{bei}_v x = \exp\left(\frac{\pi}{2} vj\right) I_v(x\,e^{j\pi/4})$$

$$= \exp(\pi vj)\, J_v(x\,e^{-j\pi/4}) \qquad (7.305)$$

$$\text{ker}_v x + j\,\text{kei}_v x = \exp\left(-\frac{\pi}{2} vj\right) K_v(x\,e^{j\pi/4})$$

$$= -\tfrac{1}{2}\pi j \exp(-v\pi j)\, H_v^{(2)}(x\,e^{-j\pi/4}) \qquad (7.306)$$

where x is real. For zero order we have simply

$$\text{ber } x + j \text{ bei } x = I_0(x\, e^{j\pi/4}) = J_0(x\, e^{-j\pi/4}) \tag{7.307}$$

$$\text{ker } x + j \text{ kei } x = K_0(x\, e^{j\pi/4}) \tag{7.308}$$

where the subscript 0 is dropped, by convention, on the Kelvin functions.

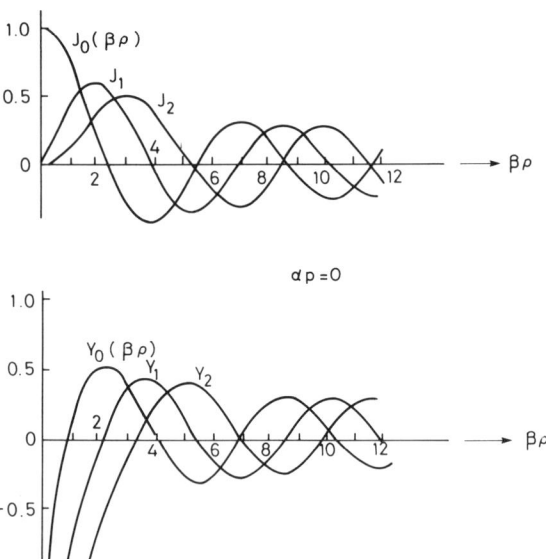

Fig. 7.36 Functions $J_m(\beta\rho)$ and $Y_m(\beta\rho)$

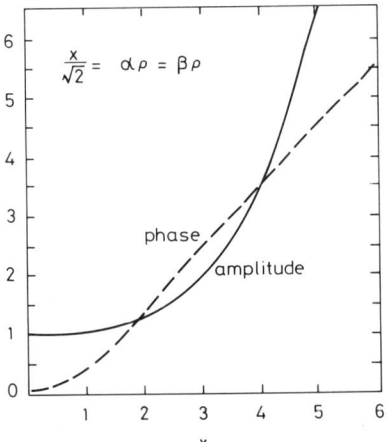

Fig. 7.37 Amplitude and phase of $I_0(j^{1/2}x)$ as a function of x

Some of the properties of Bessel functions are illustrated in graphical plots in Fig. 7.35 to 7.38 for the conditions indicated. In Fig. 7.35 we show the I and K functions, for order 0 and 1, plotted against the real

Cylindrical waves and scatter

argument $\alpha\varrho$. The singular nature of the K functions, as $\alpha\varrho$ approaches zero, is to be noted. Also we see that the I functions become exponentially large as $\alpha\varrho$ tends to infinity. The I and K functions, for real values of their arguments, are distinguished by monotonic behaviour. In Fig. 7.36 we plot the J and Y functions, for orders 0, 1 and 2, as a function of the real argument $\beta\varrho$. Here we note that the Y functions are singular as $\beta\varrho$ tends to zero but both functions J and Y are bounded as $\beta\varrho$ becomes large. Also both functions oscillate about the horizontal axis. The location of the zeros play a role in determining the cutoff frequencies, for example in circular waveguides. The other case of interest, when the argument of the I function is $\pi/4$, is sketched in Fig. 7.37, for the zero-order case only. This function characterises the skin effect of alternating currents in metallic wires of circular cross-section. Finally we give an example of the nature of the I_0 function when the argument is generally complex. These contour plots of equal amplitude and phase are shown in Fig. 7.38 (after Lösch [32]).

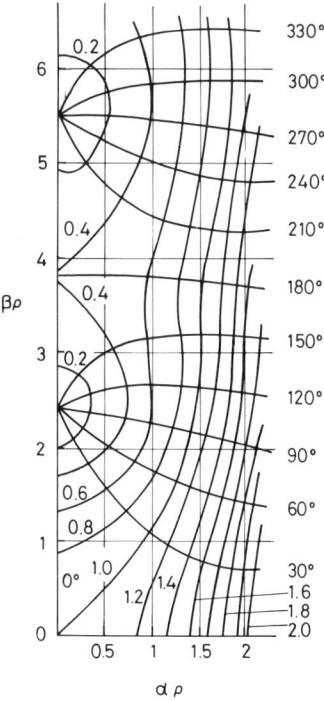

Fig. 7.38 Function $I_0\,(\alpha\rho + j\beta\rho)$ showing contours of equal amplitude and phase in degrees

7.19 References

1. J. W. STRUTT (Lord Rayleigh). On the electromagnetic theory of light, *Phil. Mag.*, 1881, **21**, pp. 81–101.
2. J. R. WAIT, *Electromagnetic Radiation from Cylindrical Structures*, Pergamon Press, 1959 (see Chapter 17).
3. D. S. JONES, *The Theory of Electromagnetism*, Pergamon Press, 1964 (see Chapter 8).
4. J. J. BOWMAN, T. B. A. SENIOR and P. L. E. USLENGHI (eds), *Electromagnetic and Acoustic Scattering*, North-Holland, 1969 (see Chapter 2).

5. M. KERKER, *The Scattering of Light*, Academic, 1969 (see Chapter 6).
6. H. C. VAN DE HULST, *Light Scattering from Small Particles* (see Chapter 15).
7. J. A. KONG, *Theory of Electromagnetic Waves*, Wiley, 1976 (see Chapter 6, Section 3).
8. J. W. STRUTT (Lord Rayleigh), On the self induction and resistance of straight conductors, *Phil Mag.*, 1886, **21**, pp. 381–94.
9. J. W. STRUTT (Lord Rayleigh), On the dispersal of light from a dielectric cylinder, *Phil Mag.*, 1918, **36**, pp. 365–76.
10. J. R. WAIT and A. MURPHY, Further studies of the influence of a ridge on low frequency ground waves, *J. Res. Natl. Bur. Stand.*, 1958, **61**, pp. 57–60.
11. D. A. HILL and J. R. WAIT, Radiation pattern of a low frequency beacon antenna located on a semi-elliptic terrain irregularity, *Arch. Elektron. & Übertragungstech.* 1973, **27**, pp. 293–6.
12. D. A. HILL and J. R. WAIT, Coupling between a dipole antenna and an infinite cable over an ideal ground plane, *Radio Sci.*, 1977, **12**, pp. 231–8.
13. J. R. WAIT, Excitation of an ensemble of parallel cables by an external dipole over a layered ground, *Arch. Elektron & Übertragungstech.*, 1977, **31**, pp. 483–93.
14. J. R. WAIT and R. E. WILKERSON, The sub-surface fields produced by a line current source on a non-flat earth, *Pure Appl. Geophys.*, 1942, **95**, pp. 150–6.
15. J. R. WAIT and K. P. SPIES, Subsurface electromagnetic fields of a line source on a conducting half-space, *Radio Sci.*, 1971, **6**, pp. 781–6.
16. J. R. WAIT, Scattering of a plane wave from a circular dielectric cylinder at oblique incidence, *Can. J. Phys.*, 1955, **33**, pp. 383–90: 1965, **43**, pp. 130–8.
17. J. H. RICHMOND, Scattering by a dielectric cylinder of arbitrary cross-sectional shape, *IEEE Trans.*, 1965, **AP-13**, pp. 335–41.
18. P. M. MORSE and H. FESHBACH, *Methods of Theoretical Physics*, McGraw Hill, 1953.
19. G. W. HOHMANN, Electromagnetic scattering by conductors in the earth near a line source of current, *Geophys.*, 1971, **36**, pp. 101–31.
20. A. Q. HOWARD, Jr., The electromagnetic fields of a subterranean cylindrical inhomogeneity, excited by a line source, *Geophys.*, 1972, **37**, pp. 975–94.
21. M. F. ISKANDER, C. H. DURNEY, D. G. BRAGG and B. H. OVARD, A microwave method for estimating absolute value of average ling water, *Radio Sci.*, 1982, **17**, pp. 11S–117S (Helsinki Symposium issue).
22. R. F. HARRINGTON, *Time Harmonic Electromagnetic Fields*, McGraw Hill, 1961, pp. 358–9.
23. J. H. RICHMOND, TE wave scattering by a dielectric cylinder of arbitrary cross-section shape, *IEEE Trans.*, 1966, **AP-14**, pp. 460–4.
24. M. CAUTERMAN, Q. T. NGUYEN and P. DEGAUQUE, Numerical modeling of geophysical structures excited by an electromagnetic ewave, *Radi Sci.*, 1982, **17**, pp. 1086–94.
25. P. F. TURNER, Regional hyperthermia with an annular phased array, *IEEE Trans*, 1984, **BME-31**, pp. 106–14.
26. G. N. WATSON, *A Treatise on the Theory of Bessel Functions*, 2nd edn, Cambridge University Press, 1944.
27. N. W. McLACHLAN, *Bessel Functions for Engineers*, 2nd edn, Oxford University Press, 1961.
28. M. ABRAMOWITZ and L. A. STEGEN (eds), *Handbook of Mathematical Functions*, Chapter 10 (by F. W. J. Olver), pp. 355–434, US Government Printing Office, 1964 (also available from Dover Publications in paperback).
29. N. N. LEBEDEV, *Special Functions and Their Applications*, Dover Publications, 1972 (see Chapter 5, pp. 98–142).
30. A. ERDELYI *et al.*, *Higher Transcendental Functions*, vol. 2, Chapter 7, pp. 1–114, McGraw Hill, 1953.
31. J. R. WAIT, *Electromagnetic Wave Theory*, Chapter 3, Section 3, 6, pp. 88–99, Harper and Row, 1985.
32. F. LÖSCH, *Tables of Higher Functions*, Chapter 9, pp. 158–299, McGraw Hill, 1960 (this is the 6th edition of the famous 'Jahnke and Emde').

Guide to reading for Bessel functions: 26 the old testament; 27 the Royal treatment; 28 the 'the standard'; 29 jewel in the Kremlin; 30 the new testament; 31 an entrée; 32 eine ausgezeichnete Ausstattung.

Other references

J. R. WAIT and D. A. HILL, Excitation of a homogeneous conductive cylinder of finite length by a prescribed axial current distribution, *Radio Sci.*, 1973, **8**, pp. 1169–76.

A. ISHIMARU, *Wave Propagation and Scattering in Random Media*, Academic Press, 1978.

J. R. WAIT, Focused heating in cylindrical targets, Part I, IEEE Trans., 1985, *MTT-33*, pp. 647–649.

J. R. WAIT, M. LUMORI, Part II, 1986, *MTT-34* pp. 357–359.

Chapter 8

Electromagnetic TM and TE spherical waves

8.1 Introduction

There is a broad range of problems in electromagnetics [1–6] where we deal with spherical regions. As such, it is convenient to represent our fields in terms of spherical waves. In this chapter, we outline the basic theory, beginning with first principles.

In what follows, we will deal with homogeneous regions that are bounded in some sense by concentric spherical boundaries. The key step in our analysis is to write the vector electric and magnetic fields $\boldsymbol{E}$ and $\boldsymbol{H}$ as the superposition of two partial fields in the fashion

$$\boldsymbol{E} = \boldsymbol{E}_e + \boldsymbol{E}_h \tag{8.1}$$

$$\boldsymbol{H} = \boldsymbol{H}_e + \boldsymbol{H}_h \tag{8.2}$$

These are chosen such that, in spherical co-ordinates (r, θ, ϕ),

$$H_{e,r} = 0 \tag{8.3}$$

$$E_{h,r} = 0 \tag{8.4}$$

The subscript e indicates that the partial vector fields $\boldsymbol{E}_e$ and $\boldsymbol{H}_e$ are associated with electric type modes. On the other hand, the partial fields designated by the subscript h are associated with magnetic type modes. Alternative descriptors are TM (transverse magnetic) when the magnetic field has a zero radial component, and TE (transverse electric) when the electric field has a zero radial component (see Chapter 4).

8.2 The Debye potentials

Now in a homogeneous medium in a region devoid of sources,

$$\operatorname{div} \boldsymbol{E} = 0 \quad \text{and} \quad \operatorname{div} \boldsymbol{H} = 0$$

Thus, it follows that

$$\frac{\partial}{\partial \theta}(\sin \theta \, H_{e,\theta}) + \frac{\partial H_{e,\phi}}{\partial \phi} = 0 \tag{8.5}$$

$$\frac{\partial}{\partial \theta}(\sin \theta \, E_{h,\theta}) + \frac{\partial E_{h,\phi}}{\partial \phi} = 0 \tag{8.6}$$

bearing in mind that

$$H_{e,r} = E_{h,r} = 0$$

We now introduce potential functions $\hat{U}$ and $\hat{V}$ as follows:

$$\sin\theta\, H_{e,\theta} = (\sigma + j\varepsilon\omega)\frac{\partial \hat{U}}{\partial\phi}, \qquad H_{e,\phi} = -(\sigma + j\varepsilon\omega)\frac{\partial \hat{U}}{\partial\theta} \qquad (8.7)$$

$$\sin\theta\, E_{h,\theta} = -j\mu\omega\frac{\partial \hat{V}}{\partial\phi}, \qquad E_{h,\phi} = j\mu\omega\frac{\partial \hat{V}}{\partial\theta} \qquad (8.8)$$

Clearly, with these choices, eqns. 8.5 and 8.6 are satisfied automatically. The introduction of the *admittivity* $\sigma + j\varepsilon\omega$ and the *impedivity* $j\mu\omega$ at this stage is for later convenience. We call $\hat{U}$ and $\hat{V}$ the *Debye potentials*.

Now we also note that scalar potentials u and v can be defined such that

$$rE_{e,\theta} = -\frac{\partial u}{\partial\theta}, \qquad r\sin\theta\, E_{e,\phi} = -\frac{\partial u}{\partial\phi} \qquad (8.9)$$

$$rH_{h,\theta} = -\frac{\partial v}{\partial\theta}, \qquad r\sin\theta\, H_{h,\phi} = -\frac{\partial v}{\partial\phi} \qquad (8.10)$$

As indicated, the transverse electric fields of the electric modes are obtained from the gradient of the potential u. A similar statement applies to the magnetic type modes.

Now from Maxwell's curl $\mathbf{H}$ equation, we can write

$$(\sigma + j\varepsilon\omega)r\sin\theta\, E_{e,r} = \frac{\partial}{\partial\theta}(\sin\theta\, H_{e,\phi}) - \frac{\partial H_{e,\theta}}{\partial\phi} \qquad (8.11)$$

$$(\sigma + j\varepsilon\omega)rE_{e,\theta} = -\frac{\partial}{\partial r}(rH_{e,\phi}) \qquad (8.12)$$

$$(\sigma + j\varepsilon\omega)rE_{e,\phi} = \frac{\partial}{\partial r}(rH_{e,\theta}) \qquad (8.13)$$

On substituting $E_{e,\theta}$ in eqn. 9.9 and $H_{e,\phi}$ in eqn. 8.7 into eqn. 8.12, we see that

$$\frac{\partial}{\partial\theta}\left[\frac{\partial(r\hat{U})}{\partial r} + u\right] = 0 \qquad (8.14a)$$

Similarly, when working with eqn. 8.13, we deduce that

$$\frac{\partial}{\partial\phi}\left[\frac{\partial(r\hat{U})}{\partial r} + u\right] = 0 \qquad (8.14b)$$

Without loss of generality we set the square bracket terms in eqns. 8.14a and b equal to zero. Thus, the scalar potential is related to the Debye potential by

$$u = -\frac{\partial}{\partial r}(r\hat{U}) \qquad (8.15)$$

An analogous procedure leads to

$$v = -\frac{\partial}{\partial r}(r\hat{V}) \qquad (8.16)$$

The next step is to use eqns. 8.7 and 8.11 to write

$$E_{e,r} = -\frac{1}{r}\left[\frac{1}{\sin\theta}\frac{\partial}{\partial\theta}\left(\sin\theta\frac{\partial \hat{U}}{\partial\theta}\right) + \frac{1}{\sin^2\theta}\frac{\partial^2 \hat{U}}{\partial\phi^2}\right] \quad (8.17)$$

Analogously,

$$H_{h,r} = -\frac{1}{r}\left[\frac{1}{\sin\theta}\frac{\partial}{\partial\theta}\left(\sin\theta\frac{\partial \hat{V}}{\partial\theta}\right) + \frac{1}{\sin^2\theta}\frac{\partial^2 \hat{V}}{\partial\phi^2}\right] \quad (8.18)$$

Now we note that, from Maxwell's curl E equation,

$$\frac{\partial(rE_{e,\theta})}{\partial r} - \frac{\partial E_{e,r}}{\partial\theta} = -j\mu\omega r H_{e,\phi} \quad (8.19)$$

and, from Maxwell's curl H equation,

$$\frac{\partial(rH_{h,\theta})}{\partial r} - \frac{\partial H_{h,r}}{\partial\theta} = (\sigma + j\varepsilon\omega)rE_{h,\phi} \quad (8.20)$$

Using eqns. 8.7, 8.8, 8.9 and 8.10 we can eliminate the transverse field components from eqns. 8.19 and 8.20 in order to write

$$E_{e,r} = -r\hat{U}\gamma^2 - \partial u/\partial r \quad (8.21)$$

$$H_{h,r} = -r\hat{V}\gamma^2 - \partial v/\partial r \quad (8.22)$$

In view of eqns. 8.15 and 8.16, the latter pair is equivalent to

$$E_{e,r} = \left(-\gamma^2 + \frac{\partial^2}{\partial r^2}\right)(r\hat{U}) = E_r \quad (8.23)$$

$$H_{h,r} = \left(-\gamma^2 + \frac{\partial^2}{\partial r^2}\right)(r\hat{V}) = H_r \quad (8.24)$$

where $\gamma^2 = j\mu\omega(\sigma + j\varepsilon\omega)$.

Then we easily deduce that the resultant transverse field components are

$$E_\theta = \frac{1}{r}\frac{\partial^2}{\partial\theta\partial r}(r\hat{U}) - \frac{j\mu\omega}{r\sin\theta}\frac{\partial}{\partial\phi}(r\hat{V}) \quad (8.25)$$

$$E_\phi = \frac{1}{r\sin\theta}\frac{\partial^2}{\partial\phi\partial r}(r\hat{U}) + \frac{j\mu\omega}{r}\frac{\partial}{\partial\theta}(r\hat{V}) \quad (8.26)$$

$$H_\theta = \frac{1}{r}\frac{\partial^2}{\partial\theta\partial r}(r\hat{V}) + \frac{\sigma + j\varepsilon\omega}{r\sin\theta}\frac{\partial}{\partial\phi}(r\hat{U}) \quad (8.27)$$

$$H_\phi = \frac{1}{r\sin\theta}\frac{\partial^2}{\partial\phi\partial r}(r\hat{V}) - \frac{\sigma + j\varepsilon\omega}{r}\frac{\partial}{\partial\theta}(r\hat{U}) \quad (8.28)$$

Finally, we equate the right-hand sides of eqns. 8.17 and 8.18 to 8.23 and 8.24, respectively, to show that

$$(\nabla^2 - \gamma^2)\begin{matrix}\hat{U}\\\hat{V}\end{matrix} = 0 \quad (8.29)$$

where ∇^2 is the Laplacian operator in spherical co-ordinates, given by

$$\nabla^2 = \frac{1}{r^2 \sin \theta} \left[\sin \theta \frac{\partial}{\partial r} \left(r^2 \frac{\partial}{\partial r} \right) + \frac{\partial}{\partial \theta} \left(\sin \theta \frac{\partial}{\partial \theta} \right) + \frac{1}{\sin \theta} \frac{\partial^2}{\partial \phi^2} \right]$$

(8.30)

We now have achieved a useful goal; as indicated by eqns. 8.23 to 8.30, the electric and magnetic field components are all expressible in terms of the two Debye potentials $\hat{U}$ and $\hat{V}$. Note that the $\hat{U}$ and $\hat{V}$ are not the same as the U and V employed in Chapter 7.

The solutions of eqns. 8.29 have the form

$$r^{-1} f(r) P_n^m (\cos \theta) e^{\pm jm\phi}$$

where P_n^m is the Legendre associated polynomial of order n and degree m [7]. The radial function $f_n(r)$ satisfies

$$\frac{d^2 f(r)}{dr^2} - \left[\gamma^2 + \frac{n(n+1)}{r^2} \right] f_n(r) = 0 \qquad (8.31)$$

Such solutions are single valued and bounded for all values of θ from 0 to π and ϕ from 0 to 2π.

It is now desirable to pin down our model in more specific terms [8, 9]. Thus we idealise the spherical target as a concentrically stratified structure. In the outermost layer of thickness d_1, the conductivity, permittivity and permeability are σ_1, ε_1 and μ_1, respectively. In terms of spherical co-ordinates (r, θ, ϕ) the layer is bounded by $r = a$ and $r = a - d_1$, where a is the earth's radius.

Now, without having to be more specific about the problem, we can deduce from eqns. 8.25 to 8.28 that, for a spectral order m and n,

$$E_{e,\theta} = -Z H_{e,\phi}, \qquad E_{e,\phi} = Z H_{e,\theta} \qquad (8.32)$$

where

$$Z = \frac{1}{(\sigma_1 + j\varepsilon_1 \omega) r \hat{U}} \frac{\partial}{\partial r} (r\hat{U}) \qquad (8.33)$$

and

$$E_{h,\theta} = -Y^{-1} H_{h,\phi}, \qquad E_{h,\phi} = Y^{-1} H_{h,\theta} \qquad (8.34)$$

where

$$Y = \frac{1}{j\mu_1 \omega r \hat{V}} \frac{\partial}{\partial r} (r\hat{V}) \qquad (8.35)$$

Now clearly eqns. 8.32 to 8.35 hold everywhere within the region $a > r > a - d_1$. In particular, they hold as $r \to a$, whence Z and Y are designated Z_1 and Y_1.

Exercise

Show that eqns. 8.23 to 8.28 may be derived in an alternative fashion if we identify $r\hat{U}$ with an electric Hertz vector with only a radial component Π_r, and analogously $r\hat{V}$ is identified with a magnetic Hertz vector with a radial component Π_r^*.

Solution

In the usual fashion, we could begin with the proposition that

$$H_e = (\sigma + j\varepsilon\omega) \operatorname{curl} \vec{\Pi} \qquad (8.36)$$

where
$$\vec{\Pi} = i_r \Pi_r = i_r r \hat{U}$$
and that
$$E_h = -j\mu\omega \text{ curl } \vec{\Pi}^* \tag{8.37}$$
where
$$\vec{\Pi}^* = i_r \Pi_r^* = i_r r \hat{V}$$
Then in accordance with Maxwell's equations,
$$E_e = \text{curl curl } \vec{\Pi} \tag{8.38}$$
$$H_h = \text{curl curl } \vec{\Pi}^* \tag{8.39}$$
Then noting that
$$E = E_e + E_h \tag{8.40a}$$
$$H = H_e + H_h \tag{8.40b}$$
we may obtain the explicit forms in spherical co-ordinates as given by eqns. 8.23 to 8.28.

8.3 External excitation of sphere

With the basic framework indicted above, we now can deal with the external excitation of the sphere. At the surface $r = a$ we stipulate, for the electric or TM modes of order n, that

$$E_{e,\theta} = -Z^{(n)} H_{e,\phi} \quad \text{and} \quad H_{e,\theta} = \frac{1}{Z^{(n)}} E_{e,\phi} \tag{8.41}$$

where $Z^{(n)}$ is the surface impedance for an individual TM mode. Correspondingly, we can write, for the magnetic or TE modes of order n, that

$$H_{h,\theta} = Y^{(n)} E_{h,\phi} \quad \text{and} \quad E_{h,\theta} = -\frac{1}{Y^{(n)}} H_{h,\phi} \tag{8.42}$$

where $Y^{(n)}$ is the surface admittance for an individual TE mode.

For the region $r > a$, we now need to postulate the required forms for the Debye potentials $\hat{U}$ and $\hat{V}$. If this external region is homogeneous with electrical constants σ_e, ε_e and μ_e, we can write for $a < r < r_s$, for r_s arbitrary, that

$$r\hat{U} = \sum_{m=-n}^{n} \sum_{n=0}^{\infty} [A_{m,n} \hat{J}_n(kr) + B_{m,n} \hat{H}_n(kr)] e^{-jm\phi} P_n^m (\cos \theta) \tag{8.43}$$

$$r\hat{V} = \sum_{m=-n}^{n} \sum_{n=0}^{\infty} [C_{m,n} \hat{J}_n(kr) + D_{mn} \hat{H}_n(kr)] e^{-jm\phi} P_n^m (\cos \theta) \tag{8.44}$$

The factors in square brackets in eqns. 8.43 and 8.44 are the radial solutions. They satisfy eqn. 8.31, where $\gamma^2 = -k^2 = j\mu_e\omega (\sigma_e + j\varepsilon_e\omega)$. $\hat{J}_n$ and $\hat{H}_n$ are two independent solutions of this equation. They are Schelkunoff's spherical Bessel functions [10], and are defined by

$$\hat{J}_n(z) = (\pi z/2)^{\frac{1}{2}} J_{n+\frac{1}{2}}(z) \tag{8.45}$$

$$\hat{H}_n(z) = (\pi z/2)^{\frac{1}{2}}[J_{n+\frac{1}{2}}(z) - jY_{n+\frac{1}{2}}(z)] \tag{8.46}$$

in terms of the conventional Bessel functions [11] J and Y of order $n + \frac{1}{2}$. The coefficients $A_{m,n}$ and $B_{m,n}$ in eqns. 8.43 and 8.44 are to be specified from the source conditions, and $B_{m,n}$ and $D_{m,n}$ are determined by the boundary conditions, as we shall indicate below. The summation over n in eqns. 8.43 and 8.44 is quite generally from $n = 0$ to ∞ through all integers, whereas m ranges from $-n$ to $+n$ through all integers including zero (i.e. note that $P_n^m = 0$ for $|m| > n$).

In writing down the forms of $\hat{U}$ and $\hat{V}$ for eqns. 8.43 and 8.44, respectively, we have restricted attention to the region $a < r < r_s$. Here r_s is chosen, at least for the present discussion, to exclude any active sources from this region.

In view of Schelkunoff's normalisation [10] we have the following very simple asymptotic behaviour for $kr \gg n$:

$$\hat{J}_n(kr) \sim \sin kr \tag{8.47}$$

$$\hat{H}_n(kr) \sim \exp(-jkr) \tag{8.48}$$

In fact for $n = 0$ these are exactly true for all kr. In any case we can identify the terms associated with $\hat{H}_n(kr)$ as outgoing waves, whereas the terms with $\hat{J}_n(kr)$ are related to the excitation of the sphere by sources at or beyond the surface $r = r_s$.

We now wish to obtain a relationship between the unknown coefficients by exploiting the boundary conditions [12]. Here, for example, we note that

$$E_{e,\theta}]_{r=a} = \frac{1}{a}\frac{\partial}{\partial \theta}\left[\frac{\partial}{\partial r}(r\hat{U})\right]_{r=a} \tag{8.49}$$

$$H_{e,\phi}]_{r=a} = -\frac{\sigma_e + j\varepsilon_e\omega}{a}\left[\frac{\partial}{\partial \theta}(r\hat{U})\right]_{r=a} \tag{8.50}$$

Thus, it is clear that the first of eqn. 8.41 is satisfied if

$$\left\{\frac{\partial}{\partial r}[A_{m,n}\hat{J}_n(kr) + B_{m,n}\hat{H}_n(kr)]\right\}_{r=a}$$
$$= Z^{(n)}(\sigma_e + j\varepsilon_e\omega)[A_{m,n}\hat{J}_n(ka) + B_{m,n}\hat{H}_n(ka)] \tag{8.51}$$

Here, we may solve for $B_{m,n}$ in terms of $A_{m,n}$ to yield

$$B_{m,n} = -\left[\frac{\hat{J}_n'(ka) - j\Delta_n\hat{J}_n(ka)}{\hat{H}_n'(ka) - j\Delta_n\hat{H}_n(ka)}\right]A_{m,n} \tag{8.52}$$

where

$$\Delta_n = Z^{(n)}/\eta \tag{8.53}$$

$$\eta = jk/(\sigma_e + j\varepsilon_e\omega) = [j\mu_e\omega/(\sigma_e + j\varepsilon_e\omega)]^{1/2}$$

The dimensionless quantity Δ_n is the surface impedance for TM modes normalised by the intrinsic impedance η of the external medium. The prime over the functions $\hat{J}$ and $\hat{H}$ in eqn. 8.52 indicates differentiation with respect to the indicated argument, i.e.

$$\hat{J}'_n(ka) = [d\hat{J}_n(z)/dz]_{z=ka} \tag{8.54}$$

It is easy to demonstrate that the second of eqn. 8.41 is also satisfied by eqn. 8.43 when $B_{m,n}$ is given by eqn. 8.52. In a similar vein, we can work with the TE or magnetic modes and apply eqns. 8.42 and 8.44 to deduce that

$$D_{m,n} = -\left[\frac{\hat{J}'_n(ka) - j\delta_n \hat{J}_n(ka)}{\hat{H}'_n(ka) - j\delta_n \hat{H}_n(ka)}\right] C_{m,n} \tag{8.55}$$

where

$$\delta_n = Y^{(n)}\eta \tag{8.56}$$

The dimensionless quantity δ_n is the surface admittance for TE modes normalised by the intrinsic admittance $1/\eta$ of the external medium.

8.4 Concentrically layered model

We now model the target itself as a concentrically layered structure as indicated in Fig. 8.1 where, for the sake of clarity, only a sector is shown. A typical layer has electric and magnetic properties σ_p, ε_p and μ_p that are constant within the region $a_p > r > a_{p+1}$. Here, p designates the pth layer, where $p = 1, 2, 3, \ldots, P - 1, P$ and where P is the total number of layers. The propagation constant γ_p and the characteristic impedance η_p of the pth region are defined by

$$\gamma_p = [j\mu_p\omega(\sigma_p + j\varepsilon_p\omega)]^{\frac{1}{2}}$$

$$\eta_p = [j\mu_p\omega/(\sigma_p + j\varepsilon_p\omega)]^{\frac{1}{2}}$$

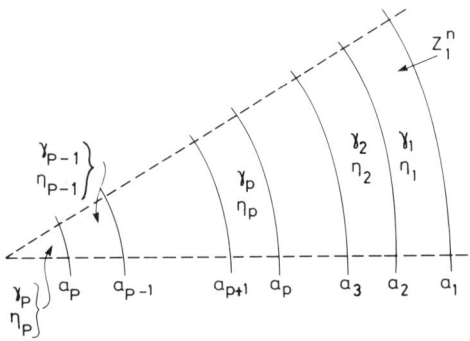

Fig. 8.1 Illustrating the equivalent nonuniform transmission line of P sections

We now consider the appropriate form of the solutions for the fields within the pth concentric region that is bounded by $r < a_p$ and $r > a_{p+1}$. This region is shown in Fig. 8.2.

To illustrate the procedure, we deal with electric or TM modes only. The corresponding results for the magnetic or TE modes follow by analogy. Within the region $a_{p+1} < r < a_p$ we may write

$$r\hat{U} = \sum_{n=0}^{\infty} \sum_{m=-n}^{n} [\alpha_{m,n} \hat{I}_n(\gamma_p r) + \beta_{m,n} \hat{K}_n(\gamma_p r)] e^{-jm\phi} P_n^m(\cos\theta) \tag{8.57}$$

The factor in square brackets is the radial solution. It satisfies eqn. 8.31 with $\gamma = \gamma_p$. Here, $\hat{I}_n$ and $\hat{K}_n$ are independent solutions and, again following Schelkunoff, they are defined by

$$\hat{I}_n(z) = (\pi z/2)^{\frac{1}{2}} I_{n+\frac{1}{2}}(z) \tag{8.58}$$

$$\hat{K}_n(z) = (2z/\pi)^{\frac{1}{2}} K_{n+\frac{1}{2}}(z) \tag{8.59}$$

in terms of the conventional modified Bessel functions of order $n + \frac{1}{2}$. The coefficients $\alpha_{m,n}$ and $\beta_{m,n}$ are as yet undetermined.

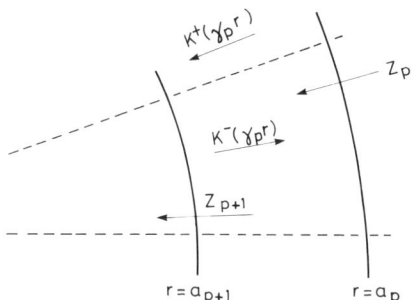

Fig. 8.2 Illustrating the input impedence Z_p at $r = a_p$ for a terminal impedance Z_{p+1} at $r = a_{p+1}$ for the pth section

When $z \gg n$ we have the simple asymptotic approximate forms

$$\hat{K}_n(z) = \exp(-z) \tag{8.60}$$

$$\hat{I}_n(z) = \sinh z \tag{8.61}$$

which in fact are exact for $n = 0$. It is also worth noting that

$$\hat{K}_n(z) = e^{-z} \sum_{s=0}^{n} \frac{(n+s)!}{s!(n-s)!(2z)^s} \tag{8.62}$$

$$\hat{I}_n(z) = \frac{1}{2}\left[e^z \sum_{s=0}^{n} \frac{(-1)^s(n+s)!}{s!(n-s)!(2z)^s} + (-1)^{n+1}\hat{K}_n(z) \right] \tag{8.63}$$

which are exact finite series representations that hold for any integer n. The summations extend over all intergers s from 0 to n. Finally, we note that the following identities hold:

$$\hat{I}_n(jz) = j^{n+1} \hat{J}_n(z) \tag{8.64}$$

$$\hat{K}_n(jz) = j^{-n-1} \hat{H}_n(z) \tag{8.65}$$

$$\hat{I}'_n(jz) = j^n \hat{J}'_n(z) \tag{8.66}$$

$$\hat{K}'_n(jz) = j^{-n-2} \hat{H}'_n(z) \tag{8.67}$$

The 'modified' spherical Bessel functions on the left sides of the above four equations are used as a matter of convenience when the region is highly dissipative. We choose these forms when dealing with the concentric spherical regions for $r < a$. (We propose here that $\hat{I}_n$ and $\hat{K}_n$ be called *Schelkunoff functions* in recognition of the tremendous contribution by Sergei Schelkunoff to the mathematics of electrical wave transmission and radiation in dissipative media.)

8.5 The iterative solution

We now return to eqn. 8.57 and note that, for the pth region, the radial solution for the nth order TM mode has the form

$$_p f_n(r) = {}_p\alpha_{m,n}\hat{I}_n(\gamma_p r) + {}_p\beta_{m,n}\hat{K}_n(\gamma_p r) \tag{8.68}$$

Now this form of the solution will hold for any concentric region. For example, in the outermost region or surface layers we merely replace γ_p by γ_1, with two new sets of coefficients ${}_1\alpha_{m,n}$ and ${}_1\beta_{m,n}$. To be explicit,

$$_1 f_n(r) = {}_1\alpha_{m,n}\hat{I}_n(\gamma_1 r) + {}_1\beta_{m,n}\hat{K}_n(\gamma_1 r) \tag{8.69}$$

for the region $a_1 > r > a_2$.

When we deal with the homogeneous core region (i.e. $0 < r < a_P$), the solution actually has the form

$$_P f_n(r) = {}_P\alpha_{m,n}\hat{I}_n(\gamma_P r) \tag{8.70}$$

which is finite as $r \to 0$. The infinite behaviour of $\hat{K}_n(\gamma_P r)$ as $r \to 0$ requires that, in effect, ${}_P\beta_{m,n}$ be identically zero.

We now note that, proceeding from the Pth region out to the first region (i.e. $p = 1$), we have $2P - 1$ unknown coefficients. Now the objective is to obtain an expression for the surface impedance $Z_1^{(n)}$ at $r = a_1 = a$. This is defined, in accordance with eqn. 8.33, by

$$Z_1^{(n)} = \frac{1}{\sigma_1 + j\varepsilon_1\omega} \frac{\frac{\partial}{\partial r}[{}_1\alpha_{m,n}\hat{I}_n(\gamma_1 r) + {}_1\beta_{m,n}\hat{K}_n(\gamma_1 r)]}{{}_1\alpha_{m,n}\hat{I}_n(\gamma_1 r) + {}_1\beta_{m,n}\hat{K}_n(\gamma_1 r)} \tag{8.71}$$

evaluated at $r = a_1$. An equivalent statement is

$$Z_1^{(n)} = \eta_1 \frac{\hat{I}'_n(\gamma_1 a_1) + ({}_1\beta_{m,n}/{}_1\alpha_{m,n})\hat{K}'_n(\gamma_1 a_1)}{\hat{I}_n(\gamma_1 a_1) + ({}_1\beta_{m,n}/{}_1\alpha_{m,n})\hat{K}_n(\gamma_1 a_1)} \tag{8.72}$$

Now the boundary conditions tell us that the tangential electric fields and the tangential magnetic fields are continuous at each of the spherical interfaces from $r = a_P, a_{P-1}, \ldots, a_p, \ldots, a_3, a_2$. These conditions require that

$$\left[\frac{1}{r}\frac{\partial}{\partial r}{}_{p+1}f_n(r)\right]_{r=a_{p+1}} = \left[\frac{1}{r}\frac{\partial}{\partial r}{}_p f_n(r)\right]_{r=a_{p+1}} \tag{8.73}$$

and

$$(\sigma_{p+1} + j\varepsilon_{p+1}\omega)_{p+1}f_n(a_{p+1}) = (\sigma_p + j\varepsilon_p\omega)_p f_n(a_{p+1}) \tag{8.74}$$

Thus we have $2P - 2$ linear equations to solve for the $2P - 2$ unknown coefficients. In a procedural sense, we would first apply the boundary conditions at the $r = a_P$ interface in order to obtain an expression for the ratio ${}_{P+1}\beta_{m,n}/{}_{P+1}\alpha_{m,n}$. The process is repeated at the $r = a_{P-1}$ interface and continued to the $r = a_2$ interface, whence the ratio ${}_1\beta_{m,n}/{}_1\alpha_{m,n}$ is fully specified. Thus the desired form for $Z_1^{(n)}$ is obtained from 8.72.

The algebraic process described above is purely iterative and requires no tedious matrix inversion. In fact, we can draw an analogy with transmission line theory. However, the one complication is that the transmission line is nonuniform in the sense that the properties of the line depend on the radial distance along the line.

To explain our analogy we need to introduce some definitions. We refer specifically to the pth section as illustrated in Fig. 8.2. The characteristic impedances of the equivalent transmission line section for $a_p > r > a_{p+1}$ for the nth TM mode are

$$K_n^+(\gamma_p r) = \eta_p \frac{\hat{I}_n'(\gamma_p r)}{\hat{I}_n(\gamma_p r)} \tag{8.75}$$

$$K_n^-(\gamma_p r) = -\eta_p \frac{\hat{K}_n'(\gamma_p r)}{\hat{K}_n(\gamma_p r)} \tag{8.76}$$

The superscripts $+$ and $-$ refer specifically to the inward and outward directions, respectively. (No confusion should exist in using the notation K_n^+ and K_n^- for wave impedances and $\hat{K}_n$ for the Schelkunoff function.) For example, the inward-looking wave function for the nth mode is

$$(rU_n) = \hat{I}_n(\gamma_p r) e^{-jm\phi} P_n^m(\cos\theta) \tag{8.77}$$

Then the associated inward-looking impedance is defined by

$$K_n^+ = -E_{e,\theta}^+/H_{e,\phi}^+ = E_{e,\phi}^+/H_{e,\theta}^+ = \frac{1}{\sigma_p + j\varepsilon_p\omega} \frac{\partial(r\hat{U})/\partial r}{r\hat{U}} \tag{8.78}$$

which leads to eqn. 8.75. An analogous development leads to eqn. 8.76.

We also introduce the following transmission ratios, for a mode of order n

$$s_{p,n} = \frac{\hat{K}_n'(\gamma_p a_p)}{\hat{K}_n'(\gamma_p a_{p+1})} \frac{\hat{I}_n'(\gamma_p a_{p+1})}{\hat{I}_n'(\gamma_p a_p)} \tag{8.79}$$

$$S_{p,n} = \frac{\hat{K}_n(\gamma_p a_p)}{\hat{K}_n(\gamma_p a_{p+1})} \frac{\hat{I}_n(\gamma_p a_{p+1})}{\hat{I}_n(\gamma_p a_p)} \tag{8.80}$$

For example, $s_{p,n}$ can be interpreted as the product of (a) the ratio of the electric field $E_{e,\theta}^+$ or $E_{e,\phi}^+$ at $r = a_{p+1}$ to that at $r = a_p$, and (b) the ratio of $E_{e,\theta}^-$ or $E_{e,\phi}^-$ at $r = a_p$ to that at $r = a_{p+1}$. The expression for $S_{p,n}$ has a corresponding interpretation in terms of the tangential magnetic fields.

Using the above defined parameters we can write eqn. 8.71 in the explicit form

$$Z_1^{(n)} = K_n^+(\gamma_1 a_1) \frac{1 + r_{1,n} s_{1,n}}{1 + R_{1,n} S_{1,n}} \tag{8.81}$$

where

$$r_{1,n} = -\frac{1/Z_2^{(n)} - 1/K_n^+(\gamma_1 a_2)}{1/Z_2^{(n)} + 1/K_n^-(\gamma_1 a_2)} \tag{8.82}$$

$$R_{1,n} = -\frac{Z_2^{(n)} - K_n^+(\gamma_1 a_2)}{Z_2^{(n)} + K_n^-(\gamma_1 a_2)} \tag{8.83}$$

and, in turn,

$$Z_2^{(n)} = K_n^+(\gamma_2 a_2) \frac{1 + r_{2,n} s_{2,n}}{1 + R_{2,n} S_{2,n}} \qquad (8.84)$$

For any value of p we have

$$r_{p,n} = -\frac{1/Z_{p+1}^{(n)} - 1/K_n^+(\gamma_p a_{p+1})}{1/Z_{p+1}^{(n)} + 1/K_n^-(\gamma_p a_{p+1})} \qquad (8.85)$$

$$R_{p,n} = -\frac{Z_{p+1}^{(n)} - K_n^+(\gamma_p a_{p+1})}{Z_{p+1}^{(n)} + K_n^-(\gamma_p a_{p+1})} \qquad (8.86)$$

but, of course, for $p = P$,

$$Z_P^{(n)} = K_n^+(\gamma_P a_P) = \eta_P \frac{\hat{I}_n'(\gamma_P a_P)}{\hat{I}_n(\gamma_P a_P)} \qquad (8.87)$$

The determination of the admittance function $Y_1^{(n)}$ for the TE modes proceeds in a similar vein. Here, it is convenient to regard the problem as the dual of the TM situation. Thus impedances become admittances. For example, the characteristic admittances for the pth section are written

$$M_n^+(\gamma_p r) = \frac{1}{\eta_p} \frac{\hat{I}_n'(\gamma_p r)}{\hat{I}_n(\gamma_p r)} \qquad (8.88a)$$

for the 'ingoing' wave, and

$$M_n^-(\gamma_p r) = -\frac{1}{\eta_p} \frac{\hat{K}_n'(\gamma_p r)}{\hat{K}_n(\gamma_p r)} \qquad (8.88b)$$

for the 'outgoing' wave. Then in analogy to eqn. 8.81 we have the following expression for the input admittance for the TE modes of order n at $r = a_1$:

$$Y_1^{(n)} = M_n^+(\gamma_1 a_1) \frac{1 + \tilde{r}_{1,n} s_{1,n}}{1 + \tilde{R}_{1,n} S_{1,n}} \qquad (8.89)$$

where $s_{1,n}$ and $S_{1,n}$ are as defined by eqns. 8.79 and 8.80, but now

$$\tilde{r}_{1,n} = -\frac{1/Y_2^{(n)} - 1/M_n^+(\gamma_1 a_2)}{1/Y_2^{(n)} + 1/M_n^-(\gamma_1 a_2)} \qquad (8.90)$$

$$\tilde{R}_{1,n} = -\frac{Y_2^{(n)} - M_n^+(\gamma_1 a_2)}{Y_2^{(n)} + M_n^-(\gamma_1 a_2)} \qquad (8.91)$$

For any value of p, $\tilde{r}_{p,n}$ and $\tilde{R}_{p,n}$ follow analogously from eqns. 8.85 and 8.86.

Exercise

Verify by a direct solution of eqns. 8.73 and 8.74, for a two-layered sphere, that the required form for Z_1 as given by eqn. 8.81 is obtained when $P = 2$.

Solution

We make an obvious contraction in notation and note that, from eqn. 8.68,

$$f(r) = A\hat{I}(\gamma_1 r) + B\hat{K}(\gamma_1 r) \qquad (8.92)$$

for $a_1 > r > a_2$, while

$$f(r) = C\hat{I}(\gamma_2 r) \tag{8.93}$$

for $r < a_2$, where A, B and C are to be determined. The boundary conditions as given by eqns. 8.73 and 8.74 are now met, at $r = a_2$, if

$$\gamma_1[A\hat{I}'(\gamma_1 a_2) + B\hat{K}'(\gamma_1 a_2)] = C\gamma_2 \hat{I}'(\gamma_2 a_2) \tag{8.94}$$

and

$$(\sigma_1 + j\varepsilon_1 \omega)[A\hat{I}(\gamma_1 a_2) + B\hat{K}(\gamma_1 a_2)] = (\sigma_2 + j\varepsilon_2 \omega)C\hat{I}(\gamma_2 a_2) \tag{8.95}$$

We can solve for B in terms of A by eliminating C. Thus, we deduce that

$$\frac{B}{A} = \frac{K^+(\gamma_1 a_2) - K^+(\gamma_2 a_2)}{K^-(\gamma_1 a_2) + K^+(\gamma_2 a_2)} \frac{\hat{I}(\gamma_1 a_2)}{\hat{K}(\gamma_1 a_2)} \tag{8.96}$$

or equivalently

$$\frac{B}{A} = \frac{1/K^+(\gamma_1 a_2) - 1/K^+(\gamma_2 a_2)}{1/K^-(\gamma_1 a_2) + 1/K^+(\gamma_2 a_2)} \frac{\hat{I}'(\gamma_1 a_2)}{\hat{K}'(\gamma_1 a_2)} \tag{8.97}$$

where

$$K^+(\gamma r) = \eta \hat{I}'(\gamma r)/\hat{I}(\gamma r) \tag{8.98}$$

$$K^-(\gamma r) = -\eta \hat{K}'(\gamma r)/\hat{K}(\gamma r) \tag{8.99}$$

Now the desired surface impedance is

$$Z_1 = \frac{1}{\sigma_1 + j\varepsilon_1 \omega} \left. \frac{\partial f(r)/\partial r}{f(r)} \right]_{r=a_1}$$

$$= \eta_1 \frac{\hat{I}'(\gamma_1 a_1) + (B/A)\hat{K}'(\gamma_1 a_1)}{\hat{I}(\gamma_1 a_1) + (B/A)\hat{K}(\gamma_1 a_1)} \tag{8.100}$$

This result is consistent with eqn. 8.81 in the special case where $P = 2$.

8.6 Scattering of a plane wave from a sphere: radar cross-sections

An important concrete case of the foregoing analysis is when a plane wave is incident on the spherical target. This situation will be considered explicitly in what follows. We will also consider the case when the surrounding medium is perfectly insulating so that k is real.

With reference to Fig. 8.3, the incident plane wave is defined by

$$E_x^{inc} = E_0 e^{-jkz} = E_0 e^{-jkr\cos\theta} \tag{8.101}$$

$$H_y^{inc} = \frac{E_0}{\eta_0} e^{-jkz} = \frac{E_0}{\eta_0} e^{-jkr\cos\theta} \tag{8.102}$$

Now the total fields in the region $r > a$ can again be obtained from Debye potentials $\hat{U}$ and $\hat{V}$ as indicated by eqns. 8.23 to 8.28, where now $\sigma = 0$. In the present case,

$$(\nabla^2 + k^2) \frac{\hat{U}}{\hat{V}} = 0 \tag{8.103}$$

for the region $r > a$.

We now make recourse to the following addition theorem [10]:

$$e^{-jkr\cos\theta} = -\sum_{n=0}^{\infty} j^{-n}(2n+1)\frac{\hat{J}_n(kr)}{kr} P_n(\cos\theta) \quad (8.104)$$

where

$$\hat{J}_n(kr) = \left(\frac{\pi kr}{2}\right)^{\frac{1}{2}} J_{n+\frac{1}{2}}(kr)$$

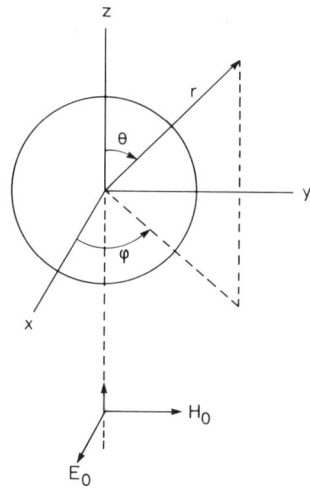

Fig. 8.3 Plane wave incident on the spherical target

is the preferred form of the Bessel function to use when kr is real. Then it follows that

$$E_r^{\text{inc}} = \cos\phi \sin\theta\, E_x^{\text{inc}} = E_0 \frac{\cos\phi}{jkr} \frac{\partial}{\partial\theta}(e^{-jkr\cos\theta})$$

$$= +j\frac{E_0 \cos\phi}{(kr)^2} \sum_{n=1}^{\infty} j^{-n}(2n+1)\hat{J}_n(kr) P_n^1(\cos\theta) \quad (8.105)$$

where

$$P_n^1(\cos\theta) = \frac{\partial}{\partial\theta} P_n(\cos\theta) \quad (8.106)$$

Now we note that

$$\left\{\frac{\partial^2}{\partial r^2} + \left[k^2 - \frac{n(n+1)}{r^2}\right]\right\} \hat{J}_n(kr) = 0 \quad (8.107)$$

and, of course,

$$E_r^{\text{inc}} = \left(\frac{\partial^2}{\partial r^2} + k^2\right)(r\hat{U}^{\text{inc}}) \quad (8.108)$$

Thus, we deduce

$$r\hat{U}^{\text{inc}} = \frac{E_0}{j\varepsilon\omega k} \sum_{n=1}^{\infty} a_n \hat{J}_n(kr) P_n^1(\cos\theta) \cos\phi \quad (8.109)$$

where

$$a_n = \frac{j^{-n}(2n+1)}{n(n+1)} \tag{8.110}$$

In a similar fashion we deduce that

$$r\hat{V}^{\text{inc}} = \frac{E_0}{j\mu\omega k} \sum_{n=1}^{\infty} a_n \hat{J}_n(kr) P_n^1 (\cos\theta) \sin\phi \tag{8.111}$$

The required form for the secondary Debye potentials for the region are now evident:

$$r\hat{U}^{\text{sc}} = \frac{E_0}{j\varepsilon\omega k} \sum_{n=1}^{\infty} b_n \hat{H}_n(kr) P_n^1 (\cos\theta) \cos\phi \tag{8.112}$$

$$r\hat{V}^{\text{sc}} = \frac{E_0}{j\mu\omega k} \sum_{n=1}^{\infty} c_n \hat{H}_n(kr) P_n^1 (\cos\theta) \sin\phi \tag{8.113}$$

where b_n and c_n are coefficients to be determined by the boundary conditions. In noting the forms of eqns. 8.52 and 8.55 it is evident that

$$\frac{b_n}{a_n} = -\left[\frac{\hat{J}'_n(ka) - j\Delta_n \hat{J}_n(ka)}{\hat{H}'_n(ka) - j\Delta_n \hat{H}_n(ka)}\right] = B_n \tag{8.114}$$

and

$$\frac{c_n}{a_n} = -\left[\frac{\hat{J}'_n(ka) - j\delta_n \hat{J}_n(ka)}{\hat{H}'_n(ka) - j\delta_n \hat{H}_n(ka)}\right] = C_n \tag{8.115}$$

where

$$\Delta_n = Z_1^{(n)}/\eta_0, \qquad \eta_0 = \mu\omega/k \tag{8.116}$$

$$\delta_n = Y_1^{(n)} \eta_0 \tag{8.117}$$

are the dimensionless immitance parameters which depend on the degree n. Here $Z_1^{(n)}$ and $Y_1^{(n)}$ are given by eqns. 8.81 and 8.89, respectively, in terms of the properties of the radially stratified target.

The secondary or scattered fields are obtained from $\hat{U}^{\text{sc}}$ and $\hat{V}^{\text{sc}}$ using eqns. 8.23 to 8.28. In the far field where $|kr| \gg 1$ and $r \gg a$ it is not difficult to deduce that

$$E_\theta^{\text{sc}} \simeq +\frac{jE_0}{kr} e^{-jkr} P(\theta) \cos\phi \simeq \eta_0 H_\phi^{\text{sc}} \tag{8.118}$$

$$E_\phi^{\text{sc}} \simeq -\frac{jE_0}{kr} e^{-jkr} Q(\theta) \sin\phi \simeq -\eta_0 H_\theta^{\text{sc}} \tag{8.119}$$

where

$$P(\theta) = \sum_{n=1}^{\infty} \frac{2n+1}{n(n+1)} \left[B_n \frac{d}{d\theta} P_n^1(\cos\theta) + C_n \frac{P_n^1(\cos\theta)}{\sin\theta}\right] \tag{8.120}$$

$$Q(\theta) = \sum_{n=1}^{\infty} \frac{2n+1}{n(n+1)} \left[B_n \frac{P_n^1(\cos\theta)}{\sin\theta} + C_n \frac{d}{d\theta} P_n^1(\cos\theta)\right] \tag{8.121}$$

To present numerical results in an economical manner, two 'cross-sections' $\sigma_e(\theta)$ and $\sigma_h(\theta)$ are introduced. They are defined in terms of the far fields as follows:

$$\sigma_e(\theta) = [4\pi r^2 |E_\theta^{sc}|^2 / E_0^2]_{\substack{r \to \infty \\ \phi = 0}} \quad (8.122)$$

$$\sigma_h(\theta) = [4\pi r^2 |E_\phi^{sc}|^2 / E_0^2]_{\substack{r \to \infty \\ \phi = 90°}} \quad (8.123)$$

These functions are conveniently normalised by dividing by the geometric cross-section πa^2. Then, in terms of $P(\theta)$ and $Q(\theta)$ as defined by eqns. 8.120 and 8.121, we see that

$$\frac{\sigma_e(\theta)}{\pi a^2} = \frac{4}{(ka)^2} |P(\theta)|^2 \quad (E\text{-plane}) \quad (8.124)$$

$$\frac{\sigma_h(\theta)}{\pi a^2} = \frac{4}{(ka)^2} |Q(\theta)|^2 \quad (H\text{-plane}) \quad (8.125)$$

The normalised cross-sections (denoted CS) for the E-plane (i.e. $\phi = 0°$) and H-plane (i.e. $\phi = 90°$) are plotted in Figs. 8.4 and 8.5. These results were computed from the power series expansions for $P(\theta)$ and $Q(\theta)$. To secure convergence, at least $2ka$ terms were used in the summation.

In presenting the numerical data, the basic parameters are the circumference ka of the sphere in wavelengths, and the value of the surface immitances (i.e. impedances or admittnces). We simplify matters greatly here by assuming that Δ_n and δ_n, as defined by eqns. 8.116 and 8.117, are effectively independent of mode number. This idealisation is equivalent to the statement that the tangential electric and magnetic fields are related by a fixed impedance Z_1 in the manner

$$E_\theta = -Z_1 H_\phi \quad \quad \quad (8.126)$$
$$E_\phi = Z_1 H_\theta \quad \Big]_{r=a} \quad (8.127)$$

For example, in the case of a two-layered sphere we may deduce directly from eqn. 8.100 that

$$Z_1 \simeq \eta_1 \frac{\eta_2 + \eta_1 \tanh \gamma_1 d}{\eta_1 + \eta_2 \tanh \gamma_1 d} \quad (8.128)$$

where $d = a_1 - a_2$ is the thickness of the upper layer, provided $d \ll a_1$ and $|\gamma_1|^2$ and $|\gamma_2|^2 \gg k^2$. Under these conditions,

$$\Delta_n \simeq 1/\delta_n \simeq Z_1/\eta_0 \quad (8.129)$$

In other words, the outer surface of the target sphere exhibits an impedance of an equivalent two-layer planar structure.

The curves shown in Fig. 8.4 actually correspond to the case where Z_1 is either zero or purely inductive. In effect, we are saying that

$$Z_1 = j\eta_0 \chi = j\,120\pi\,\chi \quad \Omega$$

where χ is the normalised surface inductive reactance. Physically, this would correspond to a perfectly conducting sphere of radius a coated by a thin layer of dielectric. Note that if $|\omega \gamma_1 d| \ll 1$ and $|\gamma_2/\gamma_1|^2 \gg 1$,

Electromagnetic TM and TE spherical waves

$$Z_1 \simeq \eta_1 \gamma_1 d = j\mu_1 \omega d \qquad (8.130)$$

so that

$$\chi \simeq \mu_1 \omega d / \eta_0 = kd \simeq 2\pi \frac{d}{\lambda_0}$$

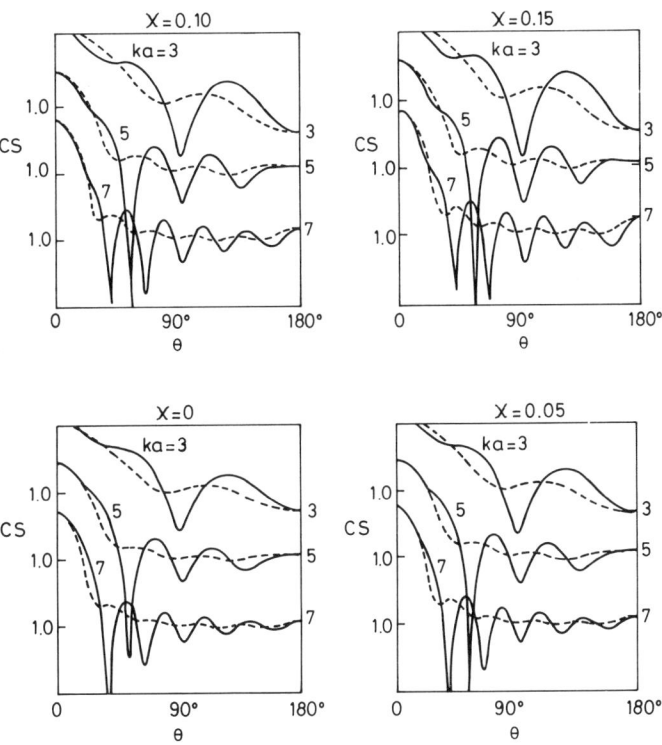

Fig. 8.4 Normalised scattering cross-sections (CS) for both E-plane and H-plane for three sphere sizes and four values of χ, the coating parameter

On the other hand, the curves in Fig. 8.5 correspond to the case where Z_1 is either zero or complex with a phase angle of $\pi/4$ or 45°. In this case,

$$Z_1 \simeq e^{j\pi/4} \eta_0 |\Delta| \qquad (8.131)$$

where $|\Delta|$ is nonzero if the sphere is finitely conducting. For example, consider a homogeneous sphere of radius a; then

$$Z_1 \simeq [j\mu_1 \omega/(\sigma_1 + j\varepsilon_1 \omega)]^{\frac{1}{2}}$$

$$\simeq (j\mu_1 \omega/\sigma_1)^{\frac{1}{2}} \qquad (8.132)$$

provided $|\gamma_1/k|^2 \gg 1$ and $\varepsilon_1\omega/\sigma_1 \ll 1$. In this case,

$$|\Delta| \simeq (\mu_1\omega/\sigma_1)^{\frac{1}{2}} \qquad (8.133)$$

The curves in fig. 8.5 show that the cross-section functions (CS) do not depend significantly on $|\Delta|$, at least for the situation where $|\Delta|$ is small.

As we see in Figs. 8.4 and 8.5, the forward scattering (i.e. $\theta \simeq 0°$) is greatly enhanced over the back scattering (i.e. $\theta \simeq 180°$). In fact, the (normalised) forward scattering cross-sections for the E-plane and H-plane cases both become approximately equal to $(ka)^2$, whereas the limiting value of the (normalised) back scattering cross-section is approximately unity. Thus in the forward direction the sphere is behaving as a 'black disc', particularly for the larger ka values, as indicated in Fig. 8.5.

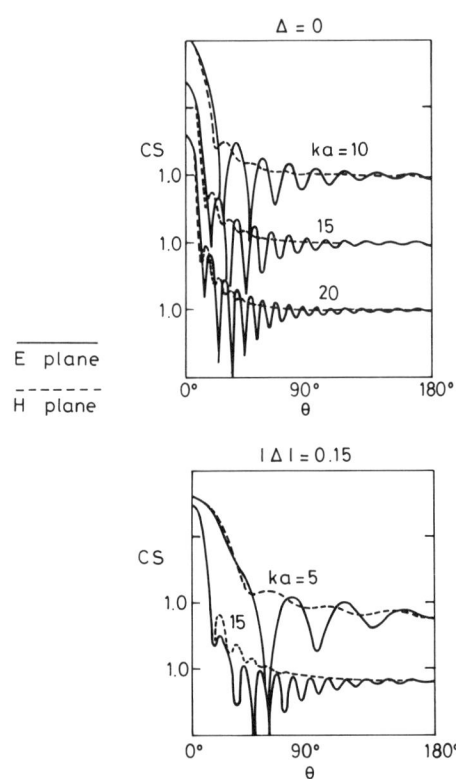

Fig. 8.5 Normalised scattering cross-section for various sizes of sphere, showing effect of finite conductivity

On the basis of Huygen's principle, it follows from van de Hulst [6] that, for small values of θ,

$$\sigma_e(\theta) \simeq \sigma_h(\theta) \simeq (ka)^2 \iint e^{-jk(x\cos\phi + y\sin\phi)} \sin\theta \, dx \, dy \qquad (8.134)$$

where the integration is over the area of the black disc. In the case of the sphere or its equivalent black disc, the integrations are readily carried out to give

$$\frac{\sigma_e(\theta)}{\pi a^2} \simeq \frac{\sigma_h(\theta)}{\pi a^2} \simeq (ka)^2 \left[\frac{2J_1(ka \sin \theta)}{ka \sin \theta}\right]^2 \quad (8.135)$$

For pure forward scattering (i.e. $\theta \to 0°$), the square bracket term in eqn. 8.135 approaches unity, which is a broad maximum. It decreases smoothly to zero at $\theta = \theta_0$ when $ka \sin \theta_0 = 3\cdot832$. For $ka = 10$, 15, and 20, the corresponding value of θ_0 are $22\cdot5°$, $14\cdot8°$ and $11\cdot0°$. They correspond quite well to the first minima in the calculated curves. The agreement actually improves for the earger ka values.

Because of the crude physical assumptions underlying the applications of Huygen's principle, it is not expected that eqn. 8.134 should be used for angles except near foward scatering. Being a scalar theory, it does not distinguish between the two wave polarisations. However, this simple physical optics model does adequately describe the shape of the main lobe.

The oscillatory behaviour of the curves in Figs. 8.4 and 8.5 is certainly evident as θ moves from $0°$ to $180°$. Here we have a good example of the 'creeping waves' contributing to the total scattered fields. Such diffracted wave contributions can be identified with the energy path that has circumnavigated the back side of the sphere. Van de Hulst [6] gives a clear physical discussion of the phenomena in a general context.

The interference between the diffracted and reflected wave contributions is most marked for E-plane polarisation. This fact is consistent with rigorous diffraction theory when such creeping waves are treated explicitly.

In the case of pure backscatter, the main reflected wave signal dominates and the effective cross-section actually approaches the geometrical cross-section πa^2. This behaviour is particularly evident for large ka values.

8.7 References

1. J. R. WAIT, On the electromagnetic response of a conducting sphere to a dipole field, *Geophys.*, 1960; **25**, pp. 649–58. Other related references are given in Chapter 7.
2. T. RIKITAKE, *Electromagnetism and the Earth's Interior*, Elsevier, Amsterdam, 1966.
3. S. P. SRIVASTAVA, Theory of the magnetotelluric method for a spherical conductor, *Geophys. J. Roy. Astrom. Soc.*, **11**, pp. 373–87.
4. G. SCHUBERT and D. S. COLBURN, Thin highly conducting layer in the moon, *J. Geophys. Res.*, 1971, **76**, pp. 8174–80.
5. G. SCHUBERT and K. SCHWARTZ, High frequency electromagnetic response of the noon, *J. Geophys. Res.*, 1972, **77**, pp. 67–83.
6. H. C. VAN DE HULST, *Light Scattering by Small Particles*, Dover, New York, 1981.
7. A. ERDÉLYI et al., *Higher Transcendental Functions*, vol. 2, McGraw-Hill, New York, 1953. On p. 250, note that

$$P_n^{(m)}(\chi) = (-1)^{n+m} 2^{-n}(n!)^{-1} (1-\chi^2)^{\frac{1}{2}m} \frac{d^{n+m}}{d\chi^{n+m}} (1-\chi^2)^n$$

where $m = 0, \pm 1, \pm 2, \ldots, \pm n$. Also,

$$P_n^{-m}(\chi) = (-1)^n \frac{(n-m)!}{(n+m)!} P_n^{(m)}(\chi)$$

is useful.

8. J. R. WAIT, Exact surface impedance for a spherical conductor, *proc. IEEE*, 1980, **68** (2), pp. 279–80. Note: $Z_{\theta\theta}$ in eqn. (9) should be $Z_{\phi\theta}$.
9. M. A. LEONTOVICH, On the approximate boundary conditions, *Investigations of the Propagation of Radio Waves*, pp. 5–15, ed. B. A. Vvedensky, USSR Academy of Sciences, Moscow, 1948.
10. S. A. SCHELKUNOFF, *Electromagnetic Waves*, Van Nostrand, New York, 1943.
11. G. N. WATSON, *Theory of Bessel Functions*, Cambridge University Press, 2nd edn, 1944.
12. J. R. WAIT, Electromagntic scattering from a radially inhomogeneous sphere, *Appl. Sci. Res.*, 1963, **10**, pp. 441–50. Note that the right-hand sides of eqns. 43, 44 and 49 should all be prefixed with negative signs. Also, replace left-hand sides of eqns. 41 and 42 by s_m^{-1} and S_m^{-1} respectively. Finally, note that on the right-hand sides of eqns. 41 and 42 the Bessel functions are all of order n rather than m.

Other references

J. R. WAIT and C. M. JACKSON, Calculations of the bistatic scattering cross section of a sphere with an impedance boundary condition, *Radio Sci.*, 1965 **69D**, pp. 299–315.

H. W. MARCH, The field of a magnetic dipole in the presence of a conducting sphere, *Geophys.*, 1953 **16**, pp. 671–84.

J. R. WAIT, Appendix A, *GeoElectromagnetism*, Academic Press, 1982.

J. R. WAIT, Complex magnetic permeability of spherical particles, *Proc. IRE*, 1953, **41**, pp. 1665–67.

S. H. WARD, The electromagnetic method, *Min. Geophys.*, 1967, **2**, pp. 224–72.

Chapter 9

Antennas mounted on smooth convex surfaces

9.1 Introduction

In the preceding two chapters we presented the formal theory of electromagnetic waves interacting with cylindrical and spherical surfaces with various applications. A common feature of the representations was the harmonic series that converged adequately provided the radius of curvature was not excessive in terms of the radio wavelength. Now clearly there are many situations where the effective circumference of the body will be a very large number of wavelengths. A good example is the earth or other planetary body when an electromagnetic wave propagates along the surface. In the early part of the century this problem attracted the attention of prominent mathematicians, who sought tractable solutions in an effort to explain the remarkable discovery by Marconi that radio waves will propagate to great distances over the earth's surface.

Here we present a short historical survey of the early work on diffraction by spheres and cylinders of large radii of curvature in the context of radiowave transmission and the related problem of radiation from antennas mounted on such surfaces. We will treat the cylindrical model in an explicit fashion, and then point out the essential equivalence to corresponding problems involving spherical surfaces. Our objective is to present universal pattern factors that can be used to characterise the radiation fields of antennas mounted on such convex boundaries, including the case where the finite conductivity of the surface material (e.g. the soil) needs to be allowed for.

The eminent mathematician G. N. Watson [1] demonstrated in 1918 that the harmonic series, for a vertical electric dipole over an idealised spherical earth, could be converted to a more rapidly convergent series in certain regions of space. This representation has become known as the residue series, as each term corresponds to a residue of a complex pole. This particular method has been exploited and refined in the late 1930s by Vvedensky [2] in Russia, Millington [3] in England, and Van der Pol and Bremmer [4] in Holland. The latter two authors in particular developed a highly accurate method to compute the values of the complex poles occurring in the residue series. Essentially, this is based on a representation for spherical wave functions by Hankel functions of order one-third, and for this reason it has become known as the Hankel approximation. It is more precise, although more involved, than the so-called tangent approximations of Vvedenskii and Millington, which were based on a second-order representation of the spherical wave functions due to Debye. Van der Pol and Bremmer have also demonstrated in an elegant fashion that a geometrical optical representation for the fields can be derived as a special case which is valid when the source and observer are well within 'line of sight'. In 1939 M. C. Gray [5] extended the Van der Pol and Bremmer treatment to a vertical magnetic dipole source.

In 1941 K. A. Norton [6] presented a systematic method for calculating the ground wave field for both vertical and horizontal polarisation. Using a number of graphs, the method was in a form suitable for use by radio engineers. In 1949 Bremmer [7] presented a comprehensive outline of the theory of ground wave propagation, listed a number of formulas for field computations, and gave certain illustrative examples.

For most applications of ground wave propagation, the methods of Norton and Bremmer are satisfactory. The exception is when a line joining the receiving antenna and the transmitting antenna just grazes the earth. Norton has proposed that this case be treated by interpolating between geometrical optics and the field computed in the shadow from the first term of the residue series. Often this is quite adequate. In this instance, Bremmer recommends that the complete residue series representation should be employed. Unfortunately, it becomes a very poorly convergent near the boundary of light and shadow, particularly if the antenna heights are large. The convergence of the residue series is also poor at short distances at low and medium radio frequencies, even when the transmitter and receiver are both on or near the ground. In the latter case, an alternative series has been developed [7, 8], it contains inverse powers of ka, and the leading term is the flat earth formula of Sommerfeld.

A more direct approach to the evaluation of the field near the light/shadow boundary is to return to the original complex integral representation which is actually the first stage in the Watson transformation. This is the line of attack adopted by Fock and Leontovich [9, 10] in the USSR. By making approximations to the spherical wave functions which were equivalent to the 'Hankel approximation' of Van der Pol and Bremmer, the integral for the field was put in a form suitable for numerical integration. Fock used Airy integrals, which are simply related to Hankel functions of order one-third (see following text). Fock [11] gave numerical results only for perfect conductivity. Fock has shown that the same integral arises in the diffraction of a plane wave by a paraboloid and any generally convex surface if the radius of curvature is large and smoothly varying.

It is our purpose to describe an application of Fock's method to the computation of the field of a source on a curved surface whose conductivity is finite. Particular attention is paid to the case where the receiving antenna is at a large distance from the surface and the transmitting antenna is on or near the surface. By utilising reciprocity, the results will also apply when the location of the receiving and transmitting antennas are interchanged. For purposes of presentation, a so-called cutback factor is introduced which is essentially the E-plane radiation pattern of a small loop antenna located on the curved surface. The term 'cutback' has been used to describe the rather pronounced reduction of the field at low angles.

9.2 Formulation

The model chosen as a basis for calculation is an infinitely long circular cylinder of radius a as illustrated in Fig. 9.1, together with the appropriate cylindrical co-ordinate system. (The surface of the cylinder is defined by $r = a$ and the z axis is directed into the paper.) The incident plane wave, which is incident normally on the cylinder, has a z component of the magnetic field only. It is given by

$$H_z^{inc} = H_0 \, e^{jk\varrho \cos \phi} \tag{9.1}$$

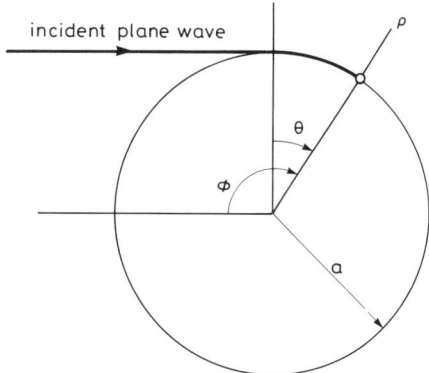

Fig. 9.1 Plane wave incident on circular cylinder

The boundary condition on the surface of the cylinder is

$$E_\phi = -ZH_z]_{\varrho=a}$$

where Z is by definition the surface impedance. For a homogeneous cylinder of conductivity σ, dielectric constant ε and permeability μ, Z can be well approximated by

$$Z \cong \sqrt{\left(\frac{j\mu\omega}{\sigma + j\omega\varepsilon}\right)} = \eta \tag{9.2}$$

where η is the intrinsic impedance of the medium. The boundary condition in this form is usually ascribed to Leontovich [12], who showed that for plane wave incidence it is valid if $|\eta/\eta_0| \ll 1$, where $\eta_0 \simeq 120\pi$ is the intrinsic impedance of free space. More recently Monteath [13] and Wait and Surtees [14] have shown a more general restriction in that the tangential fields should vary slowly in a distance equal to $|\gamma^{-1}|$, where

$$\gamma = [j\mu\omega(\sigma + j\omega\varepsilon)]^{\frac{1}{2}}$$

is the propagation constant of the conductor. Actually, a somewhat improved formula for Z for ground wave propagation is given by

$$Z \cong \eta \, (1 - \eta^2/\eta_0^2)^{\frac{1}{2}}$$

for vertical polarisation [8]. See also the discussion in Chapter 4.

8.3 Formal solution

To obtain the field scattered from the cylinder, the incident field is first written in the form

$$H_z^{inc} = H_0 \sum_{n=0}^{\infty} \hat{\varepsilon}_n \, e^{jn\pi/2} \, J_n(k\varrho) \cos n\phi \tag{9.3}$$

using a well known addition theorem in Bessel functions. In the above, $\hat{\varepsilon}_0 = 1$, $\hat{\varepsilon}_n = 2$ $(n \neq 0)$, and $J_n(k\varrho)$ is the Bessel function of the first kind of integral order n. Since the secondary field H_z^{sec} is to be a solution of the wave equation and is to give rise to outgoing waves at infinity, it is written in the form

$$H_z^{\text{sec}} = \sum_{n=0}^{\infty} A_n H_n^{(2)}(k\varrho) \cos n\phi \qquad (9.4)$$

where $H_n^{(2)}(k\varrho)$ is the Hankel function of the second kind. The secondary field H_z^{sec} is now chosen to satisfy the approximate boundary condition which can be rewritten

$$\left[E_\phi = \frac{j}{\omega\varepsilon} \frac{\partial H_z}{\partial \varrho} = -ZH_z \right]_{\varrho=a} \qquad (9.5)$$

noting that $H_z = H_z^{\text{inc}} + H_z^{\text{sec}}$. It readily follows that

$$A_n = -H_0 \hat{\varepsilon}_n e^{jn\pi/2} \frac{J_n'(x) + GJ_n(x)}{H_n^{(2)'}(x) + GH_n^{(2)}(x)} \qquad (9.6)$$

where $x = ka$ and $G = -jZ/\eta_0$. On the surface of the cylinder, the field can be written in the form

$$H_z]_{\varrho=a} = H_0 F(x, \phi) \qquad (9.7)$$

where

$$F(x, \phi) = \frac{2}{\pi j x} \sum_{n=0}^{\infty} \frac{\varepsilon_n e^{jn\pi/2} \cos n\phi}{H_n^{(2)'}(x) + GH_n^{(2)}(x)} \qquad (9.8)$$

The voltage induced in a small loop placed at $\varrho = a$ is proportional to $H_z]_{\varrho=a}$; thus $F(x, \phi)$ can be regarded as the radiation pattern of the loop in the principal or E-plane. The equivalence with eqn. 7.142 should be noted.

9.4 Complex integral representation

Unfortunately, eqn. 9.8 for the pattern function $F(x, \phi)$ is very cumbersome for purposes of calculation unless x is reasonably small. For example, something of the order of $2x$ terms are required to secure 5% accuracy. When x is large, it is desirable to represent $F(x, \phi)$ as a complex integral which can be evaluated either directly or by deforming the contour around the complex poles of the integrand. Such procedures have been used by Watson, Van der Pol and Bremmer, and Fock for treating the problem of ground wave propagation from a vertical dipole on the surface of the earth. The application of these same techniques to cylinder problems (which for large curvatures are essentially the same as the sphere problems) has been carried out by Franz [15, 16, 17], Sensiper [18], Baillin [19] and Wait [20, 21]. In most of this work (with the exception of Fock's), the complex integral was evaluated only by a saddle point method yielding useful expressions for the field in the 'lit' region of space (i.e. $|\phi| < \pi/2$.) On the other hand, the residue series representation is only useful deep in the shadow (i.e. $|\phi| > \pi/2$). In 1946, Fock evaluated the complex integral for the perfectly conducting cylinder by numerical means and thus bridged the gap between the geometric optical domain and the residue series solution deep in the shadow. The extension to imperfectly conducting cylinders is carried out in what follows. These results enable the pattern $F(x, \phi)$ to be evaluated for the whole domain of ϕ.

The complex integral representation of the function F for large cylinders can be written (see the Appendix to this chapter) in the form

$$F \cong e^{-jka\theta} g(X) \tag{9.9}$$

where

$$g(X) \cong \frac{1}{\sqrt{\pi}} \int_{\Gamma_2} \frac{e^{-jXt}}{W_1'(t) - qW_1(t)} dt \tag{9.10}$$

where Γ_2, the integration contour, runs from $\infty\, e^{-j2\pi/3}$ to 0, then out along the real axis to ∞. The outer symbols are defined as follows: $W_1(t)$ is the Airy integral, defined in the Appendix;

$$X = \left(\frac{ka}{2}\right)^{1/3} \theta = \left(\frac{ka}{2}\right)^{1/3} \left(\phi - \frac{\pi}{2}\right)$$

and

$$q = -j\left(\frac{ka}{2}\right)^{1/3} Z/\eta_0$$

Equation 9.9 for F is only valid if $(ka/2)^{1/3}\pi \gg 1$ and $|\theta| \ll 1$. This means that attention, at least for the moment, is restricted to large cylinders and to the region near the shadow boundary (i.e. the penumbra). In terms of the electrical constant ε and σ, the complex factor q is conveniently written

$$q = A\, 2^{1/6}\, e^{-j\pi/4} \left(1 + j\frac{\varepsilon\omega}{\sigma}\right)^{-\frac{1}{2}}$$

where

$$A = \left(\frac{ka}{2}\right)^{1/3} \left(\frac{\varepsilon_0 \omega}{2\sigma}\right)^{\frac{1}{2}} \tag{9.11}$$

When the conductivity is sufficiently large (i.e. $\sigma \gg \varepsilon\omega$),

$$q \simeq A\, 2^{1/6}\, e^{-j\pi/4}$$

The geometrical optics approximation for the function F can be written

$$F = (1 + R)\, e^{-jka\sin\theta} \tag{9.12}$$

where $R(\phi)$ is taken equal to the Fresnel reflection coefficient for a plane wave incident on a plane boundary whose surface impedance is Z. Therefore

$$R = \frac{\sin\theta + Z/\eta_0}{\sin\theta - Z/\eta_0} \simeq \frac{\theta + Z/\eta_0}{\theta - Z/\eta_0} \tag{9.13}$$

and consequently

$$F \cong e^{-jka\theta}\, e^{jka\theta^3/3!}\, \frac{2\theta}{\theta - Z/\eta_0} \tag{9.14}$$

for small values of θ. This equation can also be found from a saddle point evaluation of the integral $g(X)$ when X is negative and not near zero. It should be noted that if Z is replaced by $(\eta^2 - \eta_0^2 \cos^2\theta)^{\frac{1}{2}}$, R

becomes the Fresnel reflection coefficient for a homogeneous half-space with free space magnetic permeability.

In terms of X and the electrical constants σ and ε, the preceding equation becomes

$$F = e^{-jka\theta} e^{jX^3/3} \frac{2(1-j)X}{(1-j)X - 2^{2/3}A\left(1 + j\frac{\varepsilon\omega}{\sigma}\right)^{-\frac{1}{2}}} \tag{9.15}$$

remembering that $X = (ka/2)^{1/3}\theta$.

Up to this point the discussion has referred specifically to a circular cylinder of infinite length. On intuitive grounds one would expect the function F to characterise the pattern in the penumbral region of an antenna on any smooth surface whose radius of curvature is large. The quantity a is then regarded as the principal radius of curvature. It appears to be difficult to justify the preceding statements on a rigorous basis, for an arbitrary smooth surface. As shown by Fock, however, the concept is valid when applied to a perfectly conducting paraboloid, and a similar demonstration was made by S. O. Rice [22] for the parabolic cylinder.

For applications to antenna radiation over a spherical earth, it would be more logical to employ a model of a vertical electric dipole located on a homogeneous sphere with a specified surface impedance Z. This was the approach used essentially in the work of Van der Pol and Bremmer, Norton and Fock. In the present notation, the Hertz function Π (which has only a radial component) of the vertical dipole of unit strength located on the earth's surface is given by

$$\Pi \cong \frac{e^{-jka\bar{\theta}}}{a\bar{\theta}} V \tag{9.16}$$

where $\bar{\theta}$ is the angle at the centre of the earth subtended by the source dipole and the receiver at height h. In the above,

$$V = \sqrt{\left(\frac{jx}{\pi}\right)} \int_{\Gamma_2} e^{-jxt} \frac{W_1(t-y)}{W_1'(t) - qW_1(t)} dt \tag{9.17}$$

is an attenuation function and

$$x = \left(\frac{ka}{2}\right)^{1/3} \bar{\theta} \tag{9.18}$$

$$y = \left(\frac{2}{ka}\right)^{1/3} kh \tag{9.19}$$

Now when y is large, the Airy integral function $W_1(t-y)$ can be replaced by the first term of its asymptotic expansion for the important range of t in the integration. Furthermore, if x is large in such a way that $x - \sqrt{y}$ is finite, then

$$V \cong \frac{x^{\frac{1}{4}}}{y^{\frac{1}{4}}} e^{-j\frac{2}{3}y^{3/2}} g(X) \tag{9.20}$$

when $X = x - \sqrt{y}$ is a finite parameter. It is thus shown that Π, expressed as a ratio to the free space Hertz function Π_0, can be written

$$\frac{\Pi}{\Pi_0} \cong F \tag{9.21}$$

where F is identical to the cutback factor for the circular cylinder and X has the same geometrical meaning.

If the source dipole were at finite height (say h_0), then the function in eqn. 9.17 should be replaced by [9, 10]

$$\bar{V} = \frac{1}{2}\sqrt{\left(\frac{x}{\pi j}\right)} \int_{\Gamma_2} e^{-jxt} W_1(t - y_0)$$

$$\times \left[W_2(t - y) + \frac{W_2'(t) - qW_2(t)}{W_1'(t) - qW_1(t)} W_1(t - y) \right] dt \quad (9.22)$$

where

$$y_0 = \left(\frac{2}{ka}\right)^{1/3} kh_0 \quad \text{(for } y_0 < y\text{)} \quad (9.23)$$

and $W_2(t)$ is an Airy integral function. It was shown in an earlier communication [8], however, that quite generally

$$\bar{V} \cong V\left(1 + j\frac{Z}{\eta_0} kh_0\right) \quad (9.24)$$

subject to y_0 being small compared with unity and h_0 small compared with $a\bar{\theta}$.

Summarising, it can be said that the cutback factor is applicable to a transmitting vertical dipole at height h_0 and a receiving vertical dipole at height h subject to

$$kh_0 \ll \left|\frac{\eta_0}{Z}\right|$$

$$kh \gg (ka)^{1/3}$$

When jk_0 becomes comparable to $|\eta_0/Z|$, it is desirable to define a modified cutback factor F_m as follows:

$$F_m \cong \left(1 + j\frac{Z}{\eta_0} kh_0\right) F \quad (9.25)$$

which is valid if $kh_0 \ll (ka)^{1/3}$ and $h_0 \ll a\bar{\theta}$.

9.5 Discussion of the cutback factor

Using the preceding formulas for $\varepsilon\omega/\sigma = 0$ or $\alpha = 0$, the function $|F|$, described as the cutback factor, is plotted in Fig. 9.2 as a function of the angular parameter X for A ranging from 0 to 10. The geometrical optics representation is used for large negative X and the residue series representation (see the Appendix) for large positive X, whereas in the intermediate range the integral $g(X)$ was evaluated numerically. The corresponding phase factor Δ is plotted in Fig. 9.3a and 9.3b. It is defined as follows:

$$F = |F| e^{-jka\theta} e^{jX^3/3} e^{j\Delta} \quad \text{for } \theta < 0 \quad (9.26)$$

$$F = |F| e^{-jka\theta} e^{j\Delta} \quad \text{for } \theta > 0 \quad (9.27)$$

This method of normalising the phase enables one to regard Δ as a phase difference between that of the incident field and that of the resultant field in the lit region (i.e. $\theta < 0$). On the other hand, Δ is the phase difference between the electrical arc length $ka\theta$ and the resultant field in the shadow (i.e. $\theta > 0$). To facilitate the application of these curves to radiation pattern computation for a source on a spherical earth, the loss factor A is shown plotted in Fig. 9.4a as a function of the reciprocal of the ground conductivity, $1/\sigma$, for frequencies from 20 to 5000 kHz. To allow for normal atmospheric refraction the radius a is replaced by 4/3 times the actual earth radius.

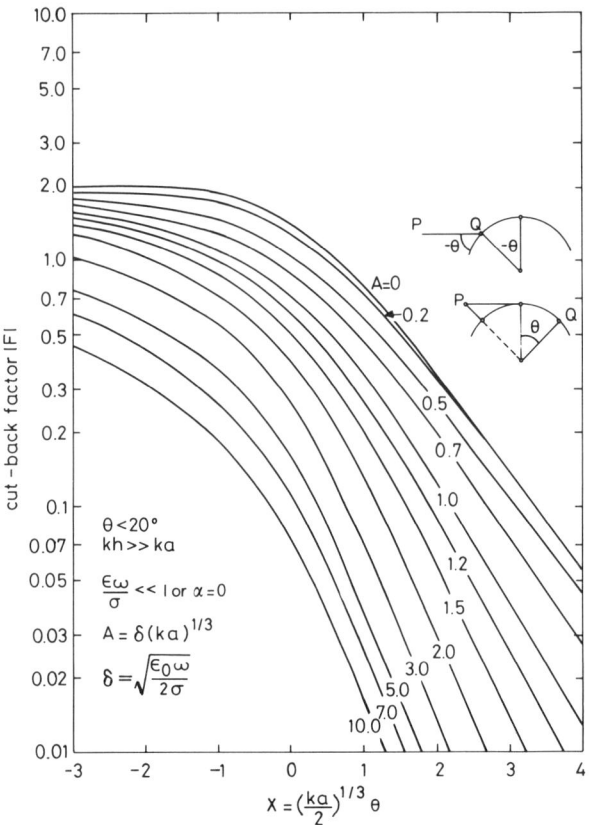

Fig. 9.2 Cutback factor $F = E/E_0$ as function of normalised grazing angle. E is field at P for source at Q on spherical surface; E_0 is field at P for source at Q in free space

At the higher frequencies and when the earth is poorly conducting, the displacement currents in the ground can be appreciable. Following Belkina [23], the complex factor q is rewritten in the form

$$q = \frac{-jp^{5/6}}{(\alpha p - j)^{\frac{1}{2}}} \tag{9.28}$$

which is equivalent to eqn. 9.11 if

$$P = (2A^6)^{1/5}$$

Antennas mounted on smooth convex surfaces

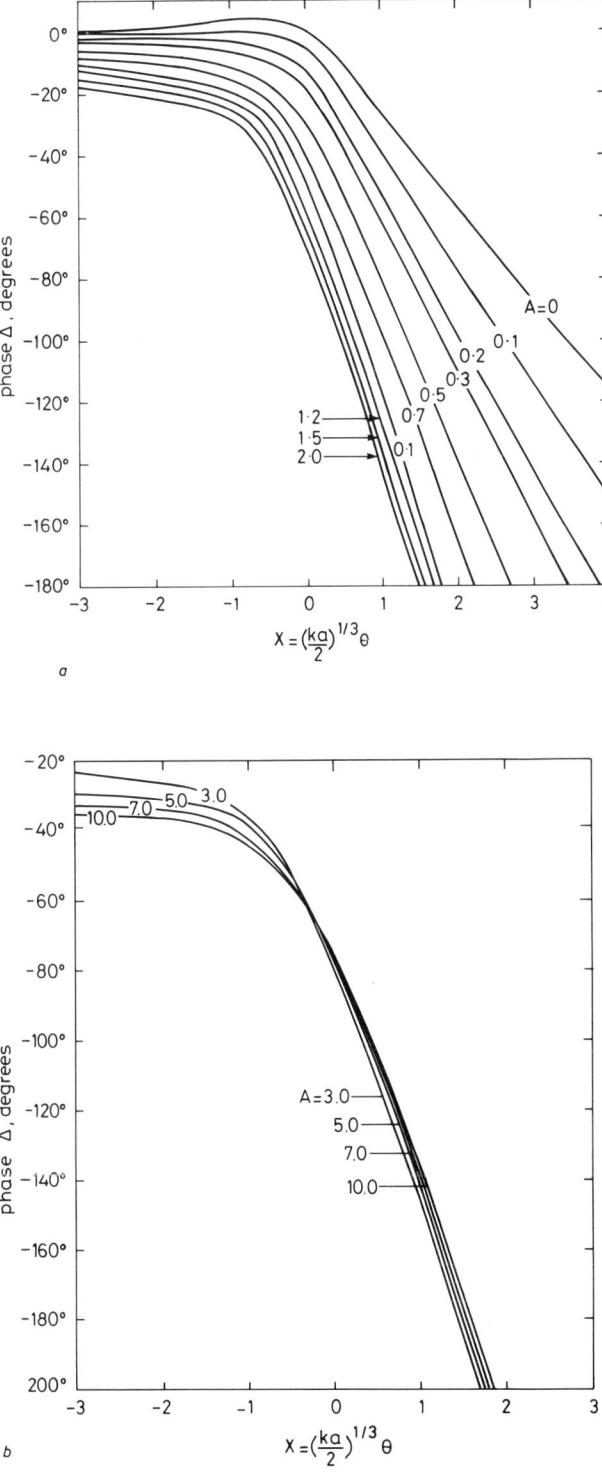

Fig. 9.3 *a, b* Phase of cutback factor ($\alpha = 0$)

$$\alpha = \frac{\varepsilon}{\varepsilon_0}\left(\frac{2}{\sigma a \eta_0}\right)^{2/5} = \frac{\varepsilon/\varepsilon_0}{(60\pi\sigma a)^{2/5}}$$

This seemingly complicated representation for q has the advantage that the displacement currents are described by the parameter α which is independent of frequency.

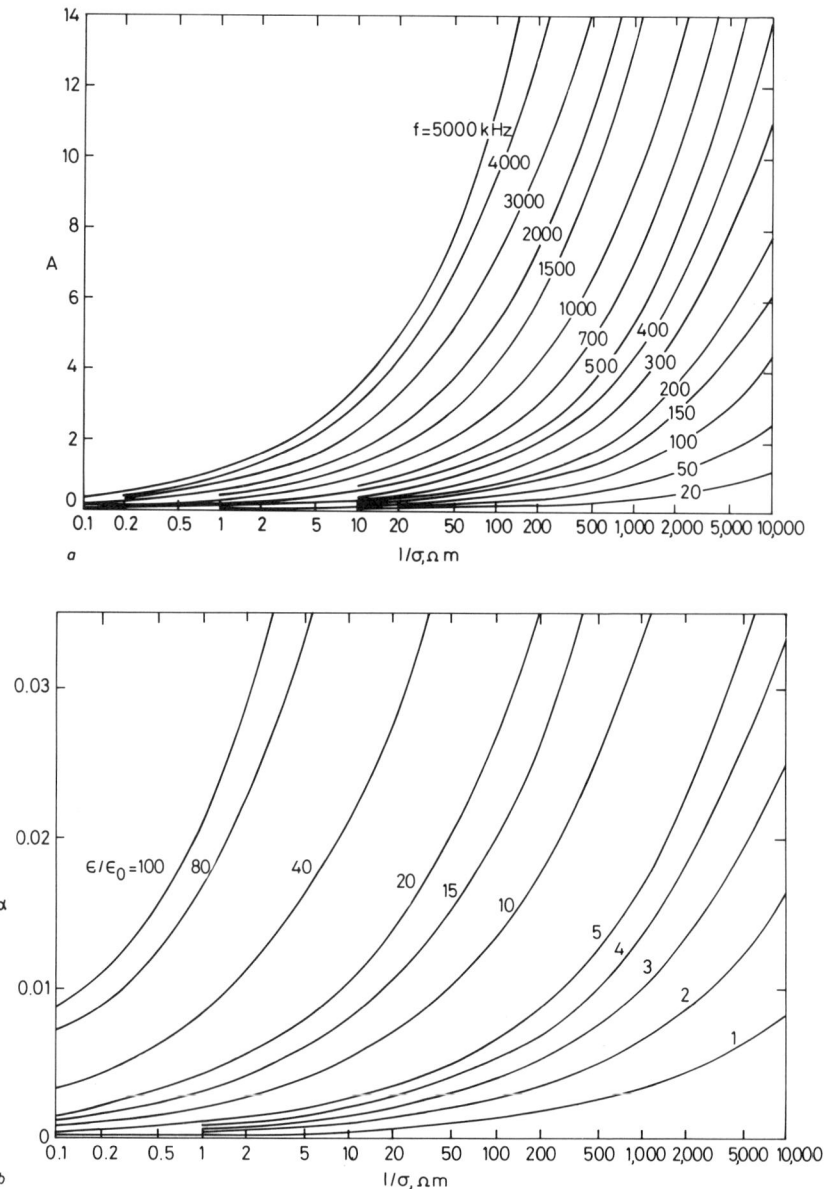

Fig. 9.4 *a* The loss factor A *b* The dielectric factor α as a function of the ground resistivity in ohm meters. $A = \delta(ka)^{1/3}$, where $\delta = (\varepsilon_0 \omega/2\sigma)^{1/2}$, $\alpha = (\varepsilon/\varepsilon_0)/(60\pi\sigma a)^{2/5}$.

Antennas mounted on smooth convex surfaces

To facilitate the conversion from the electrical constants σ and ε, the factor α is plotted in Fig. 9.4b as a function of $1/\sigma$ for various values of the relative dielectric constant $\varepsilon/\varepsilon_0$, again assuming an effective earth radius factor of 4/3. Numerical values of $|F|$ and Δ are given in Tables 9.1a and b respectively, for A ranging from 0 to 10 and α from 0 to 0·03. For the smaller A values the influence of α is negligible, and consequently only the results for $\alpha = 0$ need be shown.

The cutback factor $|F|$ is shown plotted in Figs. 9.5a and b for sea water and average land, respectively. The frequency varies from 20 to 5000 kHz. The electrical constants for sea water are taken as $\sigma = 5$ and $\varepsilon/\varepsilon_0 = 80$, and for land they are taken as $\sigma = 0·005$ and $\varepsilon/\varepsilon_0 = 15$. As an interesting illustration, $|F|$ is shown plotted in Fig. 9.5c for both land and sea at 100 kHz. The corresponding geometrical optical approximation for $|F|$ is also shown. At the larger negative values of θ, the two sets of curves merge together as they should. As θ approaches zero, however, they diverge significantly. It is noted that geometrical optics would predict that $|F|$ tends to zero at the shadow boundary (i.e. as $\theta \to 0$).

9.6 Extension to E-parallel polarisation

Up to this point, the incident wave is polarised such that the magnetic field is parallel to the axis of the cylinder. The other case of interest is when the electric field is parallel to the axis of the cylinder. The conversion of the former to the latter is simply obtained by replacing H_z by E_z and Z/η_0 by η_0/Z. Therefore, if the incident field is given by

$$E_z^{\text{inc}} = E_0 \, e^{jk\varrho \cos\phi} = E_0 \, e^{-jk\varrho \sin\theta} \tag{9.29}$$

the resultant tangential field on the cylinder is

$$E_z]_{\varrho=a} = E_0 F^* \tag{9.30}$$

where

$$F^* \cong \left[\frac{1}{\sqrt{\pi}} \int_{\Gamma_2} \frac{e^{-jXt}}{W_1'(t) - q^* W_1(t)} dt\right] e^{-jka\theta} \tag{9.31}$$

$$q^* = -j\left(\frac{ka}{2}\right)^{1/3} \frac{\eta_0}{Z}$$

Since in most applications $|Z/\eta_0| \ll 1$ and $ka \gg 1$, the quantity $|q^*|$ is a very large number and F^* can be well approximated by

$$F^* \cong -\frac{1}{a^*\sqrt{\pi}} U(X) \, e^{-jka\theta} \tag{9.32}$$

where

$$U(X) \cong \frac{1}{\sqrt{\pi}} \int_{\Gamma_2} \frac{e^{-jXt}}{W_1(t)} dt \tag{9.33}$$

The integral $U(X)$ has been evaluated numerically by Rice [22]. Noting that

$$H_\phi = \frac{1}{Z} E_z \quad \text{at} \quad \varrho = a$$

it is seen that

$$\frac{\eta_0(ka/2)^{1/3} H_\phi}{E_0} \cong -j \, e^{-jka\theta} \, U(X) \cong h_\phi$$

Table 9.1a Numerical values of $|F|$

X	$A = 0$ $a = 0$	$A = 0.1$ $a = 0$	$A = 0.2$ $a = 0$	$A = 0.3$ $a = 0$	$A = 0.5$ $a = 0$	$A = 0.7$ $a = 0$	$A = 1.0$ $a = 0$	$A = 1.0$ $a = 0.03$	$A = 1.2$ $a = 0$	$A = 1.2$ $a = 0.03$	$A = 1.5$ $a = 0$	$A = 1.5$ $a = 0.03$	$A = 2.0$ $a = 0$	$A = 2.0$ $a = 0.03$
−3·0	1·998	1·95	1·90	1·85	1·75	1·67	1·55	1·54	1·48	1·47	1·38	1·37	1·24	1·23
−2·5	2·994	1·94	1·88	1·82	1·71	1·61	1·47	1·47	1·39	1·38	1·28	1·28	1·14	1·13
−2·0	1·982	1·92	1·84	1·7	1·65	1·53	1·38	1·37	1·29	1·28	1·17	1·16	1·02	1·01
−1·5	1·948	1·87	1·79	1·71	1·56	1·41	1·25	1·24	1·14	1·13	1·02	1·01	0·869	0·868
−1·0	1·861	1·82	1·70	1·61	1·44	1·26	1·13	1·12	0·968	0·962	0·850	0·840	0·707	0·706
−0·5	1·682	1·59	1·54	1·43	1·27	1·09	0·918	0·917	0·771	0·765	0·669	0·664	0·539	0·537
0	1·400	1·34	1·27	1·15	1·01	0·901	0·711	0·710	0·579	0·570	0·490	0·485	0·386	0·384
0·5	1·059	1·00	0·952	0·904	0·801	0·659	0·500	0·499	0·401	0·389	0·317	0·316	0·301	0·300
1·0	0·738	0·722	0·692	0·661	0·563	0·446	0·316	0·315	0·251	0·248	0·181	0·180	0·119	0·117
1·5	0·488	0·477	0·466	0·444	0·384	0·290	0·196	0·195	0·152	0·149	0·102	0·0997	0·0608	0·0608
2·0	0·315	0·312	0·308	0·296	0·249	0·184	0·117	0·116	0·0838	0·0829	0·0528	0·0523	0·0282	0·0288
2·5	0·202	0·202	0·202	0·194	0·162	0·114	0·0668	0·0664	0·0453	0·0449	0·0266	0·0261	0·0122	0·0127
3·0	0·130	0·130	0·130	0·127	0·104	0·0706	0·0384	0·0376	0·0246	0·0243	0·0129	0·0127	0·00489	0·00522

X	$A = 3$ $a = 0$	$A = 3$ $a = 0.03$	$A = 5$ $a = 0$	$A = 5$ $a = 0.03$	$A = 7$ $a = 0$	$A = 7$ $a = 0.03$	$A = 10$ $a = 0.01$	$A = 10$ $a = 0.02$	$A = 10$ $a = 0.03$
−3·0	1·02	1·01	0·748	0·743	0·588	0·586	0·444	0·442	0·450	0·455
−2·5	0·919	0·916	0·651	0·647	0·512	0·510	0·379	0·382	0·388	0·394
−2·0	0·804	0·797	0·558	0·555	0·432	0·431	0·315	0·317	0·322	0·326
−1·5	0·664	0·662	0·459	0·457	0·344	0·344	0·246	0·248	0·253	0·256
−1·0	0·520	0·518	0·355	0·354	0·255	0·257	0·181	0·183	0·185	0·187
−0·5	0·685	0·684	0·249	0·249	0·175	0·182	0·122	0·124	0·126	0·127
0	0·267	0·266	0·158	0·159	0·110	0·115	0·0742	0·0760	0·0778	0·0791
0·5	0·139	0·142	0·0779	0·0788	0·0535	0·0560	0·0368	0·0375	0·0386	0·0399
1·0	0·0646	0·0661	0·0339	0·0355	0·0229	0·0246	0·0158	0·0162	0·0166	0·0176
1·5	0·0306	0·0311	0·0149	0·0156	0·0103	0·0111	0·00639	0·00668	0·00695	0·00714
2·0	0·0129	0·0132	0·00603	0·00617	0·00389	0·00426	0·00257	0·00263	0·00275	0·00288
2·5	0·00528	0·00547	0·00247	0·00261	0·00204	0·00216	0·00105	0·00108	0·00112	0·00115
3·0	0·00219	0·00225	0·000898	0·000949	0·000689	0·000693	0·000361	0·000369	0·000385	0·000407

The third significant figure is doubtful in some cases

Table 9.1b Numerical values of Δ (in degrees) (values are negative except where indicated by +)

X	$A=0$ $a=0$	$A=0.1$ $a=0$	$A=0.2$ $a=0$	$A=0.3$ $a=0$	$A=0.5$ $a=0$	$A=0.7$ $a=0$	$A=1.0$ $a=0$	$a=0.03$	$A=1.2$ $a=0$	$a=0.03$	$A=1.5$ $a=0$	$a=0.01$	$a=0.02$	$a=0.03$	$A=2.0$ $a=0$	$a=0.01$	$a=0.02$	$a=0.03$
−3.0	+0.52	0.94	2.2	3.6	6.1	8.4	10.9	10.7	13.1	13.0	15.3	15.1	14.8	18.1	17.06	17.8	17.6
−2.5	+0.85	0.91	2.4	3.9	7.0	9.1	12.6	12.4	14.4	14.2	16.9	16.7	16.4	19.6	19.5	19.4	19.3
−2.0	+1.5	0.85	2.7	4.4	8.0	10.3	14.4	14.1	15.2	14.9	18.8	18.6	18.3	21.9	21.7	21.5	21.2
−1.5	+2.4	0.74	3.0	5.3	9.4	12.9	17.0	16.6	19.2	18.8	21.5	21.3	21.0	24.6	24.1	23.5	23.0
−1.0	+3.7	0.63	4.1	7.1	12.2	16.1	20.3	19.8	23.2	22.8	25.5	25.3	25.9	29.0	28.3	27.6	26.9
−0.5	+3.9	1.2	6.3	10.3	18.1	23.2	28.2	27.5	33.0	32.6	38.4	38.1	37.8	44.5	43.0	41.5	40.0
0	0	6.0	13.7	19.8	31.5	41.8	50.9	49.7	60.8	59.3	67.6	66.9	66.2	69.0	68.1	67.2	66.1
0.5	11.4	20.7	32.0	38.8	53.5	67.3	83.0	79.0	92.3	90.8	101.1	102.2	103.4	106.2	104.1	102.7	101.3
1.0	26.6	40.3	53.9	61.3	81.2	98.4	119.0	116.0	128.4	124.4	136.0	134.8	133.1	144.1	141.8	140.0	138.9
1.5	42.7	59.5	76.9	86.3	112.4	132.8	158.8	155.5	170.0	166.8	178.2	177.0	175.5	186.3	184.6	182.7	180.9
2.0	58.0	77.9	99.0	110.7	141.1	167.4	198.5	195.0	212.0	210.0	220.3	219.1	217.9	227.1	225.5	223.9	222.4
2.5	72.9	96.5	116.0	134.8	170.5	202.0	236.8	234.8	251.0	249.2	263.9	262.5	261.1	269.8	268.2	266.6	265.1
3.0	87.6	114.9	142.7	158.7	199.9	236.3	276.9	274.5	293.2	290.5	307.6	306.0	304.4	313.5	311.6	309.4	307.5

X	$A=3.0$ $a=0$	$a=0.01$	$a=0.02$	$a=0.03$	$A=5.0$ $a=0$	$a=0.01$	$a=0.02$	$a=0.03$	$A=7.0$ $a=0$	$a=0.01$	$a=0.02$	$a=0.03$	$A=10$ $a=0$	$a=0.01$	$a=0.02$	$a=0.03$		
−3.0	23.4	22.8	22.2	21.4	29.1	27.6	26.1	24.5	33.0	30.9	28.8	26.7	35.4	31.2	27.1	23.5		
−2.5	25.2	24.4	23.6	22.9	30.7	29.1	27.5	25.8	33.6	31.5	29.4	27.3	36.5	32.0	27.8	24.1	The third	
−2.0	27.1	26.3	25.5	24.7	32.2	30.6	29.0	27.3	34.7	32.6	30.5	28.4	37.2	32.6	28.3	24.5	significant figure	
−1.5	29.2	28.6	28.0	26.3	33.2	31.5	29.9	28.2	36.1	34.0	31.9	29.8	38.0	33.2	29.0	24.6	is doubtful in	
−1.0	37.0	36.1	35.2	34.2	37.6	36.1	34.6	33.2	42.2	40.1	38.0	35.9	43.2	39.1	33.8	28.9	some cases.	
−0.5	54.2	52.5	50.8	49.1	50.8	48.9	47.0	45.0	55.3	53.1	51.0	48.0	56.0	51.2	45.7	42.6		
0	81.2	79.5	77.8	76.0	75.7	73.1	70.5	68.0	77.2	75.1	72.9	70.8	77.2	72.1	67.2	63.9		
0.5	118.0	116.0	114.0	113.7	107.9	105.2	102.6	99.9	106.8	104.7	102.5	100.4	104.9	99.9	94.5	89.5		
1.0	146.0	143.9	142.0	140.2	143.0	140.4	137.8	135.1	140.2	136.8	133.4	133.0	138.2	134.1	128.0	122.9		
1.5	187.0	184.0	181.4	180.3	181.1	178.5	175.9	173.3	178.2	174.5	170.8	167.1	176.8	171.2	165.3	159.9		
2.0	226.0	224.0	221.8	219.8	219.7	216.8	213.7	210.9	215.8	212.0	208.1	204.4	213.6	206.5	202.1	197.2		
2.5	268.1	265.5	263.0	260.0	257.6	254.3	251.0	247.8	253.9	249.8	245.7	241.6	250.8	243.9	238.5	232.2		
3.0	306.2	304.1	302.0	300.1	291.1	292.9	289.7	286.4	290.7	296.4	292.1	277.9	286.4	280.1	274.3	269.1		

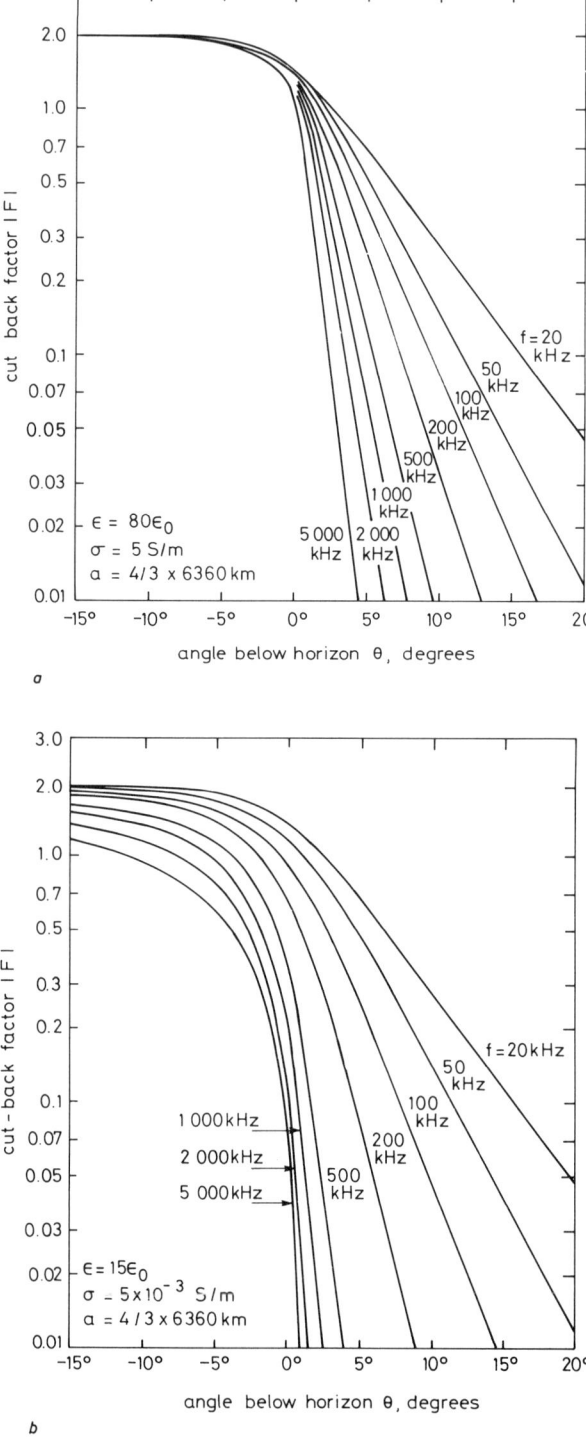

Fig. 9.5 *a* Cutback factor as a function of angle below horizon for sea water and various frequencies, *b* Cutback factor as in Fig. 9.5*a* but for average land, *c* Illustrating the inadequacy of geometrical optics for small grazing angles

Antennas mounted on smooth convex surfaces

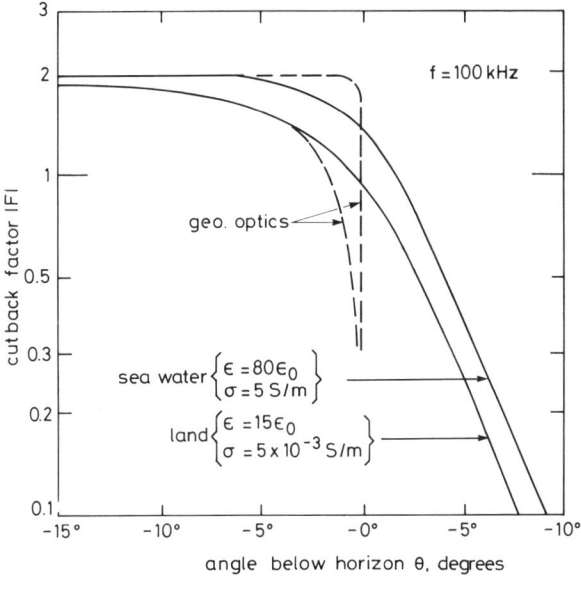

Fig. 9.5 Continued

where h_ϕ is the normalised magnetic field. For purposes of presentation, it is useful to introduce a phase factor Δ_h, in analogy to Δ, as follows:

$$h_\phi \cong |U(X)|\, e^{-jka\theta}\, e^{jX^3/3}\, e^{j\Delta_h} \qquad \text{for } \theta < 0 \quad (9.34)$$

$$h_\phi \cong |U(X)|\, e^{-jka\theta}\, e^{j\Delta_h} \qquad \text{for } \theta > 0 \quad (9.35)$$

According to geometrical optics,

$$H_\phi = -\sin\theta\, (E_0/\eta_0)\, e^{-jka\sin\theta} \qquad \text{at } \varrho = a \quad (9.36a)$$

$$\cong -\theta\, (E_0/\eta_0)\, e^{-jka\theta}\, e^{jka\theta^3/3!} \quad (9.36b)$$

which is equivalent to

$$h_\phi \cong -2X \quad (9.37)$$

The function $|U(X)|$, which is proportional to the amplitude of the tangential magnetic field $|H_\phi|$, is shown plotted in Fig. 9.6a as a function of X on a semi-logarithmic scale. The corresponding geometrical optics approximation is also shown. The function $|U(X)|$ together with the phase factor Δ_h, is plotted in Fig. 9.6b using linear scales. In this case the geometric optics result is simply a straight line and the phase is zero.

9.7 Application to moderately large cylinders

In all the preceding discussion the domain of θ and the radius of curvature is large enough that the tangential magnetic field is described adequately by eqn. 9.9. When ka is not excessively large, the formula for F should be replaced by eqn. 9.53 (see the Appendix). The leading

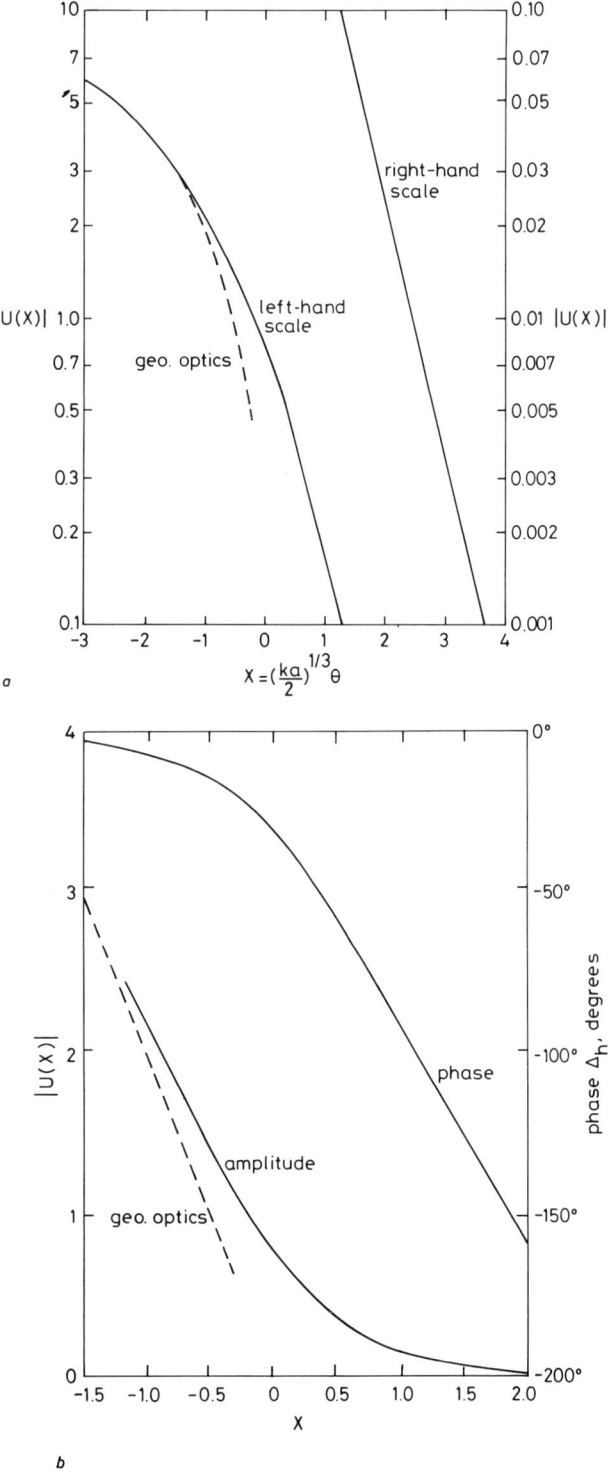

Fig. 9.6 *a, b* Amplitude and phase of cutback factor for horizontal polarization

term of this expression, which is usually adequate, reads

$$F \cong e^{-jka\theta} g\left[\theta\left(\frac{ka}{2}\right)^{1/3}\right] + e^{-jka(\pi-\theta)} g\left[(\pi-\theta)\left(\frac{ka}{2}\right)^{1/3}\right] \quad (9.38)$$

and is valid for the domain

$$\pi/2 \geqq \theta > -\Delta\theta \qquad \text{where } (\Delta\theta)^3 \ll 1$$

The second factor in the above equation can be interpreted as a wave which has creeped around from the rear of the cylinder. To illustrate the behaviour for cylinders of moderate size, $|F|$ and Δ based on eqn. 9.38 are plotted in Figs. 9.7a and b, respectively, for $ka = 8$ using the normalisation for phase defined by eqns. 9.16 and 9.17. Calculated points from the rigorous harmonic series solution (i.e. eqn. 9.8) are shown in Figs. 9.7a and b. In view of the fact that ka is not large, the agreement between the two sets of calculations is very reasonable. The corresponding curves for $ka = 21$ are shown in Figs. 9.8a and b, where the agreement between the approximate and the rigorous theory is much improved.

9.8 Concluding remarks

The numerical results presented herein can be used to predict the radiation patterns of electric and magnetic type antennas located on smooth curved surfaces of finite conductivity. For example, in certain investigations of ionospheric wave propagation it is of importance to know the amplitude of the incident wave at the lower edge of the ionosphere. The curves of the cutback factor can be used to estimate this quantity. Furthermore, the field measured at the receiving antenna on the ground can also be estimated from the cutback factor, assuming of course that the amplitude of the down-coming wave is known. It is particularly important to use these antenna cutback factors in the interpretation of long-distance ionospheric sky wave propagation at low radio frequencies. Considerable error may result if the radiation patterns are simply computed on the basis of a flat earth.

An excellent discussion of the application of cutback factors to the calculation of sky wave field strength has been presented by Norton [24]. Further applications and analytical extensions are available in the literature [25–30], which is now very extensive.

9.9 Appendix: mathematical details

The far field pattern in the equatorial plane or E-plane located on the surface of circular conducting cylinders was shown to be described by the function $F(x, \phi)$, where $x = ka$ and ϕ is the angular or azimuthal co-ordinate. This function is given by

$$F(x, \phi) = \frac{2}{j\pi x} \sum_{n=-\infty}^{+\infty} \frac{e^{jn\pi/2} e^{jn\phi}}{H_n^{(2)\prime}(x) + G H_n^{(2)}(x)} \quad (9.39)$$

where $G = -jZ/\eta_0$, Z is the surface impedance and $\eta_0 \cong 120\pi$.

The summation is over both positive and negative integers. $H_n^{(2)}(x)$ is a Hankel function of the second kind, and the prime indicates a derivative with respect to x. The first step is to express $F(x, \phi)$ as a contour integral in the complex v-plane. The contours chosen are the lines C_2 and C_1 which are located just above and just below the real axis of v, as shown in Fig. 9.9a. Some consideration shows [25] that

$$F(x, \phi) = \left(-\frac{1}{2j}\right) \frac{2}{j\pi x} \int_{C_1+C_2} \frac{e^{-jv\pi/2} e^{jv\phi}}{[H_v^{(2)\prime}(x) + GH_v^{(2)}(x)] \sin v\pi} dv \quad (9.40)$$

This integral has poles at $v = 0, \pm 1, \pm 2, \pm 3, \ldots$, as indicated in Fig. 9.9a as equispaced points on the real axis of v. It can readily be verified that the sum of residues of the integrand at these poles leads back to the series expression. (This fact is evident on applying Cauchy's theorem, which is given by

$$\oint \frac{f(v)}{\sin v\pi} dv = -2\pi j \sum_{n=-\infty}^{n=+\infty} \frac{f(n)}{\frac{\partial}{\partial v}(\sin v\pi)|_{v=n}} = -2j \sum_{n=-\infty}^{+\infty} e^{j\pi n} f(n)$$

where $f(v)$ is regular in the strip between C_1 and C_2 and the closed contour is taken in the clockwise direction.)

Replacing v by $-v$ in the integral over C_2, it is seen that

$$\int_{C_2} \frac{dv}{\sin v\pi} \frac{e^{jv(\phi-\pi/2)}}{[H_v^{(2)\prime}(x) + GH_v^{(2)}(x)]} = \int_{C_1} \frac{dv}{\sin v\pi} \frac{e^{-jv(\phi-\pi/2)}}{[H_v^{(2)\prime}(x) + GH_v^{(2)}(x)]}$$

(9.41)

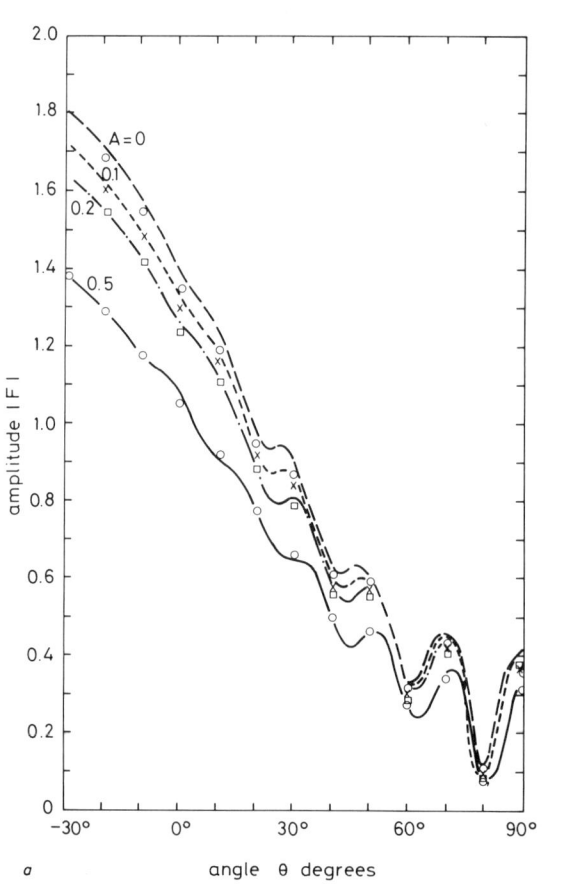

a

Fig. 9.7 *a, b* Amplitude and phase of space factor for circular model, with *ka* = 8, for various loss factors *A*, as a function of angle *θ*. The solid curves are based on the creeping wave representations, and the points are based on the harmonic series formulas

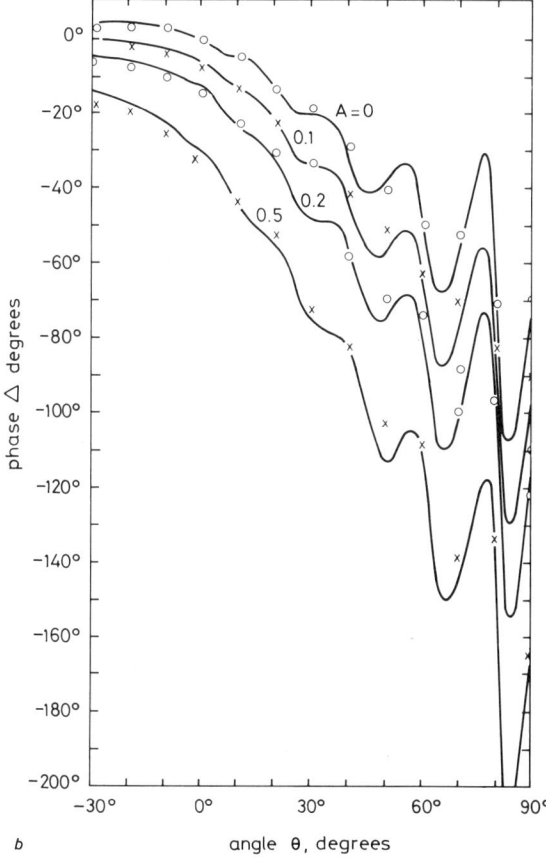

Fig. 9.7 Continued

and since
$$H^{(2)}_{-v}(x) = e^{-jv\pi} H^{(2)}_v(x)$$
the integral over C_2 is equal to

$$\int_{C_1} \frac{dv}{\sin v\pi} \frac{\exp\left[jv\left(\frac{3\pi}{2} - \phi\right)\right]}{H^{(2)'}_n(x) + GH^{(2)}_v(x)} \quad (9.42)$$

Consequently

$$F(x, \phi) = \frac{1}{\pi x} \int_{C_1} \frac{e^{jv\pi}}{\sin v\pi} \left[\frac{e^{-jv\left(\phi - \frac{\pi}{2}\right)} + e^{-jv\left(\frac{3\pi}{2} - \phi\right)}}{H^{(2)'}_v(x) + GH^{(2)}_v(x)}\right] dv \quad (9.43)$$

For this contour Im $v < 0$, and therefore it is permissible to write

$$\frac{e^{jv\pi}}{\sin v\pi} = 2j \sum_{m=0}^{\infty} e^{-jv2\pi m} \quad (9.44)$$

which leads to

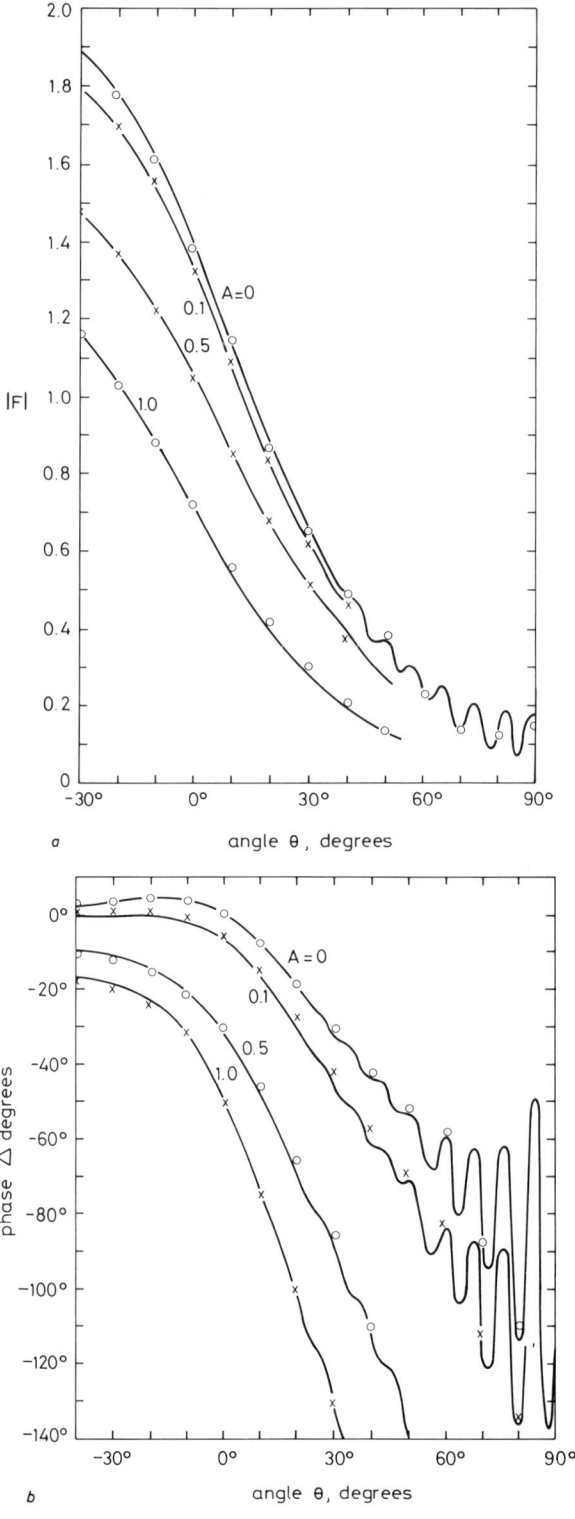

Fig. 9.8 *a, b* Same as Fig. 9.7*a* and *b* but for $ka = 21$. Note that 100 terms of harmonic series are needed to secure convergence

$$F(x, \phi) = \frac{2j}{\pi x} \sum_{m=0}^{\infty} \left[T_m\left(\phi - \frac{\pi}{2}\right) + T_m\left(\frac{3\pi}{2} - \phi\right) \right] \quad (9.45)$$

where

$$T_m(\beta) = \int_{C_1} \frac{e^{-jv(\beta + 2\pi m)}}{H_v^{(2)'}(x) + GH_v^{(2)}(x)} dv \quad (9.46)$$

with β being either $\phi - \pi/2$ or $3\pi/2 - \phi$.

The so-called third-order approximation for the Hankel functions is now introduced. This enables one to write [26]

$$H_v^{(2)}(x) \cong + \frac{j}{\pi^{\frac{1}{2}}} \left(\frac{2}{x}\right)^{1/3} W_1(t) \quad (9.47)$$

$$H_v^{(2)'}(x) \cong - \frac{j}{\pi^{\frac{1}{2}}} \left(\frac{2}{x}\right)^{2/3} W_1'(t) \quad (9.48)$$

where $W_1(t)$ and $W_1'(t)$ are Airy integrals defined by

$$W_1(t) = \frac{1}{\pi^{\frac{1}{2}}} \int_{\Gamma_1} ds \, e^{st - s^3/3} \quad (9.49)$$

$$W_1'(t) = \frac{1}{\pi^{\frac{1}{2}}} \int_{\Gamma_1} s \, ds \, e^{st - s^3/3} \quad (9.50)$$

in the notation of Fock, where the quantity t is defined by

$$t = \left(\frac{2}{x}\right)^{1/3} (v - x)$$

The integration contour Γ_1 is shown in Fig. 9.9b. (The Airy integral is related by

$$W_1(t) = e^{-\frac{2\pi j}{3}} \left(\frac{\pi}{3}\right)^{1/2} (-t)^{\frac{1}{2}} H_{1/3}^{(2)}\left[\left(\frac{2}{3}\right)(-t)^{3/2}\right]$$

to the Hankel function of order 1/3.)

In the foregoing text a function $W_2(t)$ was also employed. It has the same form as $W_1(t)$ but the contour Γ_1 is replaced by Γ_2, shown in Fig. 9.9c. The function $T_m(\beta)$ can now be written

$$T_m(\beta) \cong -j \frac{\pi x}{2} e^{-jx(\beta + 2\pi m)} g(X_m) \quad (9.51)$$

where

$$g(X) = \frac{1}{\sqrt{\pi}} \int_{\Gamma_2} \frac{e^{-jXt} \, dt}{W_1'(t) - qW_1(t)} \quad (9.52)$$

with

$$q = -j(x/2)^{1/3} Z/\eta_0$$

$$X_m = (\beta + 2\pi m) \left(\frac{x}{2}\right)^{1/3}$$

Therefore

$$F(x, \phi) \cong \sum_{m=0}^{\infty} e^{-jka[(\phi - \pi/2) + 2\pi m]} g\left[\left(\phi - \frac{\pi}{2} + 2\pi m\right)\left(\frac{x}{2}\right)^{1/3}\right]$$

$$+ \sum_{m=0}^{\infty} e^{-jka\left[\left(\frac{3\pi}{2} - \phi\right) + 2\pi m\right]} g\left[\left(\frac{3\pi}{2} - \phi + 2\pi m\right)\left(\frac{x}{2}\right)^{1/3}\right] \quad (9.53)$$

which is valid for $\pi \geq \phi \geq (\pi/2) - \Delta\phi$, where $(\Delta\phi)^3 \ll 1$. It is also readily shown that if $\pi(x/2)^{1/3} \gg 1$, only the term for $m = 0$ need be retained.

For the shadow region $(X > 0)$, it is desirable to calculate the integral for $g(X)$ by the evaluation of the residues at the poles of the integrand. The poles t_s of the integrand are determined from

$$W_1'(t_s) - qW_1(t_s) = 0 \quad (9.54)$$

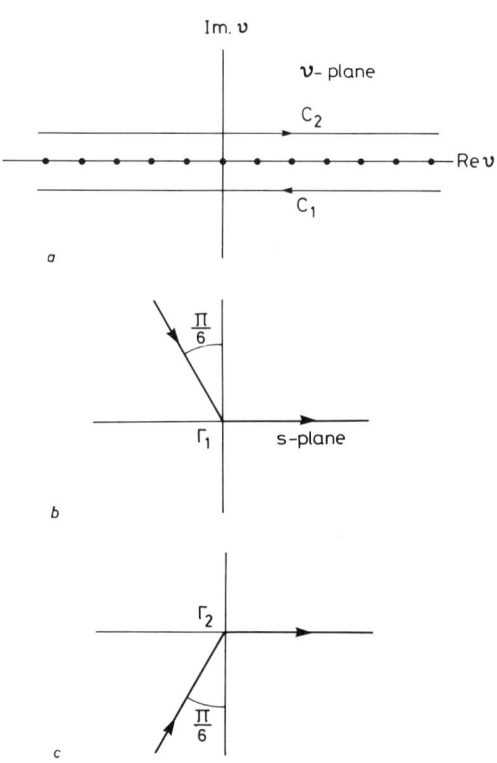

Fig. 9.9 Integration contours C_1 and C_2, which are straight lines running below and above the real axis, the points correspond to $\nu = n$, where n is any integer including zero b The integration contour Γ_1 in the s plane c The integration contour Γ_2 in the s plane

Expanding the denominator in a series around the pole $t = t_s$ leads to

$$W_1'(t) - qW_1(t) \cong [W_1''(t_s) - qW_1'(t_s)](t - t_s)$$

when terms containing $(t - t_s)^2, (t - t_s)^3$ have been omitted. But since

$$W_1''(t_s) - t_s W_1(t_s) = 0$$

$$W_1'(t_s) - q W_1(t_s) = 0$$

one may write

$$W_1''(t_s) - q W_1'(t_s) = (t_s - q^2) W_1(t_s) \tag{9.55}$$

and consequently

$$W_1'(t) - q W_1(t) \cong (t_s - q^2) W_1(t_s)(t - t_s)$$

$$+ \text{ higher-order terms in } (t - t_s) \tag{9.56}$$

For $X > 0$, C can be deformed to a path enclosing the poles which are contained (as will be shown) in the fourth quadrant of the t-plane. $g(X)$ is thus equal to the sum of the residues of the poles at $t = t_s$.

$$g(X) = -2\sqrt{(\pi)} j \sum_s \frac{e^{-jXt_s}}{(t_s - q^2) W_1(t_s)} \tag{9.57}$$

The numerical evaluation of the poles t_s, being a solution of

$$W_1'(t) - q W_1(t) = 0$$

has been discussed in detail by Fock. Actually, however, the task had been completed by Van der Pol and Bremmer a decade earlier. (The equation for the roots t_s arises also in the problem of a dipole located on the surface of an imperfectly conducting sphere.) Van der Pol and Bremmer's equation is

$$\delta \, e^{j\pi/3} \frac{H_{2/3}^{(2)}\left[\frac{1}{3}(-2\tau_s)^{3/2}\right]}{H_{1/3}^{(2)}\left[\frac{1}{3}(-2\tau_s)^{3/2}\right]} = -(-2\tau_s)^{-\frac{1}{2}} \tag{9.58}$$

where

$$\delta \cong -j \frac{(\eta_0/Z)}{(ka)^{1/3}}$$

$H_{2/3}^{(2)}$ and $H_{1/3}^{(2)}$ are Hankel functions of the second kind of order 2/3 and 1/3. But, since

$$W_1(t) = e^{-\frac{2\pi j}{3}} \sqrt{\frac{\pi}{3}} (-t)^{\frac{1}{2}} H_{1/3}^{(2)}\left[\frac{2}{3}(-t)^{3/2}\right] \tag{9.59}$$

it readily follows that

$$t_s = 2^{1/3} \tau_s \tag{9.60}$$

Although the notation of Van der Pol and Bremmer should have a historical priority, the Russian notation will be retained in the remainder of this section. When $q = 0$, the roots are determined simply from

$$W_1'(t_s) = 0$$

or from

$$H_{2/3}^{(2)}\left[\frac{1}{3}(-2\tau_s)^{3/2}\right] = 0$$

These particular roots, designated t_s^0, are well known [4]. The first five are given by

$$t_1^0 = 1\cdot01879\, e^{-j\pi/3}$$

$$t_2^0 = 3\cdot24820\, e^{-j\pi/3}$$

$$t_3^0 = 4\cdot82010\, e^{-j\pi/3}$$

$$t_4^0 = 6\cdot16331\, e^{-j\pi/3}$$

$$t_5^0 = 7\cdot37218\, e^{-j\pi/3}$$

A series expansion for t_s in terms of ascending powers of q can be obtained readily. For example, since

$$\frac{d}{dq}\left[\frac{dW_1(t)}{dt} - qW_1(t)\right] = W_1''(t)\frac{dt}{dq} - W_1(t) + qW_1'(t)\frac{dt}{dq} = 0 \qquad (9.61)$$

and

$$W_1''(t) = +tW_1(t), \qquad W_1'(t) = qW_1(t)$$

it is seen that the differential equation for t_s is

$$\frac{dt}{dq} = \frac{1}{t - q^2} \qquad (9.62)$$

The resulting expansion is

$$t_s = t_s^0 + \frac{1}{t_s^0}q - \frac{1}{2(t_s^0)^3}q^2 + \left[\frac{1}{3(t_s^0)^2} + \frac{1}{2(t_s^0)^5}\right]q^3 \qquad (9.63)$$

$$+ \text{ higher terms in } q \qquad (9.63)$$

Other expansions for t_s and a very extensive tabulation is given in a monograph by Belkina [23]. The results given here should be adequate for most conditions in practice.

9.10 References

1. G. N. WATSON, The diffraction of radio waves by the Earth, *Proc. Roy. Soc. Lond.*, **A95**, pp. 83–99; 1919, **A95**, pp. 546–63, 1919.
2. B. VVEDENSKY, The diffractive propagation of radio waves, *Tech. Phys. USSR*, 1935, **2**, pp. 624–39; 1936, **3**, pp. 915–25; 193, **4**, pp. 579–91.
3. G. MILLINGTON, The diffraction of wireless waves round the earth (a summary of the diffraction analysis, with a comparison between the various methods), *Phil. Mag.*, 1939, **27** (184), pp. 517–42.
4. B. VAN DER POL and H. BREMMER, The diffraction of electromagnetic waves from an electrical point source round a finitely conducting sphere, with applications to radiotelegraphy and the theory of the rainbow; the propagation of radio waves over a finitely conducting spherical earth, *Phil. Mag.* 1937, **24** (159), pp. 141–76; 1947, **24** (164), pp. 825–64, suppl.; 1938, **25** (171), pp. 817–34, suppl.; 1939, **27** (182), pp. 261–75.
5. M. C. GRAY Diffraction and refraction of a horizontally polarized electromagnetic wave over a spherical earth, *Phil. Mag.*, 1939, **27** (183), pp. 421–36.
6. K. A. NORTON, The calculation of ground-wave field intensity over a finitely conducting spherical earth, *Proc. IRE*, 1941, **29**, pp. 623–39.

7. H. BREMMER, *Terrestrial Radio Waves*, Elsevier, New York, 1949.
8. J. R. WAIT, Currents excited on a conducting surface of large radius of curvature, *Trans. IRE*, 1956, **MTT-4**, pp. 143–5; and Radiation from a vertical antenna over a curved stratified ground, *J. Res. Natl. Bur. Stand.*, 1956, **56**, pp. 230–40.
9. V. A. FOCK, Diffraction of radio waves around the earth's surface, *J. Phys. USSR*, 1945, **9**, pp. 256–66; *J. Theor. Exp. Phys.*, **15**, pp. 480–90.
10. M. A. LEONTOVICH and V. A. FOCK, Propagation of electromagnetic waves along the earth's surface, *J. Phys. USSR*, **10**, pp. 13–24.
11. V. FOCK, Distribution of currents induced by a plane wave on the surface of a conductor, *J. Phys. USSR*, 1946, **10**, pp. 130–40, pp. 399–409; and A. S. GORIAINOV, Diffraction of plane electromagnetic waves on a conducting cylinder, 1956, *Doklady, Akademii SSSR*, **109**, pp. 477–80.
12. M. A. LEONTOVICH, Approximate boundary conditions for the electromagnetic field on the surface of a good conductor, *Bull. Acad. Sci. USSR*, serie physique, 1944, **9**, p. 16 (in Russian).
13. G. D. MONTEATH, Application of the compensation theorem to certain radiation and propagation problems, *Proc. IEE*, 1951, **98**, IV, pp. 23–30.
14. J. R. WAIT and W. J. SURTEES, Impedance of a top-loaded antenna of arbitrary length over a circular grounded screen, *J. Appl. Phys.*, 1954, **25**, pp. 553–5.
15. W. FRANZ, The Green's functions of cylinders and spheres, *Z. Naturforsch.*, **9A**, pp. 705–16.
16. W. FRANZ and R. GALLE, 1954, Semiasymptotic series for the diffraction of a plane wave by a cylinder, *Z. Naturforsch.*, 1955, **10a**, pp. 374–8.
17. W. FRANZ and P. BECKMAN, Creeping waves for objects of finite conductivity, *Trans. IRE*, **AP-4**, pp. 203–8.
18. S. SENSIPER, Cylindrical radio waves, *Trans. IRE*, 1957, **AP-5**, pp. 56–70.
19. L. L. BAILIN and R. J. SPELLMIRE, Convergent representations for the radiation fields from slots in large circular cylinders, *Trans. IRE*, 1957, **AP-5**, pp. 374–382.
20. J. R. WAIT, Radiation characteristics of axial slots on a conducting cylinder, *Wireless Eng.*, 1955, **32**, pp. 316–23; and Pattern of a flush mounted microwave antenna, *J. Res. Natl. Bur. Stand.*, 1957, **59**, pp. 255–9.
21. J. R. WAIT and J. KATES, Radiation patterns of circumferential slots on moderately large conducting cylinders, Monograph 167 R, Institution of Electrical Engineers (London), February 1956, republished in *Proc. IEE*, **103**, Part C. 289–296.
22. S. O. RICE, Diffraction of plane radio waves by a parabolic cylinder, *Bell Syst. Tech. J.*, **33**, pp. 417–502.
23. M. G. BELKINA, Tables to calculate the electromagnetic field in the shadow region for various soils, *Soviet Radio Press*, Moscow, 1949.
24. K. A. NORTON, Low and medium frequency ionospheric propagation, Section 4.2 in transmission loss in radiowave propagation, *J. Res. Natl. Bur. Stand.*, 1959, **63D**, pp. 53–73.
25. A. D. WATT, *Very Low Frequency Radio Engineering*, Pergamon, 1967. See also J. S. Belrose, chapter 15 in *Handbook of Antenna Design* (eds A. W. Rudge, K. Milne, A. D. Olver and P. Knight) Peter Peregrinus, 1983.
26. J. R. WAIT, *Electromagnetic Radiation from Cylindrical Structures*, Pergamon, 1959.
27. J. R. WAIT and A. M. CONDA, Pattern of an antenna on a curved lossy surface, *IRE Trans.*, 1958, **AP-6**, pp. 348–359.
28. P. H. PATHAK and R. G. KOUYOUMJIAN, An analysis of the radiation from apertures in curved surfaces by the geometrical theory of diffraction *Proc. IEEE*, 1974, **62**, pp. 1438–47.
29. E. F. KNOTT and T. B. A. SENIOR, Comparisons of three high-frequency diffraction techniques, *Proc. IEEE*, 1974, **62**, pp. 1468–74.
30. G. L. JAMES, *Geometrical Theory of Diffraction for Electromagnetic Waves*, Peter Peregrinus, 1980 (see Chapter 6).

Other references

J. R. WAIT, *Wave Propagation Theory*, Pergamon, 1981 (see Chapter 9 for modal analysis of spherical earth model for azimuthally symmetric source).

S. ROTHERAM, Ground wave propagation, *Marconi Review*, 1982, **XLV**, pp. 18–48.

J. R. WAIT, Theories for radio ground wave transmission over a multisection path, *Radio Science*, 1980, **15**, No. 5, pp. 971–976.

D. A. HILL and J. R. WAIT, Ground Wave attenuation function for a spherical earth with arbitrary surface impedance, *Radio Science*, 1980, **15**, No. 3, 637–643.

D. A. HILL and J. R. WAIT, HF Ground wave propagation over sea ice for a spherical earth model, *IEEE Trans.*, 1981, **AP-29**, No. 3, pp. 525–527.

D. A. HILL and J. R. WAIT, HF Radio wave transmission over sea ice and remote sensing possibilities, *IEEE Trans.*, (*Geoscience and Remote Sensing*), 1981, **GE-19**, No. 4, pp. 204–209.

D. A. HILL and J. R. WAIT, HF Ground wave propagation over mixed land, sea and sea-ice paths, *IEEE Trans.*, 1981, **GE-19**, No. 4, pp. 210–216.

L. E. VOGLER, Radio wave diffraction by a rounded obstacle, *Radio Science*, 1985, **20**, pp. 582–590.

J. R. WAIT and A. M. CONDA, Diffraction of electromagnetic waves by smooth obstacles for grazing incidence, *J. Research of the U.S. National Bureau of Standards, Sec. D*, **63**, pp. 181–197.

Index

Admittivity, 17, 211
Airy functions, 249
Ampere's Law, 29
— general form, 37
Ampere-Maxwell equation, 37
Antenna array, 98
  broadside pattern, 100
  end-fire pattern, 101
  power considerations, 100
Antennas
  base driven, 118
  horizontal, 113
  in bore hole, 114
  near cylinders, 198
  on convex surface, 229
  vertical, 113

Bessel functions
  cylindrical forms, 199
  addition theorems, 164, 178, 231
  spherical forms, 214
Boundary conditions, 68
  for planar interface, 69, 71
  for normal components, 80

Coaxial cable, 121
Coordinate transformation, 52
Complex conductivity, 17
Complex integral representation, 232
Complex vectors, 7
Conductivity, 17
Curl operator, 11
Critical reflection, 74
Current point source, 20
  bipole, 23
  dipole, 24
  point sink, 23
Current sheet source, 44
Cylindrical conductor, 157
Cylindrical coordinates, 25
Cylindrical waves, 152
Cut-back factor, 235

Debye potentials, 154, 210
Dielectric half space, 66
Dipole current lines, 59
Dipole fields, 23, 88
electrostatic limit, 26
Dipole limit of linear wire, 31
Displacement currents, 18, 37
Divergence operator, 11
Divergence theorem, 11
Dynamic fields, 37

Earth-ionosphere waveguide, 141
Electrostatic limit, 26
Electron plasma, 73
Electric dipole, 23, 28
Evanescent fields, 74, 186
Excitation of plane waves, 44
  from planar source, 50
  of wave guide modes, 147

Far field radiation, 95
Faraday's law, 37
Faraday–Maxwell equation, 38
Fibre optics transmission, 146
Fields of linear antenna, 92
  quasi-static limit, 94
Fock theory, 230
Fresnel reflection coefficient, 70, 233

Geometrical optics, 107, 243
General wave impedances, 156, 217
Gradient operator, 10
Ground plane effects, 80, 106, 117, 123, 176
Ground wave, 169, 232
Ground topography, 171, 181

Harmonic series, 232
Helmholtz equation, 55
Hertz vectors, 154
Huygen's principle, 226

Image methods, 35, 108, 168
Integral equation method, 190

Laplace's equation, 19, 87
Laplace transform, 20
Laplacian operator, 153, 200
Legendre polynomials, 213
Line sources, 154
  in presence of cylinder, 178

Lossy capacitor, 18
Lossy half-space, 75

Magnetic induction, 27
Magnetostatics, 27
Magnetic dipole fields, 104
Magnetic permeability, 28
Magnetic potential, 28
Maxwell's equations, 37
  including source term, 38
  in cylindrical coordinates, 152
  in spherical coordinates, 212
  time dependent form, 39
Mode concept, 109
Mode equation, 138, 139, 147
Mode orthogonality, 133
Multiple scattering, 173

Non-uniform transmission line
  theory, 165, 216

Oblique excitation of plane waves, 47
Oblique incidence reflection, 64
Oblique incidence scattering, 161, 162
Ohm's law, 17, 160

Parallel plate region, 107, 147
Permittivity, 17
Phasor concept, 1
Plane wave transmission, 39
  attenuation, 41
  characteristic impedance, 46
  excitation, 44
  in material media, 42
  oblique case, 47
  phase velocity, 41
  propagation constant, 40
Plasma frequency, 73
Poynting vector, 102

Quasi-static concept, 19

Radiation patterns, 90, 101, 102
Radiation resistance, 102
Ray concepts, 109, 175
Receiving antenna pattern, 80

Reflection,
  at interface, 53
  at transmission line junction, 125
  from electron plasma, 73
  from lossy half-space, 75
  from the ground, 78
  of microwaves, 73
  of sound waves, 57
Reflection coefficient, 53
  at oblique incidence, 64
Refraction of plane waves, 64
Residue series in diffraction, 245
Resonant rays, 142

Scattering
  from arbitrary cylinder, 186
  from coated sphere, 224
  from cylindrical boss, 168
  from finitely conducting sphere, 226
  from pairs of wires, 184
  from thin cylinders, 160, 196
  from wire/cylinder combination, 182
Shadowing effects, 171
Schelkunoff functions, 219

Separation of variables, 55
Series impedance of wire, 158
Sinusoidal current assumption, 94
Snell's law, 68
Spherical coordinates, 25
Spherical target model, 216
Spherical waves, 211
Standing wave patterns, 62, 63, 66, 172
Stokes theorem, 12
Surface admittance, 72
Surface impedance, 72, 221, 231
Surface wave, 145

Tilted electric dipole, 109
Time averaging, 8
Transmission coefficient, 53
Transmission line theory, 123
Two-wire line, 123
Transmission line,
  characteristic impedance, 125
  input impedance, 126
  matching section, 127
  radial form, 216

  series impedance, 123
  shunt admittance, 123
  terminated, 126
Transverse electric mode, 129
Transverse magnetic mode, 129

Van der Pol–Bremmer theory, 251
Vector addition and subtraction, 5
Vector multiplication, 6
Vector potential, 33, 86
Very low frequency fields, 144

Watson transform, 229
Waveguide, 109
  attenuation in, 132
  cut-off condition, 131
  dielectric slab model, 145
  earth-ionosphere guide, 141
  equivalent circuit, 135
  lossy slab model, 136
  phase velocity, 131
  rectangular form, 127
Wave tilt, 80